PRAISE FOR

# FRANCONA

"Even Yankees fans are going to want to read this Red Sox book."
— *New York Daily News*

"Terry Francona's new book is not only a must-read, but it is a fascinating and entertaining look into the daily life on Yawkey Way during that memorable time period . . . full of surprising and fun anecdotes . . . There are glimpses of his relationships with the players — good and bad . . . and in most cases Shaughnessy's thorough reporting includes perspective from the players, executives, and owners."
— **MLB.com**

"Candid and compelling."
— *Chicago Sun-Times*

"A good read, well worth it for any Red Sox fan or anyone interested in the last decade of baseball. Francona and Shaughnessy tell how the Red Sox became champions and also how it all fell apart."
— **CBSSports.com**

"The former Red Sox manager collaborates with Boston columnist and sports éminence grise Dan Shaughnessy to peel layers off the mythology-prone Red Sox Nation, opining that the Sox became more about money than winning. Having led the team to its first World Series win since 1918, Francona knows what he's talking about."
— *Daily Beast*

"A great read . . . good fun and dishy."
— **NBCSports.com**

A manager at work: Terry Francona's lineup card for a May 31, 2008, game against the Orioles. Manny Ramirez hit his five hundredth career home run in this game. At the time, the Red Sox were defending World Series champions. *National Baseball Hall of Fame, Cooperstown, NY*

# FRANCONA

## THE RED SOX YEARS

**TERRY FRANCONA**

AND

**DAN SHAUGHNESSY**

MARINER BOOKS
HOUGHTON MIFFLIN HARCOURT
BOSTON  NEW YORK

First Mariner Books edition 2014

Copyright © 2013 by Terry Francona and Dan Shaughnessy

For information about permission to reproduce selections from this book,
write to Permissions, Houghton Mifflin Harcourt Publishing Company,
215 Park Avenue South, New York, New York 10003.

www.hmhco.com

*Library of Congress Cataloging-in-Publication Data is available.*
ISBN 978-0-547-92817-3     ISBN 978-0-544-22787-3 (pbk.)

*Book design by Brian Moore*

Printed in the United States of America
DOC 10 9 8 7 6 5 4 3 2 1

Terry Francona

*To my dad, Tito Francona, who managed to always make me feel like he was nearby, even when he was thousands of miles away.*


Dan Shaughnessy

*To my dad, the late William J. Shaughnessy, who saved enough S&H Green Stamps to get me my first mitt, a Tito Francona glove, in 1962.*

# CONTENTS

# CHAPTER I

# "Can't you guys do one thing right?"

**T**HEY WERE EXHAUSTED. Empty. Six weeks of spring training had given way to six months of regular season and another three weeks of thrill-to-the-marrow playoff excitement and pressure. It was autumn 2004, and the late, late nights had blended into fuzzy mornings with folks asking, "Did that really just happen?" Those four wins against the Yankees — the comeback trick that nobody'd turned in more than a century of postseason baseball — were played on four consecutive days, an aggregate 18 hours and 12 minutes over 44 intense innings. That's an average of more than four and a half hours per game. Games 4 and 5 at Fenway were actually won on the same calendar day (October 18) because Game 4 stretched past midnight and Game 5 started fewer than 16 hours later. It was something like a morning-night doubleheader.

The World Series against the St. Louis Cardinals was a blur. It was a little bit anticlimactic, if you can say such a thing about the first World Series title in 86 years for a city starving for a baseball championship. At the end, the Red Sox were an army of steamrollers. The World Series games were not nearly as exciting as those in the epic playoff against the Yankees.

Despite all the bad things that happened to the franchise after Babe

Ruth was sold in 1920, there was a sense of inevitability about the Red Sox march to the finish line after they beat the Yankees in the 2004 American League Championship Series. Major League Baseball's 100th World Series was tension-free, nothing like those Fall Classics of 1946, '67, '75, or '86, all of which ended with the Sox losing a seventh game.

The carpet in the visitors' clubhouse at Busch Stadium was still champagne-soggy when the tired, triumphant 2004 Red Sox gathered in their own ancient locker room at Fenway for the parade. The '04 Red Sox were enjoying their first days of folk-hero status in New England. All the stuff that everyone had been saying turned out to be true. These guys would never have to buy another beer in Boston. Kevin Millar was going to be able to live off the moment for the rest of his life. Dave Roberts was going to get asked about that stolen base every day until he was 90 years old. David Ortiz was the Dominican Yaz, Manny Ramirez was World Series MVP, and free spirit Orlando Cabrera would be forever worshiped as the guy who replaced the legendary Nomar Garciaparra. Lanky goofball Derek Lowe was done pitching for the Red Sox, but he was also destined to be a trivia answer as the man who got the win in the clinching game of every round of the playoffs.

Boston mayor Tom Menino and the Red Sox brass plotted the parade meticulously, embracing the concept of a rolling rally, which would maximize fans' exposure to the Hub's new heroes. Red Sox players, staff, and their families would ride in "Duckboats" up Boylston Street, then double back down the Charles River. Fans were expected to line the route. Sox CEO Larry Lucchino grew impatient, wanting the day to unfold without a hitch.

But then no one could find the sweatshirts.

And that's where our story starts.

The sweatshirts had WORLD CHAMPION RED SOX stenciled on the front. They were supposed to be worn by all the players, coaches, members of the front office, and the manager. They'd keep everybody warm and bring a measure of uniformity to this motley band of gypsy champions.

The 2004 Red Sox were not a buttoned-down bunch. They were players who passed paper cups around the clubhouse and took a nip of Jack Daniels before cold October night games in New York. Johnny Damon famously labeled his own team "Idiots," and Sox fans embraced them for their absence of conformity. They had beards, untucked jerseys, and hair . . . down to here, down to there. Damon looked like a lead guitarist from the Woodstock festival. Bronson Arroyo brought cornrows to the mound. Pedro Martinez could barely get his cap over the crop of black roughage on top of his head. Manny Ramirez looked like he was playing baseball wearing pajamas. Theo Epstein, the general manager, was a 30-year-old rock star who wore jeans to the office most days. When the unkempt Sox fell behind the Yankees three games to zero, reporters questioned their professionalism.

The sweatshirts would have helped their band-of-misfits image the day of the parade. At long last the Sox would look like a team.

But the sweatshirts were nowhere to be found as the players and their families got ready to man the Duckboats. Lucchino, the trigger-tempered smartest man in the room, was not happy.

Standing in the doorway of his office, looking out at the clubhouse filled with happy (some hungover) ballplayers, wives, and excited little kids, first-year Sox manager Terry Francona saw Lucchino engaged in an animated discussion with one of his favorite veteran clubhouse operatives. Francona went over to find out what was happening. As manager of this bombastic ball club, he was accustomed to extinguishing fires.

"What's going on, guys?" he asked Lucchino and the clubbie.

The longtime, underpaid Sox employee explained that the sweatshirts had gone missing. It didn't seem like a big deal to Francona or anybody else, but Lucchino was fuming. Francona overheard Lucchino muttering, "Can't you guys do one thing right?"

*Goddamn, Larry, we just won the fucking World Series,* Francona thought to himself. *That should be good enough. Who cares what we wear? We could go down the street naked and they'd still clap for us. It doesn't matter. This is the greatest day of our lives.*

If you go back and look at the pictures of that day, you'll notice that Francona is not wearing a Red Sox ball cap. On the day that Boston's Olde Towne Team celebrated its first World Series championship in more than eight decades, the manager of the local nine wore the cap of his college baseball team, the University of Arizona.

"It was my silent protest," said Francona.

Epstein said with a chuckle, "I think we lost those sweatshirts on purpose. They wanted these official, marketing, licensed sweatshirts to be worn. They were going to sell them. They were hideous. They looked like '80s Zubaz-type sweatshirts. They were this awful, faded blue. No one wanted to wear them, so I think we had the clubbies lose them."

Francona thought of Lucchino as something of a bully, but he knew the CEO was running on fumes—like everyone else—and was just blowing off steam. Still, the first-year manager wanted to make a point. Being a team wasn't about wearing the uniform hoodies, cropping your hair in standard fashion, or tucking in your shirttails. It wasn't about newly embossed sweatshirts that were already on sale in the team souvenir store. The '04 Sox were a wacky rainbow of diverse personalities. But they were a team. They had their differences, but they also had each other's backs. Francona knew he could count on guys like Kevin Millar, Gabe Kapler, Ellis Burks, Jason Varitek, and David Ortiz to keep things in line. The ragtag bunch cared about one another. It was the mark of every good team. Baseball is the most selfish of our team sports. It's a sequence of one-on-one battles. You don't need teammates to pass you the ball or protect you in the pocket. If you try to help yourself, you're usually helping the team. But there's a lot of human interaction and tolerance in a big league clubhouse, and the winners generally handle matters better than the losers. The special ball clubs are the ones that are able to get past the small stuff and push each other to succeed. The '04 Red Sox were one of those teams. Francona's managing style tapped into the genius of the Idiots.

He was the proverbial "players' manager." He didn't bark orders, didn't have a lot of rules, and never *ever* buried his players in public. He took all the bullets. He swallowed hard when he saw things that violated his sense of baseball etiquette or decorum. He always looked at the big picture. If mercurial Manny stopped running out his ground balls or took himself out of the lineup with mysterious hamstring injuries, Francona would leave it up to Manny's teammates to decide the best course of action.

"I always had my fingers crossed with Manny," said the manager. "Sometimes, when it really got bad, I'd meet with Damon and 'Tek and

David or Doug Mirabelli. All the veterans. I'd say, 'I see what you guys are seeing. What do you want me to do? I can bench him or suspend him, and Theo will back me up. I want to know what you guys think is the best thing for us.' And every time they'd come to the same conclusion. I think it was hardest for 'Tek, because he was so respectful of the game. But he did it. They'd always wind up telling me, 'No, we'll be a team. We'll take his numbers.' So that's what we wound up doing."

And Manny wound up being Most Valuable Player of the 2004 World Series. Of course.

Francona was never a World Series MVP or American League Manager of the Year. In the autumn of 2004, he went mano a mano with Mike Scioscia, Joe Torre, then Tony La Russa, and he came out on top in each series, but he stayed in the background. There wasn't a lot of credit for the manager of the Red Sox, nor was there much space for him on the victory platform. There were no Gatorade showers or magazine covers. The best he could do was an endorsement for Metamucil — which made him the ultimate regular guy. He was just the man who put his players in position to succeed and let their skills take over. He was the master of preparation and people management. When others told him that he probably would have made a good corporation manager, he'd deflect the praise and refer to himself as a "dumb-ass."

But deep down, he knew he could handle all of it. He knew that he was always prepared. Nobody could take away the baseball lessons he'd learned in five decades around big league clubhouses. It was his style to present himself as less than brilliant, but that was part of the ruse. Everybody who was paying attention could see that other managers never got the better of him. The Red Sox had the right guy in the dugout.

Francona managed the Sox for seven more seasons after 2004, never changing his style. He'd get to the ballpark absurdly early, pore over voluminous reports from Epstein's baseball operations department, work out in the clubhouse Swim-Ex, talk strategy with his coaches, and have all of his work done by the time the ballplayers started showing up for duty. If the owners wanted to show the clubhouse off to their friends, he'd grudgingly submit to the meet-and-greet. He'd be grateful if the suits didn't bring anyone into his office when he was half-naked. He'd patiently submit to questions from the Boston media three hours before every game, and again after every game. A lot of time was spent diffusing combustible situations. Everything the Sox did mattered. Nothing was too trivial to draw the scrutiny of talk shows and newspapers.

"I got to where I hated the traffic lights when I was driving to the ballpark," said the Sox manager. "The guy in the car next to me always had something to say and didn't feel like he had to hold back. When we weren't going good, those red lights seemed to last forever."

It isn't like this anywhere else — not even in Philadelphia, where Francona managed for four sub-.500 seasons. Boston is a smaller town, and the Red Sox are a religious experience for local fans. Columnist Mike Barnicle summed it up when he wrote, "Baseball is not life or death, but the Red Sox are." This is why years spent managing the Red Sox are like dog years. They age you disproportionately. Those before-and-after photos of US presidents looking young and vigorous on Inauguration Day, then tired and gray four years later? It was the same with the men who worked in the small corner office of the Red Sox clubhouse.

During his eight seasons in Boston, Francona occasionally allowed himself to wonder what it would be like elsewhere. He'd eavesdrop on one of Joe Maddon's sessions with the South Florida media and gasp at the easy tone of the questions. He'd see Toronto Blue Jays players almost come to blows on the mound in midgame and be amazed when the incident was buried in the local papers. He knew it would have been front-page stuff in Boston.

But those moments of longing for hardball tranquillity always passed. Managing in Boston was better, even if it was an ass ache much of the time.

What he loved most was the baseball. The games. During those three-plus hours when the team was on the field and he was in the dugout, Francona could escape the madness and immerse himself in the game he loved.

His favorite time of every day was the half-hour he'd spend in the dugout before the first pitch at Fenway. That was when the media was gone and all the preparation was over. It was just the manager, his coaches, and the band of hungry players who'd come out early to banter in the quiet time before the television camera's red light went on and quick decisions had to be made. It was a time when a manager could talk to players apart from the heat of the moment. It was a time to set up shop and get ready for the game. Francona would align three water bottles under the upper dugout bench perch — a spot where he could best watch and manage the game. He'd tape his matchup sheet to the pegboard on the dugout wall and hide his stash of Lancaster chew behind wads of bubblegum that had been unwrapped by clubhouse worker Steve Murphy. He'd sign a few autographs for folks in the rows directly behind the Sox dugout. That always ended the same way. Frenetic fans tossed baseballs toward him, and sooner or later someone would hit him in the chest while he was signing a ball. That would be the end of the signing session. The offending fan — the one who ruined it for everyone — invariably was an adult.

None of his players or coaches was required to be in the dugout in the golden half-hour before the first pitch, but Francona always knew he had a good team when there were a lot of ballplayers hanging around in those quiet minutes of pregame.

"I loved that time," he said. "That's when they haven't made an out, they haven't made an error yet. You can get a guy in the dugout 20 minutes before the game starts, and they are pretty loose. Once the game starts, you can't have that conversation anymore. It's a great time to talk to people, and I loved it."

The white noise of Boston only got louder in the years after the championships of '04 and '07, and ultimately the players stopped taking care of each other and abused their freedom. The 2011 season unfolded like many of the earlier Francona years as the Sox played 39 games over .500 for four months and came into September with the best record in baseball. But it was not like the other years. Veteran players David Ortiz, Tim Wakefield, and Jonathan Papelbon — warriors of championships past — worried about their next contract and got caught up in ancillary issues. An injured Kevin Youkilis had trouble dealing with his inability to contribute. Carl Crawford, an underachiever all year after signing a whopping, seven-year, $142 million contract, never performed like the player who tortured Boston when he played for Tampa Bay. Worst of all, pitchers Josh Beckett, John Lackey, and Jon Lester seemed to lose their focus, sometimes drinking beer and eating chicken in the clubhouse instead of staying in the dugout to encourage teammates. Players who at one time were mature enough to police themselves suddenly were in need of a managerial taskmaster. Francona opted not to change the style that had produced an average of 93 wins per season in his eight years in the Boston dugout.

"I think the chicken-and-beer stuff turned out to be more of a metaphor for our team," said Francona. "I can guarantee that these guys drank less beer than a lot of other teams. I was most disturbed by the idea that stuff wasn't staying in the clubhouse. They weren't protecting each other. If somebody was drinking, they weren't drinking a lot. I'm not saying it's right, but I was more disturbed by our lack of unity. That group, they had gained my trust. Well, they probably took advantage of it in the end. They needed a new voice."

While the Sox were unraveling like a ball of yarn bouncing down stairs, Francona was dealing with difficult personal issues. He was living in a hotel, separated from his wife of almost 30 years. His body was ravaged by more than 30 surgeries from his playing days, and he relied on pain medication to keep himself game-ready. He was worried about his son, Nick, who was commanding a sniper platoon in Afghanistan, and his son-in-law, who was dismantling homemade bombs in Afghanistan. He kept his cell phone handy in the dugout, in case there was news from Nick or one of his daughters.

But he was the same manager he'd been the whole time in Boston. He knew it would be phony to suddenly change his ways. He knew that it would send a message of panic if he started playing drill sergeant. After seven years and five months of steady success, he wasn't going to change his style. But he knew he no longer had the backing of ownership. The vaunted trio of John Henry, Tom Werner, and Larry Lucchino hadn't triggered his contract option and seemed more concerned about selling the Red Sox brand and making money than about winning championships.

Nothing could stop the September slide of 2011. The Sox, plagued by horrendous starting pitching, lost 20 of their final 27 games, blowing the biggest September lead in major league history. When they were eliminated by a wacky series of events in the midnight hour of the final night of the regular season, Francona knew it was time to go. He had little choice in the matter. Ownership was going in a new direction. It was all coming apart. Brilliant GM Theo Epstein was seduced by the Chicago Cubs, and some of Francona's trusted ballplayers didn't seem to have their heads in the right place anymore.

"There's one thing I'm going to be proud of after I'm gone," Francona said in the days after it ended. "I think they're going to find there's more shit that goes on than they realize."

A lot went on in eight years at Fenway. The sellouts, pink hats, parades, and renditions of "Sweet Caroline" were fun. Putting out fires and dealing with a complex and needy cast of characters was a daily challenge. But none of it could take anything away from what happened on the field. The baseball was always the best part.

# CHAPTER 2

# "I'm Mr. Francona's son and he wanted me to come over and say hello"

A BASEBALL LIFE is a life of interminable bus trips, tobacco spit, sunflower seeds, rain delays, day-night doubleheaders, and storytelling. There's a lot of standing in the outfield, shagging fly balls, and swapping lies. No life in sports has more downtime, more loitering, more waiting. The old salts tell the hungry young bucks not to get too high or too low. And always stay within yourself . . . whatever that means. The season is simply too long for daily reaction and analysis. It's not like football, a violent, self-important game that demands that you hit yourself over the head with a mallet for six days if you should happen to lose on any given Sunday. Baseball doesn't attach too much importance to any single game. If you lose today, you go back out there and get 'em tomorrow. There's always a chance for instant redemption. Hall of Fame skipper Earl Weaver knew what he was talking about when he said that the best part about baseball was that "we do this every day."

When you grow up the son of a major league ballplayer and dedicate your life to playing, then coaching and managing baseball teams, you appreciate the slow, steady pace of the game. You also create a worldwide network of teammates, coaches, and associates who keep finding you, sometimes years after you think you're done with them. This is how it's always been for Terry Francona.

When he was eight years old, Francona met Joe Torre, who was then a star catcher with the Atlanta Braves and a teammate of outfielder Tito Francona. Thirty-seven years after their initial meeting, Torre would come back into Francona's life as a worthy adversary in the Red Sox–Yankee rivalry of the 21st century.

When he was 11 years old, Francona met Ted Williams, the best player in the history of the Boston Red Sox, perhaps the greatest hitter who ever lived. Decades later, Francona would drive past a statue of Williams on his way to work every day at Fenway Park.

When he fulfilled a lifelong dream and played in the big leagues, Terry Francona's first manager was Dick Williams — the man who skippered the most important Red Sox team of the 20th century, the 1967 "Cardiac Kids." In the 21st century, Francona would become the greatest Red Sox manager since Dick Williams.

When his playing days were over and he became a coach and manager, Francona roomed and carpooled with a minor league lifer and cotton farmer named Grady Little. Eleven years after they were roommates, Little made a decision that altered the lives of millions of Red Sox fans and paved Francona's path to Boston.

Even some of the ballparks represented a thread. Terry Francona was a seven-year-old kid in the stands when the St. Louis Cardinals dedicated their spectacular new stadium in 1966. Sixteen years later, Francona's promising big league career was derailed when he tore up his knee chasing a fly ball on Busch Stadium's warning track. In 2004 Francona stood on the same field as manager of the World Champion Boston Red Sox. Now the place is gone, torn down to make room for a better model.

That's the baseball life. You get hired or fired by guys who played with, or against, your dad. Your college teammate, Brad Mills, is back at your side in the dugouts in St. Louis and Colorado when you win World Series for the Red Sox. Buddy Bell, your roommate with the Reds, brings you back into baseball as a coach when your playing days are over. Ken Macha, a fellow western Pennsylvanian who befriends you when you are about to be released by the Montreal Expos in 1986, rescues you from the depths of depression when blood clots almost take your life in 2002. John Farrell, another big league teammate, comes back into your life as your pitching coach and eventually succeeds you as manager of the Red Sox. Billy Beane, the man drafted one spot behind you in 1980, becomes famous as the *Moneyball* GM, then serves as your boss when you coach under Macha. When Beane turns down an offer from the Red Sox in 2002, the Sox turn to 28-year-old Theo Epstein, who hires you as the 44th manager of the Boston Red Sox. Ellis Burks, the center fielder who caught the ball you hit in your final big league at-bat, becomes one of your trusted clubhouse guys when you win the first World Series with the Red Sox in 2004.

They are baseball brothers, and they weave in and out of your life — on the diamond, in the dugout, and in the back rows of buses and airplanes.

Terry Jon Francona was born on April 22, 1959, in Aberdeen, South Dakota, where his dad, Tito, had met Roberta Jackson when he was a young outfielder in the Baltimore Orioles farm system in 1953. Roberta (always known as "Birdie") wasn't allowed to date ballplayers, but her brother-in-law, outfielder Zeke Strange, was Francona's minor league manager, and that connection allowed an exception to the rule. Tito and Birdie married in 1956 after Tito's rookie year with the Orioles. By the time their first child was born, Tito was emerging as a star outfielder with the Cleveland Indians. He hit .363 with 20 homers and 79 RBI for the Tribe in 1959. An armchair psychologist would submit that the birth of his only son moved Tito to have his best year in the bigs. In 14 other major league seasons, he never hit anything close to .363.

"He used to kill a sinker," said Tim McCarver, Tito Francona's teammate with the Cardinals in 1965 and '66. "He was a great low-ball hitter. Tito could hit anything down."

We can't go any further in this baseball story without some explanation of the name "Tito." Christened John Patsy Francona, "Tito" the elder got his lifelong nickname from his dad, Carmen Francona, a steelworker, piano tuner, and minister who raised his family in New Brighton, Pennsylvania, a borough of around 6,000 citizens, many of whom made their livings at local steel, lumber, and paper mills near the Ohio River. Tito is a popular nickname for small boys in Italian households. It comes from the ancient Roman name Titus. Carmen dubbed John "Tito" when the boy was four years old, and it is the name John Patsy Francona has answered to for his entire life.

Running around big league clubhouses when he was a small boy, Terry Francona came to be known as "Little Tito." As a grown man, a father of four, and a two-time World Champion, Terry Francona is honored to answer to the name that his dad got from his grandfather. Strangers and professional acquaintances call him Terry. His friends, ballplayers, coaches, and clubhouse confidants go with Tito.

Terry grew up in New Brighton and has fond, funny memories of his dad's parents. Carmen was 100 percent Italian, known all over New Brighton, the steelworker/preacher who could fix your piano even though he could never read a note of music. Josephine Skubis, Terry's Polish grandmother, grew up in an orphanage, met Carmen when she was only 14, and was tough enough to drive a crane during the Great Depression.

"They were the embodiment of all the Italian-Polish jokes," said Terry Francona. "My granddad ran the family. He was the patriarch. Everybody in the county knew him. When I was in college at Arizona, checking out at a Kmart, somebody heard my name and said, 'Yeah, your grandpa tuned my piano.' He was a minister for the religious services in a part of town called Hunky Alley. My mom was from South Dakota and had never seen anything like it. The family folklore is that when she came to meet my grandmother, my grandmother offered her chicken soup and set down a bowl of broth in front of her that had a whole chicken sitting in it."

Three and a half years after Terry was born, Tito and Roberta had a daughter, Amy, and the children were raised in a brick ranch in New Brighton while Tito was finishing his 15-year career in the majors. New Brighton is Steeler and Pirates country. It's near the Ohio border, close to Beaver Falls, Pennsylvania, where Joe Willie Namath was raised. At church services in the late 1960s, retired outfielder Tito Francona would nudge his son and say, "There's Joe Namath's brother." The Franconas weren't rich, but had everything they needed, including a vast basement that Terry converted into a baseball training multiplex. Big league ballplayers didn't have the cachet or the cash back then that they have in the 21st century. Tito Francona's big league salary topped out at $30,500. His big endorsement deal was $600 from Rawlings —

the company gave him three cents for every G-250 Tito Francona mitt sold. When he was done playing major league baseball, Tito took a job with the local park department, but everybody in New Brighton knew him as a former big leaguer. Among the mementos in the Francona family den were a couple of framed photographs of Tito with Ted Williams.

Another reminder that the Francona household was home to a big league ballplayer: there was a spittoon in every room. Tito Francona was not one of those ballplayers who left his chewing tobacco in the dugout.

"The inside of the driver's side door in our car was always brown," said Terry.

Early in every major league season, before school let out, young Terry went long stretches without seeing his dad. He didn't have a baseball practice partner, so Birdie assembled a contraption that would allow her son to practice throwing and catching by himself. Working out daily in his basement, Terry made himself the best nine-year-old ballplayer in New Brighton, and when Phillips 66 sponsored a regional pitching, batting, and throwing competition in nearby Beaver, Birdie drove her son to the competition.

"The other kids had their dads there coaching them and playing catch," remembered Terry Francona. "I hadn't seen my dad in three months, but my mom sat next to me on the bench and bullshitted with me the whole time."

He won the competition easily, but he never got his trophy. Event organizers disqualified Terry Francona because his dad was a big leaguer. Birdie was livid. She gathered up her son, put him in the car, and promised to drive him for a consolation ice cream. But she couldn't see the road through her tears and anger. She just kept driving and talking about the injustice of it all until her son noticed a sign that read: WELCOME TO OHIO.

A ballplayer never forgets support like that.

In the latter years of Tito's big league career, Birdie would wait until school got out in mid-June, pack up the family, and relocate to an apartment near her husband's workplace. The summer of 1965 was spent in St. Louis.

"My first baseball memory is living in apartments in St. Louis," said

Francona. "I was six or seven years old and we lived in these apartments called 'The Executive.' It's not like a place where players would live today. They were horrendous. A lot of the other ballplayers' families were there — Ray Washburn, Ray Sadecki, Bob Skinner. I used to play baseball with their kids every day. It was like a thousand degrees every day. I didn't get to the park that often, I was too young. But I was there the night they opened Busch."

"I remember Terry floating around those apartments," said McCarver. "A lot of guys used that place because it was across from the airports. Let me tell you, there was nothing 'executive' about it. I stayed there four or five years, including 1967, when Roger Maris lived there, and we used to drive to the ballpark together. I remember little Terry very well."

Little Tito got into some trouble one night when he was carpooling to the park with some of the other ballplayers' families. Birdie Francona was at the wheel, and everyone heard the bulletin over the radio that Cardinal first baseman Orlando Cepeda had been hit in the face during batting practice and would not be in the lineup. Young Terry whooped it up in the backseat because he knew that meant his dad would be starting at first base. Birdie was mortified.

Then there was the night he showed up in the clubhouse with a fistful of dollars. Curt Simmons's sons had convinced Little Tito that it was okay to sell players' game bats to fans. Business was booming until Tito asked his son to explain where all the money was coming from. Fortunately, manager Red Schoendienst and Cardinal ballplayers never knew about the enterprise.

Terry started going to the park almost every day in the summer of 1967 when his dad was playing with the Atlanta Braves. He'd hitch a ride to the park with Rita Raymond, wife of pitcher Claude Raymond, then find his dad in the clubhouse and get a dollar to spend on concession food. One dollar. Every night. It was good for a 75-cent chicken sandwich, but there was nothing he could buy with the 25 cents of change. After watching the game with Rita Raymond and the rest of the wives, he'd ride home in the backseat, listening to Tito and Claude analyze what just happened.

"I heard everything they said," he remembered. "I was the only eight-year-old who knew that you pitch guys high and tight, and low and away. I saw Bert Blyleven pitch when I was 11, and when I told my dad, 'That guy has the greatest breaking ball I've ever seen,' that's when my dad figured out that I was really paying attention."

In Atlanta he introduced himself to Joe Torre, the hot-hitting catcher who always had a five-o'clock shadow by lunchtime. Years later, when Francona and Torre were the two managers in the greatest rivalry in the sport, Torre would break the ice at the beginning of every series by greeting Francona with a handshake and the question, "How's your dad?"

The final year of Tito Francona's career was one of the best years of Terry's life. Tito went to spring training in Mesa with the Oakland A's, and Birdie came out with the kids for a three-week vacation. Terry was ten. It was hardball heaven in the Arizona desert. He got to be batboy every day, hanging around with Sal Bando and pitcher Al Downing. He played catch with a sculpted young outfielder named Reggie Jackson, who'd been a star at Arizona State. He took a road trip with the team on a day when his dad stayed back in Mesa. Tito had to have his knee drained and was scheduled to play in a B game. At the road game with the big leaguers, Little Tito spilled a bunch of pine tar in the middle of the game and was too embarrassed to tell anybody. He rode back home to Mesa with super-sticky fingers.

Most of the time he was comfortable around the big league ballplayers, comfortable enough to gawk at Rick Monday's attractive young wife and tell the outfielder, "You're my idol."

To this day, when Monday sees Terry Francona, he laughs and says, "You're my idol!"

"Rick says he doesn't even remember me doing that, but I told him, 'That's my story and I'm sticking to it,'" said Francona.

Little Tito got an authentic green satin Oakland A's jacket for Christmas in 1969. He wore it to school every day.

"It was the best present I ever got," said Terry Francona.

When Tito was traded to the Milwaukee Brewers early in the 1970 season, it worked out well for his only son. Milwaukee wasn't as hot and humid as St. Louis and Atlanta, and the Brewers in those days played in the American League, which opened up a new world of teams for Terry. Plus, he was finally old enough to go to the ballpark with his dad every day. Brewers manager Dave Bristol didn't like kids hanging around the clubhouse, but this was Tito's last gasp in the bigs, so nobody complained about the 11-year-old boy. Downing helped Terry hide from Bristol. The kid shagged fly balls with the big leaguers while the Brewers were taking batting practice. When the visitors took their turn in the cage, Little Tito went up into the stands to snag foul balls with the fans. After batting practice, Little Tito would make one more visit to the clubhouse to line his pockets with candy bars — like a rube traveler stuffing his luggage with the contents of a big-city hotel mini-bar. Major league clubhouses are well stocked with all forms of sweets, snacks, and beverages. Ballplayers support this bounty in the form of tips to the clubhouse workers, but it looks like free stuff to an 11-year-old, and Tito Francona never said a word about Little Tito raiding the candy rack. He took care of the clubbies when his boy wasn't around. Years later, Terry Francona's generosity toward the clubbies would become well known inside big league clubhouses.

In late July 1970, the Washington Senators, managed by Ted Williams, came to County Stadium. Two months away from retirement, Tito Francona made sure his only son didn't miss an opportunity to meet baseball's last .400 hitter. "Teddy Ballgame" had taken time to meet with rookie Tito Francona when Terry's dad made his big league debut in the spring of 1956. A mutual friend asked Williams to visit with the young Orioles outfielder, and when the kid from western Pennsylvania walked into the visitors' dugout at Fenway before his first big league game, the "Splendid Splinter" was waiting for him. If you wanted to talk hitting, Ted was your friend. The pitchers were the enemies, even the ones on Ted's own team. Like most ballplayers of his era, Francona believed that Williams was the greatest hitter who ever lived, an opinion shared by the louder-than-life Boston batting champ. In the den of his New Brighton home, Tito Francona keeps a photo of his debut day meeting with Ted Williams.

When the Senators were taking batting practice at County Stadium in 1970, Tito Francona took his son aside and pointed across the diamond toward the big man in the visitors' dugout.

*That's Ted Williams. Go introduce yourself.*

Young Terry didn't need his dad taking him by the hand. Father knew best. It would make a better impression if the kid walked over there by himself.

Wearing his ball cap and carrying his glove, 11-year-old Terry Francona walked across the field, behind the batting cage, and down the steps of the visitors' dugout, where Williams was sitting.

"Mr. Williams, I'm Mr. Francona's son and he wanted me to come over and say hello."

Williams loved to make parents look good in front of their own kids and was impressed by the manners of Little Tito.

"Well, you are a great-looking kid!" bellowed Williams. "And your dad is one helluva ballplayer. I just want to know one thing, young man. Can you hit?"

"He was great to me," Terry Francona remembered. "He took a minute and said hello and shook my hand. It meant a lot to my dad."

It was "bucket list" time in Tito Francona's career. He knew these were his final days in the bigs, so — pushing his luck a little — he went to Bristol and asked if he could take his son on a ten-game road trip through Minnesota, Chicago, and Kansas City in early August. Bristol said okay.

That was it. Birdie took Terry to buy a sport coat, combed the kid's hair, and sent him on his way. Her 11-year-old son was going to live the big league life for a week and a half.

"I had a ball," Terry Francona said more than 40 years later. "I'd be in the hotel room with my dad and get up early and go down to that lobby while my dad slept. All the players and coaches were coming and going. If somebody needed a newspaper or a cup of coffee, I'd get it. To this day I love hotel lobbies. I love watching the people. I think it always reminds me of those first days on the road with my dad."

At the ballpark on the road, players would dress him in the smallest Brewer uniform they could find, then roll tape around him to tighten and tuck the billowing parts. Tito's son was polite, respectful, and appreciative. He made it a point to talk to everyone, a quality that stayed with him throughout his baseball life. *Be nice to all the workers. Try to remember their names.* For an 11-year-old, he knew a remarkable number of people in ballparks across America.

Tito wanted to shield his young son from some things about the baseball life. Late one night, long after a game in Kansas City, father and son were sitting in the middle of the dark Brewer bus when both became aware of some X-rated talk coming from the back of the Greyhound. The Brewers had won their game, and no doubt a few post-game beers were consumed. Everybody forgot there was an 11-year-old kid sitting toward the front of the bus. Trying to drown out the racy stuff coming from the back of the bus, Tito quickly shifted into protective dad overdrive, talking loud and steadily — about anything and everything. Twenty years later over Thanksgiving dinner, Tito related the story of the night he adroitly protected his son from the nasty conversation, and only then did Terry admit that he'd been listening to the ballplayers and had heard every word. He'd respected his dad's effort, but he wasn't going to miss the racy tales of the young ballplayers.

"My dad rarely played that year, but I watched every game and got to make that trip. It was the best summer of my life."

"I'll always remember him on that trip," said Tommy Harper, the Brewers' best player in 1970 and later a spring training instructor with the Francona Red Sox. "Little Terry was all spiffed up at the start, but by the end of the trip he looked just as rumpled and tired as the rest of us. It was an eye-opener for a little kid like him."

That was the final season of Tito Francona's 15-year career. He hit .231 in 52 games with the Brewers, but the season was a great success because he finally got to spend time at the park with the boy who was born the year he hit .363 for the Cleveland Indians.

Little Tito played organized Little League in New Brighton. His teams were sponsored by Bachman's Garage and Mill Creek Electric. Early in his baseball career, he sometimes missed part of the season when he was off to whatever city his dad was playing in, but he'd come back for the Little League playoffs and stay with his grandparents.

"I loved those visits," he said. "It always meant I was in for a pair of red Chuck Taylor high-top sneakers. They cost something like $11, and my parents thought they were too expensive, but my grandpa would buy me a pair every summer when I went there for Little League playoffs."

Wearing the orange-and-white Bachman's Garage uniform, he pitched, played short, and sometimes worked behind the plate. Every couple of innings, nervous Little Tito would sneak behind the left-field fence to pee in the creek that ran alongside the outer wall. It was a nervous habit that stayed with him throughout his professional career.

"One of the greatest nights of my life was playing a Little League tournament a half-hour from home in Leetsdale," he recalled. "They had lights, and I was so excited about playing a night game, I had to run out behind the fence and take a piss about four times because I was so nervous.

"I never once thought that I wasn't going to play in the big leagues. That's what I was going to do. I had a guidance counselor say, 'Terry, you have to put down a profession.' I'd tell them, 'Major league baseball player.' There was no plan B, and that used to frustrate people. Now that I've gone through it, I can certainly understand why. You can get derailed so easily, but I never felt for one second that anything would stop me."

Tito, who never yelled at his son, was short on instruction. Even though he'd been a big leaguer for 15 years, he wasn't going to push baseball on the boy. He'd hit Terry fungoes when he came home from work, but that was all. Terry could always tell when his dad was getting tired because the balls started coming at him quicker and harder. On Terry's game days, Tito watched from a distance, sitting on a lawn chair, rarely commenting. If Little Tito was going to get anywhere in this game, he'd have to figure things out for himself.

The kid figured it out. By his sophomore season in high school, he was starting in center field, playing first base, and putting up an ERA of 0.33 for Greg Fazio's "Fighting Lions." Mr. Fazio was a New Brighton history teacher and baseball coach, and he made no attempt to tinker with the teenager's perfect hitting mechanics. Terry hit .550 and threw a no-hitter in his first season of varsity baseball. In his junior season, he hit .769, making only nine outs all season. The competition wasn't the best, and there weren't many games. Western Pennsylvania weather kept the boys indoors for most of the spring. There were seasons when the first outdoor baseball was the day of the first game. The Lions never worried about sunblock, and there weren't a lot of college scouts in the stands, but word got around about the young slugger from New Brighton. He was selected by the Cubs in the second round of the amateur draft in his senior season of 1977, even though he'd batted only ten times his senior year because of a shoulder injury. (Though Francona went 7-10 for the Lions in '77, he fretted about his drooping batting average.)

Like every high school athlete, he was anxious to play professional ball, but the family decided that he would go to college unless the Cubs offered $40,000. When the Cubs came in at $18,000, Francona packed his bags for Tucson. He was off to play for University of Arizona coach Jerry Kindall, who had played with Tito Francona in Cleveland.

Kindall ran an excellent program, long on instruction. Four-hour practices were the norm. Francona loved the weather and the baseball-centric college experience. Away from the field, he was somewhat ill prepared for college life. He wasn't much of a student and spent a lot of time on the dormitory pay phone, calling the folks back in New Brighton. Tito and Birdie worried that their son might not last at Arizona. When Terry learned that Tito was at a work conference in Las Vegas, the homesick freshman "borrowed" a friend's orange El Camino car and drove eight hours for a visit.

"I told my friend I was going to Phoenix, but wound up going to Vegas, and I lost his gas cap and he wanted to kill me," said Francona.

He developed a crush on one of the girls in his mathematics class, but was too bashful to ask her out.

Baseball saved him — that and the emotional support from his new best friend, Brad Mills. Mills was a junior college transfer, a neat freak, and a workout warrior.

"I remember meeting Terry for the first time, and he was not impressive," said Mills. "Jerry Kindall had told me all about us getting one of the best players in the country, and I was excited about that. Then I went to a team event and started meeting everybody, and here's this guy laying on the couch. He was wearing ragged Levis, cut just below the knee, and red high-top Chuck Taylors. He had hair down to his shoulders and a ragged T-shirt. As I came over, I stuck out my hand and said, 'Brad Mills.' He didn't sit up or anything, he just put out his hand and said, 'Terry Francona.' I said, 'What? Are you kidding me?' That was all I said. The next day Jerry Kindall called him in and talked to him, and by our team meeting that afternoon he was cleaned up and his hair was cut."

It was the beginning of a long, strong friendship. After their days as college roommates, Francona and Mills went on to become big league teammates with the Montreal Expos in the early 1980s. Their wives and children became friends. Mills would serve as Francona's first-base coach when Francona managed the Phillies from 1997 to 2000 and as Francona's bench coach in Boston for the glory years. They would win two World Series together.

"I could tell him anything," Francona said later. "When I became a manager and he was a coach, I used to ride him so hard. But every now and then, if I was teasing, he'd give me a look and I'd know that was it. I'd leave. It was like he was saying, 'Enough.' I always knew. That's how close we were."

Playing college baseball thousands of miles from home, Francona remained tethered to New Brighton. Every time he hit a home run for the Wildcats, he'd get a $10 bill in the mail from his grandfather. In his third season, he used some of Carmen's reward money to take Jacque Lang on a date. She was the knockout girl from freshman math, and by junior year Francona had worked up the nerve to ask her out. Their first date was at the Lunt Avenue Marble Club. They were engaged a little over a year later, and patriarch Carmen Francona performed the wedding ceremony.

By the end of his junior season, Francona was America's top amateur baseball player, winner of the 1980 Golden Spikes Award. He hit .401 and was named College Player of the Year by *The Sporting News*. In the College World Series at Omaha, Nebraska, he hit .458 and was named MVP. The Expos drafted him with the 22nd pick of the first round, immediately before the New York Mets used their top selection on high school sensation Billy Beane.

Baseball lives intersected in the days after the draft when Montreal general manager John McHale called to sign the Expos' first-round pick. It was an exciting moment for proud father Tito Francona. In 1958 McHale had been general manager of the Detroit Tigers and refused a request for a raise from spare outfielder Tito Francona.

"I hadn't played much with the Tigers that year, and I was only making about $7,500," recalled Terry's dad. "I called John McHale to tell him my wife was pregnant and to ask for a $1,000 raise. I first called collect, and he wouldn't accept my call. When I called back and asked for the $1,000, he said, 'No way!' Then all that time went by, and there we are in 1980 and the Expos draft Terry and John McHale is calling to get him signed. I said, 'John, how good is your memory? Remember when my wife was pregnant and you wouldn't give me the extra $1,000? Well, you're going to have to pay now. This is the baby.'"

Francona signed for $100,000, more than five times what the Cubs had offered him three years earlier.

Sending their son off to a life in professional baseball, the Francona parents had sound advice.

Sounding a little like Manny Ramirez, Tito told his son, "See the ball and hit the ball."

"You are now a piece of meat," said Birdie.

They were both right.

Francona's first encounter with Red Sox legend Dick Williams was underwhelming for both. Williams was the manager of the Montreal Expos, and Terry Francona's status as a first-round pick qualified him for spring training with the big league club in West Palm Beach. The young outfielder gained considerable weight after his first summer in pro ball and made a poor impression when he showed up in Florida.

"In 1980 I played from January to Thanksgiving, playing every day, and I was worn out," said Francona. "I didn't have much power, and at the end of the year they told me to go home and get away from baseball. I did that. I must have gained 25 pounds. I showed up at spring training, and I was all excited being in the major league camp, but I was so fat my shirt wouldn't stay tucked in. I knew a lot of the guys — Tim Wallach and Brad Mills were there. I was probably more comfortable than I should have been. Bob Gebhard was running the minor leagues then, and he saw me and called me over and started in on me, saying, 'You fat fuck!' Everybody in the dugout was laughing. They killed me. That was Dick Williams's first impression of me. I wasn't around for a lot of spring training, but he never talked to me while I was there — not until my last day. I got into one of the early spring games and took a called third strike on a curveball. I walked into the dugout, put my bat in the rack."

Williams finally spoke.

"That's it, kid," said the terrifying manager.

"Does that mean I'm done for the day?" Francona asked.

"No, you're going to Daytona," snapped Williams.

Five months later, Little Tito was in the big leagues. And Dick Williams was waiting for him in the visitors' dugout in the Houston Astrodome.

Many Red Sox fans born before 1960 contend that Williams was the best manager in franchise history. Like numerous dugout geniuses, Williams was a marginal major league ballplayer. He cut his teeth in the vaunted Brooklyn Dodger farm system and played with five teams from 1951 to 1964. Williams was hired by Red Sox GM Dick O'Connell after a second-straight ninth-place finish in 1966. The Sox had lost 100 games in 1965 and hadn't been over .500 in eight seasons, but O'Connell knew he had a lot of young talent and needed a taskmaster to change the country club atmosphere at Fenway. Thirty-eight-year-old Williams was the perfect candidate for the seismic shift. He wore his hair in a Marine flattop and made it clear that he was the boss when he arrived at spring training in Winter Haven, Florida, in 1967. The rookie manager's first move was to strip young Carl Yastrzemski of his captaincy. There would be only one chief to command 25 Indians. Yaz was secretly relieved to forfeit the captaincy, but Williams's gruff style made it look like an insult. Williams didn't worry about hurting ballplayers' feelings. He told the media that talking to George Scott was like talking to cement. He made the lethargic Red Sox do calisthenics and run laps. He benched Yaz for not running out a ground ball. He went months without speaking to players who fell out of favor.

"He had everybody doing everything," Yaz recalled. "There was no downtime. He had been a utility player, and I think that was one of the things that made him such a great manager. He had the extra men go above and beyond what they would normally do. He had them out there for two hours instead of just 20 minutes a day."

Boston fans and media members thought Williams sounded a little crazy when he predicted, "We'll win more than we lose."

The 1967 Red Sox permanently changed the culture of Boston baseball. The Sox went from losers to winners overnight. They captured the town like no team before them, or since. They were the Cardiac Kids, and they won the greatest pennant race of all time (four teams separated by only a game and a half going into the final weekend) on the final day of the season before losing the World Series to the Cardinals in seven games. Years after he was fired by the Red Sox, Williams won two World Championships as manager of the Oakland A's and took the San Diego Padres to their first World Series in 1984. He won 1,571 games over 21 big league seasons. He was inducted into the Hall of Fame in the summer of 2008, and his name was still gold at Fenway when he died in the summer of 2011.

On August 19, 1981, Francona and Williams — the two best managers in the long history of the Boston Red Sox — intersected in the dugout of the Houston Astrodome.

Francona shredded Triple A pitching at Denver in the summer of '81, and the Expos called him up to see how he'd do against major league pitching. Because of an air traffic controllers' strike (which ended when President Ronald Reagan fired all the striking controllers), American air travel was chaotic in August 1981. Getting from Denver to Houston was not easy. Francona had to make a pair of connecting flights and didn't land in Houston until after the game had started. His cab didn't pull up at the Astrodome until the fifth inning.

"The driver left me off outside the center-field entrance," Francona recalled. "I had my bats, my clothes, all my stuff. I talked my way in, and I was lugging stuff all around the ballpark. Nolan Ryan had a no-hitter going, and I thought maybe I should get some popcorn and watch the game. I worked my way around to the clubhouse with all my stuff. I was sweating like a pig, and they told me to get my uniform on real quick. I got myself through the maze and into the dugout, and there was Dick Williams, barking, 'Get your bat!' I was thinking, *Damn, give me a break, let me enjoy this for a minute.* I was glad that Nolan was taken out by the time I got in."

Francona made his big league debut, leading off the ninth against Dave Smith. He grounded to Houston first baseman Cesar Cedeno. Dick Williams was not impressed. A couple of days later, Little Tito led off, went 0-5, and got thrown out at the plate in a 1–0 loss. Williams liked the kid's enthusiasm, but never shed the veneer of toughness. When Francona failed to get a bunt down, Williams let him have it.

"I was thinking, *This guy hates me.*"

But Williams did not hate Little Tito. After days of silence, the kid saw some approval when he hit a routine single down the right-field line, noticed All Star outfielder Dave Parker jogging after the ball, and stretched his hit into a double. As he got up from his slide into second, Francona looked out of the corner of his eye and saw Williams jumping out of his seat. Williams immediately gathered himself and sat back down, but he'd made a statement.

Those who spend time around baseball come to appreciate the difficulty of making it to the major leagues. Millions of young boys dream of playing big league ball, and most of them play some form of Little League, where the process starts of kids moving ahead and kids being left behind. In America the best players advance to Babe Ruth, Legion, high school, AAU, maybe college ball. If you are the best of the best, you get signed to a professional contract, then face more years of trying to move up the system. It takes talent and mental toughness. Some prospects can make it on sheer ability; others move past more talented kids because of their mental makeup. By the time a young man makes it to the major leagues, he's vaulted past hundreds of thousands of ballplayers who started off thinking they would play baseball for a living.

Terry Francona is self-deprecating when he talks about his baseball career. It's the way he was raised. But it's a lie. A gap-hitting corner outfielder, equally effective against righties and lefties, he was a superstar at every level of amateur baseball. He swung at almost everything, and he almost never struck out. ("As soon as I saw anything straight, I swung — I was the anti-poster-child for *Moneyball*.") He was the best player in the NCAA and later named a member of the College World Series Legends Team. In parts of ten seasons in major league baseball, he hit .274 with 16 homers and 143 RBI, playing 708 big league games for the Expos, Cubs, White Sox, Indians, and Brewers.

"There's one other stat that nobody knows about," said Dustin Pedroia, who was born in California in 1983 and would become Francona's favorite player as a Red Sox. "He is the only player with a minimum of 1,000 plate appearances to never work an 0-2 walk. How awesome is that? He had no fight in him. None! That's unbelievable."

Kidding aside, Francona might have evolved into a star at the big league level, but his career was derailed in his first full season in the bigs in 1982 when he crashed into a wall chasing a fly ball hit by Julio Gonzalez at Busch Stadium in St. Louis.

"I was playing left field," he said. "I went back on the ball and went to jump and planted on the warning track, and it gave way and I felt my right knee explode. My momentum took me into the base of the wall. When I hit that wall, I thought my life was over. It looked like a cartoon. They had to peel me off. I'd never felt pain like that. It was completely torn. Every time I went to Busch Stadium after that, we'd stay at the Marriott, which had a view of the field. I could look out my hotel window and see where I left part of my knee."

Francona was 22 years old, playing every day, and hitting .321 when he tore his anterior cruciate ligament and meniscus. He was never the same player. In June 1984, he blew out his other knee, running down the first-base line, avoiding a tag by John Tudor.

"From then on, my career was trying to hang on," he said. "I couldn't hit for power, and I couldn't run. As much as I wanted to be a good player, I wasn't. I realized at an early age that you have to produce, and I couldn't. I wasn't helping the team. I think maybe I started to get the message in 1989 when Rawlings sent me a right-hander's glove. I'm left-handed!"

No one tried harder. Francona needed as many as ten rolls of tape just to get himself on the field. But his legs were gone, and so was most of his game.

His ball-playing days were almost over by the time he first saw Fenway Park when he was with the Indians in 1988.

"The first day of that series I came to the ballpark with Bud Black," he remembered. "The cab let us off, and we worked our way in from some gate in right field. We walked under the stands, go to the tunnel that connects the clubhouse to the dugout, across those clattering boards. I always remember those clattering boards. Everybody who ever played at Fenway knows what I'm talking about. They fixed a lot of things in the ballpark over the years, but those boards always sounded the same. You could hear somebody coming from the dugout. I bet Babe Ruth walked on those same boards. It was a thrill for me.

"When I got to the dugout and looked out and saw the park, I thought I'd died and gone to heaven. I hit against Mike Boddicker in that series. And against Roger Clemens. I pinch-hit, and I fouled off a ton of balls. He finally got tired of it and threw one about 100 miles an hour, and I struck out. A year later, when I came back with the Brewers, I went out to left field early in the afternoon when we were taking early hitting and climbed up that ladder that they used to use to retrieve the home runs that landed in the net. This was before the Monster Seats were up there. I just wanted to go to the top, and it was stupid. The thing is like 37 feet high. You know how high that seems?

When I was climbing down, Robin Yount and B. J. Surhoff were throwing baseballs at me. I was like, 'Hey, fuck you guys, man. I'm scared up here.'"

"He was a great teammate, always positive and witty," said Black. "He always had something that would lighten the mood, and that's important in a 162-game season. Plus, no matter how bad his knees were, you could see he was a hitter. He always put the bat on the ball."

"It was always an easy conversation with Tito," recalled John Farrell, a pitcher with the '88 Indians who would be the Red Sox pitching coach under Francona for four seasons (2007 to 2010). "There was a lot of commonality with us. For some reason, it seemed like he gravitated towards pitchers. There was no direct competition on the roster, and it just clicked. We'd grab lunch before games and have a beer after games. I never saw him before he was hurt. With us, he was a bench player who could hit righties and lefties. All of our conversations centered around the game. He knew the game. He taught me a lot. A lot of things were new to me then, and Terry was always helping someone. He was a guy you felt comfortable being around because of his experience and his willingness to share that. He has a keen, intuitive feel for the game, and he had it even then."

Francona enjoyed his final days in the majors playing for young Brewers manager Tom Trebelhorn. He knew he was the 25th man on the roster, so he did all the little things. He pitched an inning against Oakland ("I struck out Stan Javier on a knuckleball") when Trebelhorn ran out of bullpen arms. He strapped on catcher's gear and volunteered to work behind the plate, which would have made him one of the few left-handed catchers in baseball history.

He played in 90 games for the Brewers in 1989, including an early-season game in which he got himself ejected after he was intentionally walked.

"We'd had a dispute with Ken Kaiser early in the season," Francona remembered. "About two months later, he was working home plate for one of our games against the White Sox. I was sent up to pinch-hit, and first base was open, so they ordered an intentional walk. Kaiser and I were yelling at each other the whole time. Every pitch. Carlton Fisk was catching, and he couldn't believe it. Finally they threw ball four, and as I was starting down the line to take my base, Kaiser said, 'I make

3,000 calls a year!' and I hollered back, 'And 2,000 of them suck!' That did it. He tossed me."

Trebelhorn didn't mind. Late in the season, he used the oft-limping Francona as a pinch runner.

"That was Trebelhorn being nice," Francona remembered. "With about ten games left, I'd mentioned to a couple of guys that I needed to get in eight more games to trigger a $25,000 bonus. The next day I pinch-ran. And again for another bunch of games. I remember thinking that Trebelhorn was crazy, but then I pinch-ran in the next-to-last game of the season and got my number. I mentioned something to Trebelhorn about it, and he said, 'It's my way of saying thanks.' That went a long way with me."

The non-stars notice the small things. That's why they often become the best managers.

In April 1990, Francona stepped to the plate in the eighth inning of an 18–0 Brewers rout of the Red Sox. It was Marathon Day in Boston, the only 11:00 AM start in baseball every year. Facing right-hander John Leister, Francona went out on a sinking liner to center. A few days later, Trebelhorn called him into his office and told him he was being released.

The fly ball in Boston was Terry Francona's final major league at-bat. He'd finished his playing career on the same diamond where his dad made his big league debut after meeting Ted Williams in 1956.

But that harmless fly ball to center in an 18–0 ball game was not Terry Francona's final deed at Fenway Park. He was just getting started. He was going to earn a place in Boston baseball history.

# CHAPTER 3

# "They're not going to fire a guy over one mistake"

JUST A FEW MONTHS removed from his final big league game at Fenway Park, Terry Francona was climbing his way down the ladder of professional baseball. He still enjoyed the game, but knew there weren't many more days of seeing his name in any lineup. In the summer of 1990, Terry and Jacque had three small children and lived in a townhouse in Tucson while he made $15,000 playing right field for the Louisville Cardinals. There were plenty of signs that his playing days were almost over, like the time he wiped out his first-base coach as he tried to beat out an infield hit, pulling a hamstring. Francona's minor league skipper, Gaylen Pitts, pinch-hit for him on multiple occasions, a slight that would have provoked a tantrum from most players with almost ten years of big league experience. Instead, Francona did everything he could do to help the team. He warmed up pitchers and counseled slumping prospects. When Louisville ran out of pitchers, Francona toed the rubber, compiling an ERA of 1.17 in seven and two-thirds innings on the mound.

There were no complaints from the ex-big-leaguer. He was happy to be playing baseball every day, and he was learning new skills. He was also starting to think like a manager, evaluating the young players who were hoping to make the big leagues.

Rare is the player who knows when it's time to quit. You always think there's one more season in your broken-down body. After playing the 1990 season in the minors, Francona went home and had surgery on his knee, shoulder, and wrist. He was going to give it one more shot in 1991.

He went to spring training with the Cardinals minor leaguers in 1991, but it was obvious that he was through. Playing in a "B" game on a back field in Dunedin, Florida, he hit an RBI single and a fluke triple ("The ball bounced off a pole and rolled forever," Francona explained) before coming to bat one last time. Batting with the bases loaded in the sixth inning, he fouled a ball off his right knee and felt it balloon instantly. He hit the next pitch over the fence and could barely stagger around the bases before coming out of the game.

Cardinals GM Ted Simmons called him into his office the next day to notify him that he'd been released.

"I was pissed," Francona said. "They gave me a ticket to Phoenix, and my home was in Tucson. And they knew they were going to release me before I played in that last game and I hurt my knee badly. I never would have been able to play again after that contusion."

The cold manner in which Francona was released stuck with him. He pledged to do it more gently and generously if he ever found himself in a position of baseball power.

Unlike many athletes whose careers are cut short by injuries, Francona didn't waste time thinking about "what might have been."

"I always felt everything was an opportunity for me," he said. "Even when I had to go back to Triple A. I always felt like, if I was good enough, I would have found a way to be good. In Montreal, I stood in the training room with Andre Dawson, and we used to get our knees taped in an identical way. He went out and played and was an All-Star and a Hall of Famer. I'd sit on the bench and get released. I never felt like it was unfair. I changed my goals and tried to be valuable enough to be on a team, but I never felt bitter. When something else happened, I'd move on and look at it as an opportunity. At the end, I looked around and realized I really wasn't that good. There were guys there that I thought were better than me that hadn't sniffed the big leagues. I figured I should just count my blessings."

Done with major league baseball, he went home to Jacque and the kids, flopped on the couch, and started watching reruns of *Gilligan's Island*.

"I had $80,000 in the bank," he said. "I figured we were all set."

Jacque did not think they were set. She talked Terry into taking a real estate course. It was a bad fit. Francona is not a salesman, and he is not a negotiator. He enrolled in night school, a six-week course that met for three hours twice a week.

"I was just busying myself," he admitted. "I had no intention of ever being a real estate agent. I was prepared to do something else when it was over, but one night during a break I called home and Jacque said, 'You'd better call Buddy [Bell]. It's pretty important.'"

Bell offered him a job as hitting instructor for the White Sox Gulf Coast Rookie League team. He didn't ask Bell how much money the Sox were paying. He went back to real estate class, dumped his books on the teacher's desk, and recommended he give the texts to another student.

"No one's going to buy a house from me anyway," Francona added.

He flew to Florida the next morning. The White Sox paid him a total of $9,000 for the next three months.

He loved being back in baseball, teaching hitting to enthusiastic young ballplayers. He learned organizational skills.

"Up until then, I thought you just played the game," he said. "That was all I knew."

A year later he was managing in South Bend. That was the season his mom was losing her long battle with breast cancer.

"She called me the day before the season was over. I could tell her memory was going. I told her I was coming home the next night. Our season ended and I drove home, and by the time I got there her memory was gone. I stayed home four or five days with my dad. He told me not to stay because he didn't want me to see her like that. She went back to the hospital, and I got in my car and drove back home to Arizona, and that's when she died.

"She was the perfect mom. She was a saint. I am still trying to figure out how she got pregnant. She was both mom and dad for a lot of years because my dad was gone so much."

The young manager was promoted to Double A Birmingham in 1993, making a whopping $32,000 per year in the Southern League.

Francona's Birmingham Barons won the Southern League Championship in his first season, and *Baseball America* named him its Minor League Manager of the Year. That winter Jacque gave birth to their fourth child and third daughter, baby Jamie.

There was nothing "minor," however, about Terry Francona's second summer in Birmingham. It was the year Michael Jordan took a leave of absence from the NBA (he "retired" in October 1993) to play baseball and was assigned to play for the White Sox Double A affiliate in Birmingham.

"The Summer of Michael" taught Francona everything he would need to know about dealing with the mass media and the sports star culture of the late 20th century. Traditionally, minor league baseball is a place where mistakes can be made and repaired without fanfare. No one cares if your star outfielder is pouting, or if he invents a hamstring injury to get himself a day off. Fistfights on buses and in clubhouses go unreported. Everything is managed behind closed doors, far from daily tabloids and the glare of ESPN.

Jordan was the most famous athlete in the world. He was accustomed to a life of luxury. The World Champion Chicago Bulls traveled in a private jet, and the 1992 Olympic Dream Team — Jordan was the top attraction — had trained for the Barcelona Olympics in Monte Carlo.

When Francona had his first meeting with Jordan in Birmingham, Jordan's first question for his new manager was, "Do we fly?"

"That wasn't the question I was expecting," said Francona. "We had major bus trips everywhere. The shortest ride was three and a half hours. It was 16 or 17 hours from Memphis to Orlando, and we did that. I told him we bused everywhere. He came back later and said, 'What if I can get us a better bus?' The next day there were four buses in the parking lot. It was a bus audition. One of the buses was for a touring rock band. We ended up riding in a new bus, but I'm sure he didn't have to pay for any of it. It wasn't even the best bus in the league. Greenville's bus had beds. We just had a newer bus. Michael signed the door, so they called it 'the Jordan cruiser.'"

Where to sit was another matter.

Bus seating is part of the routine of baseball. The manager goes up front. Before 1992, most managers sat in the first row on the right side, opposite the driver. That changed for a lot of skippers in May 1992 when Angels' manager Buck Rodgers was badly injured when the lead team bus veered off the New Jersey Turnpike on a New York–to–Baltimore trip. Rodgers survived the crash, but his ordeal reminded managers that the person in the front right was the most exposed individual on the bus. When he became a manager, Terry Francona preferred the window seat in the second row behind the driver. From there he could talk to the driver without being too close to him, steer clear of disgruntled players, and stretch his ever-swollen legs toward the front or across the side — often across the ubiquitous cooler of beer.

Already a three-time NBA champ (he would win three more after baseball), Jordan was making $30 million per year in endorsements when he joined Francona's bus-riding Barons in the Southern League in '94. The Barons paid Jordan $850 per month and 35-year-old Terry Francona told Michael he had to run out his pop-ups.

Jordan was in the White Sox system because White Sox owner Jerry Reinsdorf also owned the Chicago Bulls. The White Sox made just a couple of requests of Francona: Don't bat Michael in the number-nine spot — too embarrassing. And don't refer to the experiment as a "circus."

It could have been a circus, but Francona managed to keep things relatively normal. *Hard Copy* loved the Barons. Tom Brokaw would appear, and Francona would say, "Michael shows up on time, he works hard, and he's a great guy." And that was that. It was an adjustment for the rest of the minor leaguers — who are generally happy if their games are broadcast on the radio — but the manager was not going to be starstruck. Francona's biggest concession to Jordan was to provide an all-access pass for Michael's bodyguard, George Koehler. Jordan's main contributions were the luxury bus and a club record 467,867 fans. The Barons finished 65-74. Jordan hit .202 with 51 RBI and 30 stolen bases.

"Michael respected what we were doing so much, and that made it work," said Francona.

After the season, the White Sox asked Francona to manage Jordan again with the Scottsdale Scorpions in the Arizona Fall League. It was there that Francona first encountered Nomar Garciaparra, a young shortstop out of Georgia Tech who'd signed with the Red Sox.

Francona enjoyed his time with Jordan. They played golf together in Scottsdale with White Sox GM Ron Schueler and professional golfer Billy Mayfair. The manager beat Jordan out of $800 on the 18th hole when Mayfair got Francona out of a sand trap by throwing the Titleist onto the green when Jordan wasn't looking.

A couple of times Francona even played in pickup basketball games with Jordan. Michael got angry once when an exhausted Francona took the last shot in a best-to-11 game.

"I always take the last shot," said Jordan.

"Now you know how I feel when I watch you try to hit a curveball," said the manager.

After the 1995 season in Birmingham, Francona returned to the big leagues as third-base coach of the 1996 Detroit Tigers under manager, and old friend, Buddy Bell. The '96 Tigers were one of the most buffeted teams in hardball history, losing 109 games and providing nightly fodder for the wiseguys on ESPN. Francona was just happy to be back in the majors. The money was better. No more five-hour bus rides or Comfort Inns.

One of his favorite memories from the 1996 season was throwing batting practice to 11-year-old Prince Fielder in Tiger Stadium. Prince was the son of Tiger wideload first baseman Cecil Fielder, and the little big man put a couple of balls into the upper deck. Francona teased Prince when the youngster failed to pick up the balls in the cage after hitting. Picking up the balls in the cage after hitting is a universal practice in professional baseball — even if you are an 11-year-old future big league millionaire named Prince.

After the 1996 season, Francona got an important phone call from Tigers GM Randy Smith as he was driving from Detroit back to Tucson to join Jacque and the four kids.

"I was in Albuquerque and about up to my neck in Taco Bell wrappers when Randy called me," Francona recalled. "I thought I was getting fired because we'd lost all those games. He said, 'Do you know anybody with the Phillies? They want to interview you.' I told him I didn't know anyone with the Phillies. He said, 'Well, give them a call.' I immediately called Buddy Bell. He knew why I was calling. I asked what I should do, and he said, 'Go interview. You won't get the job, but it will be good experience.' About three weeks later, I was the manager of the Phillies. I hadn't had any major league interviews. I'd only been a major league coach for one year."

The Phillies had fired Jim Fregosi after the '96 season and were looking for a young talent to steer them through a rebuilding phase. They'd been in decline since losing the 1993 World Series against the Toronto Blue Jays.

Francona was surprised to be considered. Phillies GM Lee Thomas had played many years against Tito, but Terry remembered only one encounter with the Phillies boss. Back in '94, when he was managing Jordan in the Arizona Fall League, Thomas had asked Francona for a baseball autographed by Jordan.

"I'd have given him a whole autographed bat if I knew it would have given me a chance to manage," said Francona.

Thomas checked with many of his baseball friends. He got a strong recommendation from Joe Torre, manager of the World Champion New York Yankees.

Francona was 36 years old when the Phillies named him manager. He was only five years removed from watching Gilligan, the Skipper, and Thurston Howell III on his townhouse couch in Tucson.

Armed with his first multi-year contract and a $50,000 mortgage loan from Phillies owner Bill Giles, he moved Jacque and the kids from Tucson to Yardley, Pennsylvania, just 32 miles from Veterans Stadium in downtown Philadelphia.

Francona's four seasons as manager of the Phillies produced an aggregate record of 285-363. The Phillies never contended and never finished above .500. The media was tough, and the fans were worse. Former Phillies World Series hero Larry Bowa had been the scrappy favorite to replace Fregosi. Shortly after he was hired, Francona was booed at a Philadelphia 76ers game when his image appeared on the Jumbotron above center court. His tires were slashed on Fan Appreciation Day at Veterans Stadium. He was ridiculed for giving Scott Rolen a day off on Scott Rolen Bobblehead Day. He was ripped when he let Bobby Abreu sit out on opening day against Randy Johnson because Abreu didn't want to face the fearsome southpaw. There was a popular notion that star pitcher Curt Schilling was running the team. All the experts said Francona was too much of a players' manager. He didn't have enough rules. There wasn't enough discipline. He played cards with his ballplayers. Philadelphia sports talk jock Angelo Cataldi crushed Francona almost daily.

"They started moving my parking spot so I could sneak out the door," Francona remembered. "People were getting aggressive."

"It was hard seeing him get kicked around like that," said Mills, who coached first base for Francona in Philadelphia. "He was doing a damn good job with what he had. The wiseguys on the radio didn't know him and didn't know what he was trying to do."

Francona had a roster peppered with players who struggled at the big league level. His closer, Wayne Gomes, was rumored to be a fan of eating hot dogs in the bullpen during games.

"I brought him into a game one night, and he had mustard on his uniform," Francona recalled. "I told him he had to cut that out, and he claimed the mustard got on him when a fan threw a hot dog at him. The worst part of the whole story was — we were playing at home! I remember walking back to the dugout thinking, *Boy, this is where I'm at in my career. My closer has mustard on his jersey.*"

Managing the Phillies was nothing like coaching third base for Buddy Bell in Detroit. He took the losses personally. Many a night a fretful Francona, lost in thought, would zip past the Yardley exit on Route 95 North, then turn back when he started seeing signs for New York. After losing 97 games in 2000, Francona was fired by Phillies general manager Ed Wade.

"I was fired on the last day of the season, before the game," Francona remembered. "After the game, I went golfing with two of my coaches, Chuck Cottier and Millsie. It was weird because I knew the firing was coming. I thought, *This will be good. I'm done. I can take a deep breath. This will be a relief.* But it wasn't. I realized that I had spent all this time with these people and I had all these emotions. It was hard for me. I didn't handle it very well."

He quickly landed another baseball job, one that didn't require him to wear a uniform. In 2001 Francona served as a special assistant under Cleveland executives Mark Shapiro and (GM) John Hart. It gave him a new look at the inside operations of a baseball team. He traveled to minor league affiliates. He sat in on the draft. He scouted other major league teams, searching for a new center fielder for the Indians. (The Phillies provided citizen Francona with a security guard when he sat behind home plate at Veterans Stadium.) He ultimately recommended that the Indians acquire Milton Bradley, prophetically reporting, "Nobody comes close with tools, but my inability to connect with him throws up a little bit of a red flag."

Francona's first year out of uniform was a good learning experience, and more fun than he expected it to be.

He returned to the big league dugout as Jerry Narron's bench coach with the Texas Rangers in 2002. This was somewhat awkward because a bench coach is generally a best friend and aide-de-camp of the manager and Francona was neither. He didn't know Narron particularly well and resisted the opportunity when first approached by then–Texas GM Hart. Narron said he was comfortable with Francona, so they worked the '02 season side by side in Arlington, Texas. The Rangers finished last, Narron was fired, and when Buck Showalter was hired, the entire staff went looking for new work.

Francona felt he was ready to manage again. Mets general manager Steve Phillips called him to interview for the job that came open when Bobby Valentine was fired. (The irony of this would not surface until 2011.) Francona knew he had no shot at the Mets job, but he went anyway and enjoyed his give-and-take with Phillips. The Mets boss explained that sometimes you have an interview for "down the road." Francona appreciated Phillips's candor and generosity with his time. Art Howe got the job — the same Art Howe who won 102 and 103 games his last two seasons working for Billy Beane.

Seattle was a more realistic option for Francona. The Mariners needed a manager, and Francona believed he had a chance. But something happened the night before he was scheduled to meet with Seattle boss Pat Gillick. Making notes, preparing thoughts in his Seattle hotel room, Francona felt pain in his chest. Then there was pain in his arms. Sweating, he told himself, *I came all the way out here to have a heart attack. I should have just done this at home.*

He stayed up all night. In the morning he rode on an exercise bike in the hotel fitness room and did a few push-ups. He arrived at the interview 90 minutes early, but the pain and the flop-sweats came back when he started answering Gillick's questions. Several times he asked Gillick to repeat himself.

"He interviewed all day, half-dead," said Jacque.

"To this day I tease Gillick that his questions almost killed me," said Francona.

It was no laughing matter. Francona's multiple knee surgeries had left him with blood clots, staph infections, and internal bleeding. When he had both knees scoped to alleviate staph infections in October after the 2002 season, the procedure led to a pulmonary embolism on each side of his lungs. Within a month of the Gillick interview, he was in the hospital and doctors were considering the amputation of his right leg. After a major surgery in which his leg was split open to reduce pressure and allow fluid to drain, he spent three weeks at home in bed, unable to go downstairs or to make it to the bathroom by himself. There was massive clotting. He stopped worrying about the Mets and the Mariners. He worried about walking with a limp for the rest of his life, losing his leg, or dying.

"I think I probably should have died with all that happened," he said. "There were a couple nights in the hospital where I was thinking, *I can't take this anymore.* The nurses would come running in because I'd stop breathing. I was in bad shape. There were people around who did not think I was going to make it. I know I came real close to losing the leg. Sometimes stories get enhanced, but that one actually gets downplayed."

Pain medication was a critical part of recovery.

"I lived on it at that time when I was in the hospital," he said. "I learned every trick in the book, getting a lot of help from the nurses. You keep giving yourself a blot to get through the next 20 minutes. When I left the hospital, I was on heavy-duty drugs, and it was tough."

In recovery after Christmas, able to walk around a little again, he heard from new Oakland manager Ken Macha. Francona and Macha had first crossed paths in Francona's final spring training with the Expos in 1986. Macha was a roving coach/instructor with the Expos and was sympathetic toward Francona, a fellow western Pennsylvanian. Francona was cut loose before the start of the season and eventually signed with the Cubs, but Macha never forgot about him. When they managed against one another in the Arizona Fall League in 1994, Macha told Francona that he'd keep him in mind if he ever had a chance to manage in the majors. Francona didn't think much about it at the time. A lot of friends talk that way. It seemed unlikely. But when

Macha got the job in Oakland, he called Francona and offered him a job as bench coach.

It was a good offer. Francona knew he wasn't going to be managing in the majors in 2003. Sitting alongside Macha with the playoff-bound A's looked a lot better than lingering in last place in Arlington, Texas. But Francona wasn't healthy, and his recovery was slow. A couple of weeks before spring training, Francona called Macha to remove himself from the Oakland coaching staff. He told Macha it wouldn't be fair. He was unable to walk to his car. How was he going to help Macha in his first managing gig with the contending A's?

Macha told Francona not to worry. The A's didn't need Francona to throw batting practice or hit grounders to infielders. Macha wanted Francona for his baseball mind and his positive attitude.

"He saved me," said Francona. "He said, 'I don't give a shit. Just come out and be my bench coach.' You have no idea how that made me feel. I hung up the phone and told my wife, 'I've got to get healthy enough to go out there.' When I flew to Phoenix to join them, the walk to the rental car was the longest walk I'd made in three months. Nobody had any idea how sick I'd been."

At the start, he had trouble bending over to pick up a baseball. He wore a helmet while pitching batting practice. His legs would swell and stretch his pants by the end of the workouts. At night he'd go back to his hotel room and elevate his legs until morning. He was miserable, and it was hard to hide his limitations. Games at the Metrodome in Minnesota were particularly difficult because of a series of steps that led from the dugout to the clubhouse. Everybody in baseball knew about the Metrodome steps. Cal Ripken Jr. made them an Olympic event, trying to get to the clubhouse in a minimal number of strides. For Francona, those steps were Kilimanjaro. It would take him as much as five minutes to get from the dugout to the visitors' clubhouse in the Metrodome. He'd walk three or four steps, then sit and rest. He was embarrassed.

"I was on so much pain medication," he remembered. "Oxycontin. I weaned myself off that year, but it took almost half the season. I needed it. Anybody who had my body is going to have to have a certain amount of medication. I never took oxy again after that. It scared me. But I could see why people get hooked.

"I learned how to maintain. My right leg is so damaged, so many clots went through there. Sometimes it gets so swollen, my leg barely fits in my pants. I've got a degenerative hip, but that's so far down the list I can't even get to it. I'm cold all the time because of the blood thinners. It's all uncomfortable, but it's not going to make me die now. It just pisses me off."

Francona's health improved as the 2003 baseball season played out. Macha gave him a lot of responsibility, and the A's of Tim Hudson, Mark Mulder, and Barry Zito were good. Francona handled charts, dealt with players, and found himself enjoying baseball again. This was the year when Michael Lewis's *Moneyball* hit the national bestseller list, and Francona took delight in holding a copy of the tome in front of his face every time Macha and Billy Beane came aboard a team bus or charter aircraft.

Francona wasn't with the A's when Lewis did his reporting on the 2002 Oakland season, but he saw what it was like for Macha to work for Beane in 2003.

"I don't think the portrayal in the book was an exaggeration," said Francona. "There were times when I'd walk by Macha's office after a game and Billy would be sitting there with a lot of strong opinions flying. I'd come in after and tell Ken, 'I know what I'm supposed to be doing is cheering you up. I wish I could do a better job of it.'"

Francona's place alongside Macha on the A's bench gave him a first-hand look at the team he would manage in 2004. The Red Sox and A's were first-round opponents in the 2003 playoffs, a series that unfolded in spectacular fashion, with the Red Sox winning a fifth-and-deciding game, 4–3, on a Monday night in Oakland. The A's won the first two games at home and looked ready to wrap up the series at Fenway when bad things started happening to the Western Division champs. Base-running blunders crushed Oakland in Game 3. (Francona was one of a group in the Oakland dugout who tried to get Eric Byrnes's attention after Byrnes failed to touch home plate in the game's most crucial play.) The A's looked like they had things in hand the next night, but Oakland closer Keith Foulke couldn't hold a 4–3 lead in the eighth, and the series went back to Oakland for a fifth-and-deciding game.

The future manager of the Red Sox got an inside look at Boston's backdoor operations after the A's dropped Game 4 in Boston on Sunday. It was a quick turnaround, with both teams flying coast to coast to get back to the Bay Area for a winner-take-all game scheduled to start 3,000 miles away late Monday afternoon. The A's trip turned out to be more difficult than that of the Red Sox. Oakland's team buses left Fenway for Hanscom Field in Bedford, Massachusetts, where they encountered extraordinary security measures. No one will ever admit it, but Francona learned later that the A's delay might have been due to the influence of Boston's longtime traveling secretary, Jack McCormick. McCormick is a former Boston police officer and deeply connected with greater Boston's security and aviation networks.

"Everybody in our party was almost cavity-searched," said Francona. "We sat on that plane for three hours. I guarantee you the Red Sox got to the West Coast two hours quicker than we did. After I came to Boston, I realized what kind of pull Jack has. He was pretty proud of himself about that one."

"No comment," said McCormick. "Just a random security crackdown, I'd guess."

Boston won Game 5 in Oakland on the strength of Manny Ramirez's prodigious three-run homer in the sixth inning. There were some hard feelings at the finish when Derek Lowe grabbed his crotch after fanning Terrence Long with the bases loaded. Francona didn't even notice. Francona couldn't believe the A's lost to the Red Sox, and it was going to hurt him in the wallet. Losing Game 5 meant his bonus check was around $20,000 instead of something in the neighborhood of $80,000.

By the time the Red Sox charter left the Bay Area, bound for an epic, fateful seven-game series with the Yankees, Francona was in his blue Mercedes sports utility vehicle, starting a four-and-a-half-day drive home to Yardley, Pennsylvania.

The soon-to-be Red Sox manager was in a cold car, outside his daughter's volleyball practice, when he first started listening to the Red Sox and Yankees playing Game 7 in Yankee Stadium on Thursday night, October 16. The big game was not appointment television for the Oakland bench coach and father of four. He was still mad about losing to the Red Sox and hadn't followed the American League Championship Series very closely. He was blissfully unaware of the tension back in Boston or the nail-biting Lucchino sitting with eccentric Sox owner

John Henry in the lower bowl at Yankee Stadium. This was their fight, not his.

Late in the night, the Sox were five outs from a trip to the World Series. They were set to win the American League pennant on Yankee soil. They led the Pinstripes, 5–2, going into the bottom of the eighth with Pedro Martinez on the mound. Martinez almost never spit up a lead of three or more runs. This game was in the bag. Back in Boston, members of the Fenway Park grounds crew had already stenciled the World Series logo into the grass behind home plate. Throughout New England, fans were preparing to celebrate the ultimate victory. After all the pain inflicted by New York and the Yankees — going back to the sale of Babe Ruth in 1920 — winning a pennant at Yankee Stadium would be the sweetest of victories. Lucchino had dubbed New York "the Evil Empire," and George Steinbrenner had responded: "That's bullshit. That's how a sick person thinks. I've learned this about Lucchino: he's baseball's foremost chameleon of all time."

Francona was happy for Grady Little. He had roomed with Grady's brother, Bryan "Twig" Little, when the two played for Double A Memphis. Francona and Twig Little had met Fidel Castro while playing for a college all-star team touring Cuba in 1979. During their Expo years, Francona would stop in Texas and pick up Twig for the long drive to spring training in West Palm Beach. Francona and Grady Little were minor league managers at the same time. They both won *Baseball America*'s Minor League Manager of the Year Award, Little with Greenville in '92, Francona with Birmingham a year later. In the fall of 1992, Francona and Grady Little had lived together when Terry served as Grady's bench coach with the Grand Canyon Rafters of the Arizona Fall League. They spent a lot of hours driving to and from the ballpark, always stopping at Circle K so Grady could buy some lottery tickets. Little had an easygoing personality and told a lot of stories about farm life, but he never scored big on any of his lottery tickets. He was not a particularly lucky guy.

The Fall League is not about winning baseball games. It's about identifying and developing individual talent. Pitchers become shortstops, and first basemen become outfielders. There's not a lot of strategy, and no one uses Bill James spreadsheets to prepare for the opposition. ("We didn't even know who the hell the guys on the other team were," remembers Francona.) Francona liked the way folksy Grady Little treated his players. He gave his young players a lot of room, and they liked playing for him.

Terry Francona had returned from volleyball practice and was padding around his house in Yardley, Pennsylvania, when Pedro struck out Alfonso Soriano with his 100th pitch to end the seventh in Yankee Stadium. Francona wasn't glued to the game, but he liked to keep a TV on in every room in the house. Francona was at his computer, playing online cribbage, not paying much attention to the game, when Pedro Martinez walked off the mound and pointed to the heavens after fanning Soriano to end the seventh.

Everybody who watched the Red Sox knew what it meant when Pedro came off the mound pointing toward the sky. It meant he was through for the night. Lights out. Case closed. Crack open an El Presidente.

Grady Little had another notion — a gut instinct that defied the mountain of data on his desk. He was afraid of Scott Williamson, Mike Timlin, Alan Embree, and the rest of the arms in his bullpen. He wanted Pedro to give him another batter or two in the eighth, or maybe another full inning.

On paper, this was a bad idea. The numbers were clear: In 2003 Pedro turned to dust after his 105th pitch. Batters hit .370 off him after he passed the magic number. But Little didn't like the numbers, and he resented the young executives who never played the game telling him what to do. He was amused by Henry, the stat-driven owner, and Lucchino, the hard-driving CEO, but he never confronted any of them. He just did the job the way he'd learned from managing almost 2,000 minor league games. Seasons in Bluefield, Hagerstown, Durham, and Richmond had taught Grady Little more than Bill James's *Baseball Abstract*. All the data supplied by baseball ops was insulting to scouts and baseball lifers who beat the bushes and trusted their eyes. Little didn't like the geeks telling him what to do.

He also didn't think much of the alleged "pressure" of a major league baseball game. Before Game 7 Little told the media, "When you're standing out on your porch and watching that storm coming and you know what danger your crop is in, that's pressure."

The cotton farmer let Pedro come out for the eighth. The storm was coming.

Pedro got Nick Johnson to pop up for the first out, then surrendered a double to Derek Jeter and an RBI single to Bernie Williams as New York cut the lead to 5–3. Pedro was up to 115 pitches when Little came out of the third-base dugout. The conversation was brief, and then, to the surprise of everyone in Red Sox Nation, Little patted Pedro on the back, turned, and went back to the Red Sox dugout.

In his box seat in the lower bowl of Yankee Stadium, a furious Henry turned to Lucchino and asked, "Can we fire him right now?"

It got away quickly after that. Hideki Matsui hit a ground-rule double to right, and Jorge Posada tied the game with a bloop double to center on Pedro's 123rd pitch. Little finally came out to get Martinez.

The Red Sox lost in the bottom of the 11th when Aaron Boone launched a Tim Wakefield knuckleball over the wall in left.

Two nights later, with the 2003 WORLD SERIES logo still embedded in the grass behind home plate at Fenway, David Wells threw the first pitch for the first game of the 99th World Series, played between the Florida Marlins and the New York Yankees at Yankee Stadium.

While the Marlins and Yankees played in the Bronx, Terry Francona was at Georgetown University in Washington, DC, on a recruiting visit with his son, Nick, a star left-handed pitcher at Lawrenceville Prep. Georgetown coach Pete Wilk annually invited four or five top recruits and their parents to the Hoyas' fall-ball fund-raiser dinner. Boston multimedia personality Mike Barnicle was at the event with his three sons, including Nick Barnicle, a catcher on the Georgetown varsity. In between speeches, Barnicle approached Francona and suggested that Oakland's bench coach should give the Red Sox a call.

"It's a good fit," Barnicle told Francona. "You've managed in the big leagues, you coached with the A's, and you know the Red Sox. Your name ends in a vowel — Larry Lucchino will like that. You should give them a call."

"The Red Sox already have a manager," Francona told Barnicle. "They're not going to fire a guy over one mistake."

# CHAPTER 4

## "He kind of blew us away. . . . Is the guy too nice?"

**B**ASEBALL COMMISSIONER BUD SELIG has an unofficial rule prohibiting teams from making major announcements during the World Series. Selig doesn't like teams making moves that detract attention from the Fall Classic.

On Monday, October 27, the first business day after the conclusion of the 2003 World Series, the Red Sox fired Grady Little, who had won 188 games in two seasons.

"I didn't realize what was coming," Francona said. "I actually thought Grady handled it pretty well that night. When he left Pedro in the game and they lost, I thought, *Okay, that'll go away.* Now that I've been through it, I know."

Theo Epstein was the man in charge of replacing Grady Little. In the autumn of 2003, Epstein was already a rising front-office star who breathed youth and charisma into the stodgy offices on Yawkey Way. He was part of a new generation of baseball executives grounded in knowledge and data rather than experience on the playing field. Many of them were educated at elite universities and grew up reading about baseball, worshiping at the altar of Peter Gammons, the Hall of Fame

*Boston Globe* scribe who invented the Sunday notes columns that became a staple of newspapers across the country in the 1970s.

Theo and his twin brother, Paul, were born on December 29, 1973, children of brilliant, liberal parents who taught their sons to think independently and never root for the New York Yankees. The Epsteins are a family of letters and service. Theo's grandfather, Oscar-winner Philip Epstein, wrote *Casablanca* with his twin brother Julius. Theo's dad, Leslie, is a novelist who for many years served as director of the creative writing program at Boston University. Along with their older sister, Anya, the Epstein twins grew up in a roomy apartment building on Parkman Street in Brookline, Massachusetts. With no lawn of their own to mow, the boys played baseball and soccer at the nearby Amory Street playground and made a makeshift Fenway out of the concrete floor and walls of the parking lot behind the Holiday Inn along the Beacon Street trolley line. The Epstein twins were city Pony League champs as members of the Brookline Yankees, but were more accomplished as soccer players at Brookline High School, where the Warriors would make it to the state tournament in Theo's senior season.

When Theo and Paul were 12, they nervously watched the Red Sox play the Mets in the 1986 World Series. Fenway was just a couple of stops down the Green (C) Line, and the Red Sox were a passion. The twins were home alone on Saturday night, watching the World Series on television, when the Sox were set to finally win a championship. The boys hatched a plan for the magic moment: they wanted to be suspended in midair — not of this world — when the Sox finally clinched it. Every time the Sox got to within one strike of winning, the boys would leap off the couch. It was exhausting and frightening as Calvin Schiraldi, then Bob Stanley, delivered a series of pitches, none of which delivered the long-awaited grail. Over and over, the boys leapt into the air, only to crash to the floor, disrupting the neighbors downstairs and carving more pain into their preteen souls. They saw three consecutive Met singles, a passed ball that was ruled a wild pitch, then a hideous, unspeakable Little League error — the ground ball between Bill Buckner's wickets. There went Game 6. When the Sox lost Game 7 two days later, the '86 World Series had christened and damaged a new generation of Sox sufferers. It broke Theo's 12-year-old heart and made him a card-carrying member of a group that would come to be known as Red Sox Nation.

Theo's brother Paul grew up to be a caregiver for troubled youths at Brookline High School. His sister Anya married Oscar-nominated screenwriter Dan Futterman (*Capote*) and went to Hollywood, where she enjoys a career as a film and television producer and writer (*Homicide, In Treatment*). Theo mapped out a career in baseball that may land him a plaque in Cooperstown.

After graduating from Brookline High, Epstein entered Yale in the fall of 1991, landed a gig with the school's daily newspaper, and plotted to get himself into a baseball front office. Sitting in his dorm room in New Haven, it was easy to see that the Baltimore Orioles might be a good place to start. In '91 the Orioles were owned by Yale grad Eli Jacobs, and former Bulldog running back Calvin Hill was working in the Baltimore front office. Hill is famous for multiple reasons. He was Yale's running back in the infamous 29–29 Harvard-Yale game of 1968, and one of his fraternity brothers was George W. Bush. He played 12 seasons in the National Football League, made four Pro Bowls, and won a Super Bowl with the Dallas Cowboys. He is a featured character in Garry Trudeau's long-running *Doonesbury* cartoon strip. His wife was Hillary Rodham's roommate at Wellesley, and his son is NBA legend Grant Hill. That's a lot of celebrity for one individual, but Red Sox fans should embrace Calvin Hill as the man who got Theo Epstein his first baseball job.

Knowing his Yale roots might get him noticed, Theo wrote a letter to Hill in 1992, searching for an internship. Hill put the letter in front of Dr. Charles Steinberg, the Orioles' director of public affairs. Steinberg knew what it was like to get into baseball at a young age: he had grown up in Baltimore and as a 20-year-old was assigned the task of arranging Earl Weaver's "matchup" index cards — a skill that Theo and other Gammons Youth would turn into hardball science in the 21st century. Steinberg invited freshman Theo for an interview during Yale's spring break, and thus was born a relationship that had enormous impact on the World Champion Red Sox of the 21st century — before everything soured inside Fenway Park's ancient walls.

Epstein spent the summers of 1992 and 1993 in the front office of the Baltimore Orioles, assembling a project to pay tribute to baseball's

Negro Leagues, which had been shamelessly dismissed, even in the years after Jackie Robinson integrated the major leagues by coming aboard with the Brooklyn Dodgers in 1947. Epstein's Negro League project was a featured attraction when baseball's All-Star Game came to Camden Yards in the summer of 1993, and Steinberg insists that the exhibit was responsible for Negro Leaguer Leon Day's plaque in Cooperstown. The project also launched the career of the man who hired Terry Francona to manage the Boston Red Sox.

In Baltimore, young Epstein caught the eye of Larry Lucchino, the man who ran the Orioles and built Camden Yards. Soon after Camden was built, Lucchino went to the Padres, taking Steinberg with him. In San Diego in 1995, Steinberg called for Epstein, who was graduating from Yale. Twenty-one-year-old Epstein started at the bottom. He was responsible for the messages that appeared on the Jumbotron ("Julie, Will You Marry Me?") and monitored the whereabouts of "Flag Man," a Padre mascot. He also handed out press notes to baseball writers in the press box. He didn't have his driver's license and relied on Steinberg for rides to and from work. At the urging of Lucchino and Steinberg, he attended the University of San Diego law school.

Theo moved into the Padres' baseball operations department in 1997. He also graduated from law school, passed the bar, and was immediately offered a $140,000 position with an Anaheim law firm. He was making less than $30,000 with the Padres. Reacting to the offer, San Diego GM Kevin Towers bumped Epstein to $80,000 and made him director of baseball operations in 2000. In February 2002, after the Red Sox sale to John Henry, Tom Werner, and Lucchino was formalized, the Red Sox fired GM Dan Duquette, hired veteran company man Mike Port as interim GM, and hired Epstein as Port's assistant.

It was always understood that Port was a short-term solution for the new Red Sox owners. During the 2002 Sox season, Epstein reported directly to Henry and Lucchino, but he was not the GM-in-waiting. Henry had his eye on Oakland superstar GM Billy Beane. When he owned the Florida Marlins, Henry had been wildly impressed with a presentation Beane made to baseball executives. Henry loved the way Beane was able to get the cash-strapped A's into the playoffs for three straight seasons. Henry's entire life was rooted in mathematics, and he saw Beane as a kindred spirit.

When the Sox finished out of the playoffs in 2002, while Oakland won 103 games with a $41 million payroll, Henry went after Beane. Henry offered Beane $12.5 million over five seasons, and when Beane agreed, Henry uncorked a bottle of champagne. A day later, the Sox owner was shocked when Beane changed his mind and decided to stay with Oakland. The Sox had few realistic options. Toronto's J. P. Ricciardi, a central Massachusetts native, had also taken himself out of the running. Somewhat apprehensive because of Epstein's age, Henry and Lucchino agreed to take a chance on their boy wonder. On November 25, 2002, 28-year-old Theo Epstein became the youngest general manager in the history of baseball. He promised to build a "scouting and player development machine." In anti-Duquette fashion, he also said, "I'm not standing here thinking I have all the answers."

Theo's first game as GM was a disaster. With no established closer, the Sox were forced to go with a "bullpen by committee" — a collection of kids and veterans who'd never demonstrated they had what it took to finish games in the big leagues. It smacked of new-age Bill James arrogance, the Sox insisting that they knew more than anybody else. The "committee" coughed up a 4–1 lead in the ninth on opening day, losing when Tampa leadoff hitter Carl Crawford hit a three-run walk-off homer off Chad Fox.

"That was a kick in the gut," said Epstein. "Guys in the media were licking their chops because the whole bullpen story was so easy to write."

The 2003 Sox discovered David Ortiz, won 95 games, and came back from 0–2 against Oakland to make it to the seventh game of the ALCS. They came within five outs of making it to the World Series.

After coming so close, Epstein came out swinging in the winter of 2003–2004. The Sox successfully traded for stopper Curt Schilling and signed Oakland's closer Foulke (no more committee). They openly courted the Rangers' Alex Rodriguez. They placed Manny Ramirez on waivers. They made a three-team deal with Texas and Chicago in which Ramirez and Nomar Garciaparra would have been replaced by A-Rod and Magglio Ordonez. The deal was voided by Selig, however, and A-Rod wound up going to the Yankees on Valentine's Day, right before the start of spring training.

Epstein also made his first managerial hire: 44-year-old Terry Francona.

Francona had been on a couple of job interviews by the time Epstein first called. Two days after Grady Little's disastrous decision in Game 7 of the ALCS not to relieve a fatigued Pedro Martinez, Francona flew to Chicago to meet with White Sox general manager Ken Williams at a restaurant near O'Hare.

"It was a difficult interview," Francona said. "Every time I started to say something meaningful, the waiter would come by and ruin my momentum. You get a feeling of how things are going right away in a situation like that. I knew I wasn't going to be their manager."

The White Sox hired Marlins coach Ozzie Guillen and two years later won the 2005 World Series.

The day after the Sox fired Grady Little, Francona met with Jim Beattie and Mike Flanagan, co-general managers of the Baltimore Orioles. Flanagan and Beattie had already interviewed local favorites Eddie Murray, Rich Dauer, Rick Dempsey, and Sam Perlozzo. The Oriole GMs began the Francona interview by asking him how he'd feel about inheriting the coaching staff left behind by Mike Hargrove. This didn't sit well with Francona. He'd been "assigned" to Jerry Narron in Texas. He believed a manager should have input into the naming of a coaching staff. It sounded as if the Orioles just wanted to change managers and stick with the same losing culture. The Baltimore interview did not go well. It got worse when the Orioles had him meet with the Baltimore-Washington media after the session.

Oriole beat reporters quickly asked about Francona's four sub-.500 seasons in Philadelphia.

"Philadelphia may not want to hear this, but it's almost like having a mulligan," he answered. "I had my chance to make my mistakes, to learn from them, and to gain confidence, just like a player does."

Francona was right about folks in Philadelphia not wanting to hear about his "mulligan." His phone lit up on the drive back to Yardley.

"I kind of put my foot in my mouth on that one," he admitted. "What I meant was that you learn from your mistakes, but it didn't come out that way. I knew I really misspoke."

He wasn't going to make the same mistake when the Red Sox called.

He knew his name had been bounced around in Boston since before the start of the World Series. He also knew the Sox were strongly considering Angel pitching coach Bud Black — the same Bud Black who had accompanied him from the Sheraton Boston on his first visit to Fenway Park when the two were teammates with the Indians in 1988.

Black had already spoken with Epstein about the Boston job, but wasn't sure he wanted to move his family to the East Coast. He'd had numerous conversations with Francona about the Sox job when Theo called Francona.

It was awkward. Francona had been urging his friend to pursue the coveted Sox job, but now they were calling him. He called Black to explain the conflict. Black told him to go for it.

"I was in a good situation with the Angels," said Black, who was named manager of the San Diego Padres in 2007. "The timing wasn't right for me personally or professionally. When you make that step, you have to be ready, and I wasn't ready."

"I knew if he interviewed with Boston, he'd get it because he's a sharp guy," said Francona. "In my mind, part of the reason he took himself out of it was because the Red Sox called me."

Only young Theo Epstein seemed to see Francona's time in Philadelphia as a potential building block to a long managerial career.

"I was working in the National League [with the Padres] when Tito got the job with the Phillies," said Epstein. "I remember thinking, *Holy shit, that guy got the job really young. He must be really good.* And then watching those teams play, I remember thinking, *Boy, they really fucked that one up.* But when we had our manager search, I thought there must have been something there to make them hire him in the first place. I wondered if it just couldn't come out in Philadelphia or if it might come out now because he had failed. I called Billy Beane about him, and Billy was really strong in his endorsement for Tito, so we brought him in. He fit a lot of the criteria we were looking for. We wanted someone with experience and someone who was open to fresh ideas and could relate to the contemporary player."

Francona was the only candidate on Epstein's list who had major league managerial experience. With Black out of the running, the Red Sox list was pared to Francona, Angels bench coach Joe Maddon,

Dodgers third-base coach Glen Hoffman (a former Red Sox infielder), and Rangers first-base coach DeMarlo Hale, a minority candidate who'd been a manager in the Red Sox minor league system for seven years. Selig insisted that teams grant at least one interview to minority candidates for any managerial position, and the John Henry Red Sox were particularly beholden to the commissioner; he'd delivered the team to Henry in a backroom deal in 2001.

Francona knew he wasn't going to manage the Orioles or the White Sox, but this was a much better job opportunity. The Sox were a high-payroll team, ready to win. They'd just gotten to within five outs of making the World Series, and they were going after another starter, a closer, and maybe another superstar or two. They filled their ballpark for every game, and the fans followed their team with the same passion as fans in Philadelphia. Best of all, the Sox had a GM and owner who were not obsessed with hiring a "big name" candidate.

"Do your homework" is a way of life for Terry Francona. It's the way he was raised by Tito and Birdie. He was not a superior student — baseball always took priority over books — but he was never less than prepared in any situation. He was smart enough to know that it didn't take Ivy League brainpower to be on time or to prepare more thoroughly than the next guy.

He did his homework before coming to Boston for his interview. He sat down and composed his baseball manifesto, nine pages of notes on how he felt about managing. He wrote about the respect he felt for everyone who played the game. He wrote about being on time and busting your ass. He wrote about the little things you notice when you watch players when they don't know you are watching. He got his hands on a Red Sox press guide that featured photographs of every person who worked in the front office. He already knew what Theo Epstein looked like, but he was going to be meeting a lot of people, and he knew how good it makes people feel when you remember their names.

In addition, he called old friend Mark Shapiro in Cleveland. Shapiro knew Epstein well.

"Don't try to bullshit him," said Shapiro. "He'll tie you in knots."

Twenty-nine-year-old Theo and his 32-year-old assistant Josh

Byrnes were waiting for Francona at Fenway on Thursday morning, November 6. Byrnes had been Cleveland's scouting director when he was only 27, and he had come to the Red Sox from the Rockies.

Epstein and Byrnes had already interviewed Hoffman, so they were prepared when Francona sat down in the third-floor conference room near the Red Sox executive offices. In addition to Beane's tout, the Sox had a strong recommendation from Lee Thomas, who had hired Francona in Philadelphia and most recently had served as a special assistant to Epstein. Thomas resigned from the Red Sox before the Francona interview, but he'd had only good things to say about Francona. He explained that the 1997–2000 Phillies were rebuilding. He told Epstein that Francona did a good job in a difficult situation.

Early in the interview, Theo handed Francona a multiple-choice quiz. It was not a joke. He told the candidate to take his time filling it out, adding, "There's no right answer."

Some of the questions were amusing. Option D for the question "What's most important?" was "Making sure your uniform looks good in the dugout."

"We threw in a couple of disarming ones," said Epstein. "Some were *Sophie's Choice* questions, and some were off-color."

"It was to show things that were a priority versus things that were not a priority," reasoned Francona. "Like, how you look in your uniform versus how you feel if someone doesn't run out a ball. How would you react? Theo said not to worry about the answers, but they formed the interview a little from that. He'd look at the answer and say, 'This seems like it is important to you — can you elaborate?' That was how the interview started, and I thought it was a really good way to do it. There was no awkwardness. I had already committed to the answers."

Epstein and Byrnes gave him a chance to explain and defend his multiple-choice responses. They talked for about two hours, which gave Francona a chance to spill all the content from his nine pages of notes. It went well. He got more comfortable as the afternoon went on.

"As we went through the interview it was really clear that Tito was a perfect candidate," said Epstein. "In part, because of what he went through in Philadelphia, but also because of the incredibly important characteristics and endearing qualities that he had. Those things were allowed to come out because of his experience in Philadelphia."

It was no secret that Francona had a well-earned reputation as a players' manager. Tobacco-spitting hardball old-schoolers scoff at the notion of a manager who gets along with his players. Dick Williams certainly never worried about hurting anyone's feelings. Tito Francona played 15 years in the bigs in an era when managers never bothered to explain anything to ballplayers. But the days of "my way or the highway" went out with Billy Martin and Earl Weaver. The evolution of free agency and multimillion-dollar, long-term, guaranteed contracts changed how managers go about their business. Big league managers of the 21st century need to major in communication and minor in psychology. The ballplayers acquired the hammer long before 2003. Epstein and Byrnes were young and smart enough to embrace and celebrate new-school managerial styles, but they needed to know that Francona would not be a pushover.

Yes, he told them. He'd played cards with his ballplayers in Philadelphia. He got along well with Curt Schilling, Scott Rolen, Doug Glanville, Pat Burrell, Rico Brogna, and Wayne Gomes, among others. He'd given Bobby Abreu a day off when fans were demanding to see Abreu in the lineup. He'd rested Rolen on Scott Rolen Bobblehead Day at Veterans Stadium. But he'd also challenged six-foot-seven, 250-pound pitcher Bobby Munoz to a fight in front of the whole team after Munoz embarrassed a young catcher. He'd tried to send Abreu home when Abreu was repeatedly late. (Management balked at the punishment.)

Theo already knew many of the stories. Thanks to Thomas, he knew a lot about things that went on behind closed doors in the old Phillies clubhouse.

Francona made no apologies for his Philadelphia years. No mulligan this time. He told the Red Sox that he would care about his players more than anyone else. He would respect them more than anyone else. In return, he would ask more of them than anyone else.

"When you get fired, your self-esteem takes a hit," he said. "They wanted to know if I'd learned anything from all that. I told them that of course I had learned, but it only reinforced to me how I felt about things. As a manager, I think you're supposed to take responsibility and shield people that need it. I kind of wore it a little bit, but I thought that was my job."

Next up was game simulation. Moving to a room with a big-screen television and a brown leather couch, Francona was electronically transported to the seventh inning of a 2003 game between the Angels and the A's. Pretend you are managing the A's, they told him. Barry Zito is on the mound and is at 105 pitches (the magic number that turned Pedro Martinez into a pumpkin). The Angels have X, Y, Z due up, and you have A, B, C in your bullpen.

They told him he had two minutes to make a decision on what to do with Zito.

"Bring in Keith Foulke," said Francona.

Theo checked the simulation. Foulke was not available.

"I told you Macha was a dumb-ass," teased Francona.

They all laughed. Then Theo checked the roster and realized that Foulke had been on the disabled list during the game in question.

"It was fun, but a little crazy," Francona remembered. "I remember telling them, 'We can do this, but this isn't how I'm going to manage.' You can't just throw a bunch of stats at a situation. I am always prepared, but I can't be prepared in a situation where you're just throwing a bunch of numbers at me. What they had there was good, but it wasn't perfect."

In the middle of the afternoon, Sox CEO Larry Lucchino met with Francona. It was an important meeting. There's been considerable debate and confusion about the governance of the Red Sox since Henry's group bought the team in December 2001, and the balance of power shifted several times in Henry's first decade of ownership, but Lucchino has been the single steady force. A graduate of Princeton, Lucchino came into baseball with his mentor, trial attorney Edward Bennett Williams, when Williams purchased the Baltimore Orioles in August 1979. Ever the advocate — "Larry is always someone's lawyer," Charles Steinberg said — Lucchino had made a fair share of enemies in three-plus decades running baseball teams. He was the top advocate for Grady Little when Little became the Sox ownership's first managerial hire in 2002, and his stamp of approval was crucial if Francona was going to succeed Little in 2004.

"I only spent a half-hour with Larry that first day," said Francona, who had talked to his friend Bill Giles about Lucchino. "For the most part, he wanted to talk about the psychological aspect of baseball. He kept asking me to assign a percentage to the psychological side of the game, as opposed to the physical side. I hesitated and said, 'I'm not sure I can give you a number,' but that was not an acceptable answer. He wanted a number. I'm not even sure what number I gave him, but I know that's most of what we talked about that day.

"It was intense and challenging, and it was long. But I knew they were serious, and I appreciated that."

"He kind of blew us away," said Epstein. "It was a fantastic interview."

After six hours (twice as long as his session in Baltimore), Francona's first Boston interview was over, but the day was hardly done. The "new" Red Sox wanted to see how their managerial candidate handled the media. A Boston manager spends more than an hour of each game day answering questions for local and national media outlets. There's a formal session three hours before each game, a thorough Q-and-A after each game, and assorted one-on-ones with the flagship radio and television rights holders. A team publicist sorts through additional special requests. The manager becomes the face of the franchise, no small role in a baseball-crazed market like Boston. Media skill ranks right behind game strategy and player relationships.

Francona was well equipped to deal with the voluminous and sometimes bloodthirsty Boston baseball news media. He'd grown up in big league clubhouses. He'd managed in Philadelphia, perhaps the only market more vicious than Boston. He'd spent a year managing Michael Jordan, making him no doubt the only minor league skipper trailed by Tom Brokaw, Ted Koppel, and *Hard Copy*. Besides his experience, Francona was anecdotal, self-deprecating, funny, and polite.

Meeting local media when you don't yet have the job is awkward, but Francona was comfortable.

"I think the Red Sox are going to interview a lot of tremendous people," he told the assembled reporters. "But I think I can be an asset to this ball club. In Philadelphia I was very young, and I was learning on the run. I had a goal back then to be a major league manager. Now I have a goal to be a successful major league manager. Because of the situations I've been in in the past, I think that's possible."

When he was asked about working with the *Moneyball* A's, he gave a very un-Grady-Little-like response: "Some of the things I believed and maybe used in layman's terms got explained to me a little bit better.

Some people taught me how to make it more applicable, and for that I'm grateful."

After the session, he went to dinner at the Atlantic Fish Company with Epstein, Byrnes, and Jed Hoyer.

"Jed ate his ass off that night," said Francona. "I swear, the whole night, all I saw was the top of his head."

The diminutive Hoyer went on to become general manager of the Padres and Cubs.

Dining with the young baseball executives who held his future in their hands, Francona threw back a few beers . . . just a few. He'd heard of instances where a candidate had a few too many beers and got a little silly. This was not the time for that. He was auditioning, and everything was a test.

"It went great," said Epstein. "He was the guy, clearly, for us. We just had to go through a bit of a process."

"When I left Boston that night, I really wanted the job," remembered Francona.

The Sox were impressed, but the search was only beginning. Maddon and Hale were still on tap to be interviewed, and Hoffman was under consideration.

"I came in after Tito," said DeMarlo Hale, who would eventually join the Sox as part of Francona's staff in 2006. "I definitely got the sense that Tito was the front-runner. I knew I wasn't ready. You could sort of tell by the way everybody was talking, even the Red Sox players."

Francona had claimed the inside track. He knew he was the frontrunner when the Sox invited him back to work in their offices for a day. And then again for another day. It was a little strange. The Sox were still interviewing candidates, still had no manager, but Francona kept flying to Hanscom Field in Bedford, where he'd be picked up by traveling secretary Jack McCormick (the same guy who made sure the A's were delayed getting out of Boston a month earlier) for a day of work in the third-floor baseball operations offices near the New England Sports Network (NESN) studio, the club's television station. While the managerial search was allegedly ongoing, Francona returned to Boston multiple times.

He got comfortable. He played more simulation baseball ("Diamond Mind") against 32-year-old "baseball operations assistant" Brian O'Halloran.

"It was basically a computer version of Strat-O-Matic," said O'Halloran.

In a light moment, Epstein told Francona he could have the job if he consistently beat O'Halloran. Francona knew Theo wasn't serious, but the comment encouraged Francona to bend the rules in the pretend baseball games.

O'Halloran had worked as an unpaid intern under Epstein in San Diego. He had graduated from Colby, had an MBA from UCLA, and spoke fluent Russian, but he did not play baseball after high school.

Alone in the office shared by Josh Byrnes and Jed Hoyer, Francona and O'Halloran sat side by side in front of a computer and started a three-game series between the 2003 Red Sox and the 2003 A's. O'Halloran managed the Sox. Francona took the A's. O'Halloran manipulated the controls, following Francona's instructions when it was time for the A's "manager" to make decisions. O'Halloran's Red Sox routed the A's, 6–1, in Game 1. When the Sox went out to an early lead off Barry Zito in Game, 2, Francona went to his imaginary bullpen and summoned Rich Harden.

"You know that's your Game 3 starter, right?" said O'Halloran.

"I've got to win this fucking game," said the suddenly engaged Francona.

Francona's A's rallied to win Game 2. Using Ted Lilly as a substitute starter, Francona won Game 3 and copped the series.

When they asked him what he would have done in the Grady-Pedro situation, he joked and said he'd have taken Pedro out, but he didn't really feel that way. A lot of managers would have done what Grady Little did, including Francona and Bobby Valentine, the man who would ultimately succeed Francona. (Valentine, then the ex-manager of the Mets, got a little airtime when he claimed he'd been informally interviewed by Lucchino and took himself out of the running by admitting he'd have done the exact same thing as Grady Little.)

Epstein involved everyone in the process, including veteran Bill Lajoie, a "special assistant" who had been general manager of the Detroit Tigers. The Lajoie-Francona interview did not go well.

"It was over the phone, and he was hammering me about the way

I'd handled pitching, and it kind of aggravated me," said Francona. "I called him back and gave him a list of names. I asked him if he knew the names, and he said, 'Some of 'em.' And I said, 'Well, that was my fucking bullpen. You don't even know who they are, and you're questioning how I use them.'"

Speaking with the Sox regularly while still not officially an employee was somewhat uncomfortable for Francona. He was still a member of the Oakland A's coaching staff, but he was unofficially working for the Red Sox. The Sox asked about Schilling, who was considering waiving his no-trade clause for a chance to pitch for the Red Sox. They asked about free agent Foulke, who had led the American League with 43 saves for the A's in 2003. Francona was conflicted. He didn't want the Red Sox to sign Foulke if he was going back to be Macha's bench coach in Oakland.

In mid-November, he spoke to Epstein.

"Am I going to get this job here or what?" he asked, half-joking.

"Hang in there, you'll be fine," said Theo.

By the fourth week of November, Francona was comfortable enough about his chances to tell the *Globe*'s Gordon Edes, "I think they're going to name me manager. I believe that. I hope they do, but I don't think it will be tomorrow."

The day after making those comments, he returned to Boston, yet again, this time for a physical with Sox team physician Dr. Bill Morgan. Epstein told him not to bring a sport coat. There wasn't going to be any announcement of his hiring, not on this trip.

There wasn't much of a physical either.

"I loved Dr. Morgan," said Francona. "He had me put my arms out, then he pushed down on my hands to see how much resistance I had. I guess I did okay, because after he did that he said, 'You're good,' and that was it."

Francona also went to meet with Henry at the owner's mansion in Fort Lauderdale. Henry wasn't spending much time in Boston in the off-season of 2003–2004, but he wasn't about to make this important hire without meeting Epstein's favorite candidate. Francona flew into West Palm Beach, where the Sox had arranged for a car service to take him to Henry's estate.

Nervous about the interview—Epstein warned him that Henry

was soft-spoken and sometimes quiet around new people — Francona chuckled when the Town Car passed the Palm Beach International Polo Club on the way to Henry's estate. Francona's mind flashed back to a charity event in the 1980s when he'd been reprimanded along with Expos teammates Larry Parrish and Tim Wallach. Part of a charity tennis tournament during spring training, the three young players offended club officials by betting on one another's croquet strokes. Two decades later, he was on his way to a job interview with a man who lived near the stodgy club. He felt a little like Groucho Marx, who famously said he would never want to belong to any club that would have him as a member.

Francona's anxiety increased when the car went through not one but two security gates on the Henry property. When he finally got inside Henry's home, he was seated in a front room with a lot of windows, then waited nervously for almost an hour. He was all alone except for the servant who brought him an iced tea and told him that Mr. Henry would be right down.

Francona kept checking the crib sheet in his pocket. He wanted to be prepared. But he kept thinking, *What's up with this guy? Is this some kind of test?*

Finally, Henry came into the room, looking typically frail and pale, and he was also hunched over. Henry explained that he'd recently suffered a back injury.

The conversation was brief. Ten minutes. Francona didn't know what to make of it. He wondered if he'd offended Henry. Then, as Henry was walking Francona to the car, he almost committed a major faux pas. Henry's six-year-old daughter ran up to the two men, and Francona started to say, "What a cute grandchild," then caught himself. He was no longer the young wiseguy who got admonished at the Palm Beach International Polo Club. In a quarter-century of professional ball, he'd learned never to ask a dome-bellied young woman, "When's the baby due?" So he held back on the "granddaughter" compliment, perhaps changing the course of Red Sox history.

Still, he wondered why the interview was so brief.

"I wasn't sure if he was hurt or if I offended him, but Theo called me later and said, 'I heard it went great.'" Henry later told people it was the single best interview he'd ever had in baseball.

Francona had cleared every hurdle, but the announcement was still on hold while Theo attended to the matter of bringing Schilling to Boston.

The Red Sox did not hire Francona as their 44th manager in order to lure Schilling to Boston. It doesn't work that way. A manager is too important to the overall well-being of a franchise. It's not like college basketball where a school might hire a coach if said coach can induce a once-in-a-lifetime recruit to come with him to study and play. This happened to Holy Cross College in the mid-1960s when the small Catholic institution hired Power Memorial High School coach Jack Donahue, hoping that Donahue could convince his star center — Lewis Alcindor — to follow him to Worcester. It never happened, of course. Donahue went to Holy Cross, where he never won an NCAA tournament game, and Alcindor went to UCLA, where he won three national championships for John Wooden and the Bruins.

By late November 2003, Henry, Lucchino, and Epstein all knew they wanted Francona to be their manager, and having him waiting in the wings certainly did not hurt their chances of getting Schilling to agree to a trade to Boston. Schilling, a World Series co-MVP with Arizona in 2001, had veto power over any trade, and when the Diamondbacks were looking to deal him, the big righty at first said he would only accept a deal to the Yankees or the Phillies. Schilling didn't know that Epstein was already working on a deal with D-Backs general manager Joe Garagiola.

Epstein told Schilling that Terry Francona was going to be the Red Sox manager. He also wrote (with Lucchino) a letter to Curt and Shonda Schilling.

"There is no other place in baseball where you can have as great an impact on a franchise, as great an impact on a region, as great an impact on baseball history, as you can in Boston," wrote the Sox executives. ". . . The players who help deliver a title to Red Sox Nation will never be forgotten, their place in baseball history forever secure."

It worked. Schilling immediately started lobbying for Francona to get the Sox job.

On November 24, the day Selig gave Schilling a 72-hour window in which he could waive his no-trade clause and come to terms with the Red Sox, Schilling responded to a media question about the possibility of coming to Boston by saying, "Terry is a huge part of this. Terry is the number-one attraction there for me. If he's not the manager there, my interest in going to Boston would diminish drastically. I love playing for him, I enjoyed playing for him. I knew where I stood with him when I walked through the door. He didn't get dealt a full deck when he was in Philadelphia. He's up there with the people I played for in terms of the respect factor. That will play a big part in my decision."

Aware of how his remarks would be interpreted, Schilling said, "People are going to put a real bad spin on this. I only made myself available [for a trade to Boston] when I understood Terry was the number-one candidate there after a lot of interviews. It would be disrespectful to insinuate otherwise that I'm the reason he was going to get the job."

The next day Epstein and Hoyer flew to Arizona to court Schilling. Lucchino also made an appearance, arriving from San Diego the day before Thanksgiving. Schilling and his wife were impressed with the Sox presentation and invited Epstein and Hoyer to Thanksgiving at their home in Paradise Valley. (Lucchino went back to his family in California.) The day after Thanksgiving, the day of the commissioner's deadline, Schilling agreed to a three-year, $37.5 million deal with the Red Sox and announced to the world, "I guess I hate the Yankees now."

"I think Schill was trying to help," Francona said later. "But then it came out that the reason I got my job was because of Schill. I don't think that's true. I was pretty deep into the process. They weren't interviewing any more people. But I really didn't care what people thought."

"Schill couldn't have been stronger with his praise for Tito," said Epstein. "With Schill's personality, I think he thought, *Hey, they want me so bad they'll hire this manager just to get me,* but the reality is that it was done."

Early in the next week, as November turned to December, Francona got the call he'd been hoping for.

"I was at my daughter's volleyball practice, in a car, freezing my ass off," he remembered. "Theo had told me Larry was going to call and for me to act all surprised. So Larry called and said, 'We'd like you to be the manager.' I said, 'Great,' and he said, 'We'll pay you just over a million for two years.' I asked if he was serious about the money, and he said, 'Yes, take it or leave it, because if you don't want it, we'll get somebody else.' I said I'd have to think about it."

The hard-line contract offer was typical of Lucchino, ever-mindful of demonstrating that he was the boss. Epstein understood this, but Francona was just beginning to learn. Francona called Epstein to complain about the proposed salary. By major league standards, it was terrible. Epstein said he understood and bumped Francona to three years for a total of $1.55 million.

This is how things would work for the next eight years. Francona would register his complaints to ownership through Epstein. Theo would be the man in the middle, right up until the day Francona was fired.

On Thursday, December 4, six days after the Schilling announcement, Francona was introduced to the Boston media as the new manager of the Red Sox. The first question he was asked was about Manny Ramirez not running out ground balls.

He handled the queries with ease, even when WBZ's ubiquitous Jonny Miller asked if every Sox manager was just a man waiting to be fired.

"Think about it for a second," Francona said. "I've been released from six teams. I've been fired as a manager. I've got no hair. I've got a nose that's three times too big for my face, and I grew up in a major league clubhouse. My skin's pretty thick. I'll be okay. . . . This is the most exciting day of my baseball life."

When the press conference was over, Francona and Epstein retreated to the baseball operations offices on Fenway's third floor.

"What do you want to do now?" asked Epstein.

"I'd like to make some phone calls," said Francona.

They set him up in a tiny office, and the new manager started dialing. His first call was to Dallas Williams, who had been a coach with the Red Sox in 2003. Williams was one of Francona's best friends, but the Sox brass had informed Francona that Williams would not be part of the coaching staff in 2004. It was a tough phone call for Francona, but he knew it was always better to let people know so they could start making other plans.

Then he started calling Red Sox players, all the members of the 2004 Red Sox. He didn't reach every player, but he left them all messages.

*Hey, Manny, it's Tito. Just wanted to say hello. I'm looking forward to it. I can't wait. If you don't want to call back, I get it, but I just wanted to let you know how excited I am.*

He never heard back on that one.

"I knew Manny wasn't going to call me back," he said. "I'm not an idiot. But I wanted to make that call. When I was a player, I always wished the manager did something to make me feel important."

After making his calls, Francona went to dinner at Davio's with Epstein and Lucchino. They called Mike Barnicle, and Lucchino told the columnist, "In spite of your efforts to help Terry get this job, you can't be his third-base coach."

Riding the train home to Philadelphia the next day, Francona thought back to a moment he'd shared with his best friend Brad Mills in their final year in Philadelphia when the Phils were playing out the string, en route to 97 losses and unemployment.

"Millsie and I were driving down Central Avenue in Arizona, and the Diamondbacks were fighting it out for the pennant, and we were something like 30 games out. We were just getting our ass handed to us. As we were pulling up to this beautiful new ballpark, we were so jealous. I said, 'Millsie, these guys have a chance to win every day. Someday that could be us.'"

Back in Boston, Epstein composed a memo to Henry, Werner, and Lucchino, explaining the managerial search and its conclusion: "We were looking for a manager who would embrace the exhaustive preparation that the organization demands.... By using video and computer simulations, we attempted to discover how each candidate would react to game-speed strategic decisions.... Given the demands of the media and our players, we sought a manager who would be able to communicate with all constituencies in a positive and intelligent manner. We were looking for a 'partner' not a 'middle-manager.'... Terry Francona quickly emerged from the applicant pool. His experiences (Philadelphia manager, Cleveland front office, Texas/Oakland bench coach) gave him a remarkable understanding of our vision. His preparation, energy, integrity, and communication skills are exceptional."

The only question Theo still had was the one he expressed at the press conference.

"Is the guy too nice? Does he treat people too well?"

---

# CHAPTER 5

# "We'd better win"

THE NEW MANAGER of the Red Sox had a lot on his mind when he drove south for spring training in February 2004. He was excited to be managing again, and the big-payroll, bigger-expectations Red Sox had acquired Curt Schilling and Keith Foulke over the winter.

But he knew there were going to be some superstars bent out of shape when the Sox gathered in Fort Myers.

There were no secrets. In December 2003, Theo had worked out a deal with the White Sox that would have sent Nomar Garciaparra and Scott Williamson to Chicago for outfielder Magglio Ordonez and a pair of pitching prospects. At the same time, Manny Ramirez had been placed on waivers, then traded to Texas for Alex Rodriguez. The Garciaparra and Ramirez deals were quickly killed when Commissioner Selig and the Major League Baseball Players Association blocked the A-Rod–Manny trade because of creative contract restructuring that would have lowered the annual value of Rodriguez's contract from $27 million to $20.75 million. Henry and Lucchino were furious with the commissioner and with the players association, but they were not the ones who were going to have to deal with the superstars in the clubhouse. That would be Francona's job.

It was impossible to anticipate Manny's reaction. He was not a creature of cause-and-effect. Garciaparra was a different case. Francona
knew Nomar. He'd managed Garciaparra in the Arizona Fall League
before Nomar hit the big time. Everyone knew Garciaparra was hurt
and embarrassed that the Sox hadn't made more effort to sign him to
a contract extension. He was furious that Henry had met with Rodriguez, made deals with the Rangers and White Sox, then asked him to
come back and play for the Red Sox as if nothing had happened. It
didn't help that teammate Kevin Millar had spoken publicly about the
Sox making an "upgrade" with Rodriguez. Over the winter, when reporters asked Nomar how the trade reports made him feel, he turned
the question around and asked the scribes how they would feel if they
woke up one day and read in the paper that their replacements were
being interviewed.

"I called Nomar during the winter when all this came up," said Francona. "He was on vacation in Hawaii. I told him, 'I know you got some
shit going on, but it's not me and you.' He said he understood that. He
was great about it. But it turned out to be like a lot of other things. I
thought I knew what was going on, but I don't think I quite realized
'Boston' yet — how nothing was a little story in Boston."

Francona had another enlarged, wounded ego in Pedro Martinez.
Pedro's Hall of Fame skills were in slight decline by 2004, and he was
also going into the final year of his contract. Martinez was ever-mindful of the current status of his contract. In the spring of '04, he was
additionally uncomfortable with the widespread notion that Schilling
was going to be the man to save the Red Sox and break the team's
86-year championship drought. Pedro was aware that Schilling had
called him a "punk" after Martinez shucked ancient Yankee coach Don
Zimmer to the ground during a Yankees–Red Sox dustup in the '03
playoffs. Pedro never forgot a slight.

"I think I had him pretty much figured out," said Francona. "I knew
he was on his own program. I did my homework on everybody."

"Homework" is no exaggeration. With help from baseball operations assistant Brian O'Halloran, Francona had mailed a letter to every
manager, coach, scout, and player development person in the Red Sox
system, requesting information and opinions on all players in the organization.

"That was really helpful," said the manager. "I had info on all the players and the people evaluating them."

The Red Sox equipment truck pulled out of Van Ness Street, adjacent to Fenway, on Monday, February 16, two days after the bombshell trade that sent Alex Rodriguez to the Yankees for Alfonso Soriano. The A-Rod trade perfectly bookended Boston baseball's nuclear winter, which had started with the Grady Little Game, and also played to the fears of Red Sox fans convinced that the Yankees would always get the better of the Sox. In the aftermath of the trades and broken deals, Henry and Steinbrenner traded barbs — Henry claiming, "Baseball doesn't have an answer for the Yankees," while George responded with, "He chose not to go the extra distance for his fans." Selig ordered both franchises to cease and desist.

Driving south from Yardley, Pennsylvania, Francona heard the noise, but couldn't focus too much on Manny, Nomar, Pedro, A-Rod, George, or the low-talking owner of the Red Sox. He had to get ready for baseball.

"The drive south is one of my favorite things every year," he said. "It signifies getting warm and a new chance to start over. You start out wearing a jacket and you shed that, and you eat a bunch of tacos and you see all those signs for South of the Border, and then you take off your sweatshirt, and by the time you get to Florida you're in a T-shirt and it's sunny. I always go down at least a week early."

He exited Route 75 South at Daniels Boulevard, drove another five minutes, then wheeled his blue Mercedes SUV into the Homewood Suites adjacent to the Bell Tower shops and restaurants in cluttered Fort Myers. Homewood was nowhere near Fort Myers Beach, and it also wasn't especially close to City of Palms Park or the Red Sox sprawling minor league complex at the dead end of godforsaken Edison Road. Most big league ballplayers and managers rent condos near golf courses or beaches during spring training. That was too swanky and isolated for Francona. He liked the convenience and the company at Homewood. The Hilton property was always filled with polite, elderly snowbird Sox fans and pasty college teams making their spring baseball/softball trips. Homewood served breakfast in the common area every day, and there was a wine-and-cheese hour on weekday afternoons. Red Sox relic Johnny Pesky always seemed to be holding court with elderly Sox fans, talking about a train ride when Jimmie Foxx boosted him into the upper bunk. Luis Tiant was another Sox legend housed at Homewood, and spring residents grew familiar with the cigar smoke wafting from Tiant's SUV, which was always parked in front of the hotel.

Nobody bothered Francona when he smoked a cigar on the patio outside his first-floor, poolside room.

"That place was perfect for me," he said. "They make your bed, and they have the best coffee in the league in the lobby, and I'm never there. I never liked coming home to a condo by myself. The Sox got me a place at one of those golf courses one year, but I gave it to the clubhouse guys and went back to Homewood. It was just comfortable. My dad loved coming down and staying there. I'd try to take him to dinner, but he loved those Swedish meatballs they put out at five in the afternoon."

Armed with a cup of hotel coffee and a banana from the not-yet-set-up breakfast spread, the new manager of the Red Sox was up and out of Homewood every morning by 5:30 AM.

"I was anxious, nervous, all of those things," said Francona. "I was really ready to get started. I couldn't wait."

He set up shop in the manager's office at the Sox minor league clubhouse and greeted pitchers, catchers, and anxious veterans who arrived before the required reporting day. He got to know Jonny Miller, the longtime radio reporter from WBZ in Boston. Miller was a press box–clubhouse legend around the Red Sox. He was born with cerebral palsy, grew up in well-to-do Newton, graduated from Boston University, and built a successful career gathering sound from professional sports locker rooms. Miller covered all of Boston's professional sports teams at one time or another and was a favorite of Larry Bird and Pedro Martinez. By the time Terry Francona arrived in Fort Myers in 2004, Miller was exclusively a baseball reporter, one who worked longer hours than anyone else on the beat. In his midfifties, Miller suffered from back issues that made it difficult to stand for long periods. Polite players sometimes offered him their seat, but Miller usually refused. He was a workaholic and asked brutally tough questions ("Grady, can it get any worse than this?"). He got to the ballpark long before any other reporter, sometimes even before the early-bird manager.

Francona: "The day of our first pitchers and catchers' workout, I got there at about 5:45 in the morning, and it was pitch-black, and the front door to the complex was locked. Millsie had the key, and I'd beaten him there, so I walked around to the side and fucking Jonny Miller scared the shit out of me. He came around the corner, and he was carrying a bunch of books in his hands, and we bumped into each other, and I was like, 'What the fuck?' His answer was, 'I had to get some coffee.' I said, 'That's not what I meant. What the hell are you doing here?'

"I learned pretty quickly that Jonny always got to ask the first question," said Francona. "He asked things that other people wouldn't ask. I got to know him pretty quick. I think the writers sometimes get a kick out of him, and sometimes get irritated, but they know that he can ask stuff that maybe they want the answer, but they don't feel like getting yelled at. There was something with Jonny almost daily. I do think he was a big Red Sox fan. You could tell if somebody asked a question he didn't like. He'd be shaking his head, getting mad, thinking, *That dumb-ass.* He'd always be first in the room after games, and you had to let him get all of his stuff in order. He'd shuffle in and lean his cane against my desk, and his tape recorder would usually make a buzzing sound at the beginning. Most of my press conferences started with me saying, 'You all set, Jonny?'"

Francona loved the early days of spring. He'd make the rounds in the clubhouse every morning, finding a new player who'd arrived or introducing himself to a minor leaguer who was in his first big league camp. He told the pitchers and catchers to do their work and not worry about any formal meetings until the full squad arrived. At the end of each workout, the manager sat on a picnic table and met with reporters outside the Sox clubhouse. In the spring of 2004, Francona noticed a lot of reporters from New York. The Sox were a hot topic in Gotham, and at least five representatives from New York outlets stalked the Sox daily. *New York Newsday* included a daily Boston item on its spring training pages, entitled "The Misery Index."

Nomar and Pedro made it to camp a day before they were required to report. Nomar made little attempt to hide his contempt for the organization — even when he smiled.

"That smile is more of a sneer," observed Red Sox chairman Tom Werner. ". . . I'm sure he didn't like to read in the papers over the offseason that he was part of the Rodriguez trade, but this is a business that these people are playing."

Meanwhile, Pedro seemed to be in a good mood. Veteran Sox watchers considered it a blessing that the talented righty arrived before March 1. Through his Boston years, Martinez had made it a habit to arrive late, or just barely before he was contractually obligated to join the team. Hard-core Sox fans knew that Pedro's father's birthday was in February, and more than once the Cy Young winner was late getting to Fort Myers because he had to stick around the Dominican Republic for his dad's birthday celebration.

It was big news back in Boston on February 26 when Manny wheeled into camp, wearing a New York Giants number 80 Jeremy Shockey jersey. Manny entered the complex through the front door with his agent, Gene Mato, and the two were quickly whisked into a meeting with Henry, Lucchino, and Werner. Ramirez submitted to the routine club physical and took swings in the indoor batting cages adjacent to the media trailers, but did not speak to the media.

He didn't have much to say to his new manager either.

"I went up to him and introduced myself, and it wasn't good," said Francona. "He wouldn't talk to me, and he wouldn't shake my hand. I tried to talk to him, and he said, 'You just want me to like you.' I said, 'No shit. You're right.' It's not what I expected.

"There wasn't anything I could do about that, but to me it was important that they all know that it was a clean slate for everybody," said Francona. "I had all the information, but I wanted them to know that everybody was starting fresh with me. It's my job to know things, but it's also my job not to hold something against somebody."

By any measure, Ramirez is one of the greatest right-handed sluggers in the history of baseball. Red Sox general manager Dan Duquette told Boston fans they were getting the next Jimmie Foxx when he signed Ramirez to an eight-year, $160 million contract in December 2000, and Duquette was correct. Ramirez was a rare big-money free agent who turned out to make good on the club's investment. Failed drug tests in 2009 and 2011 tarnished Ramirez's significant accomplishments, but there can be no dispute about what he did with the bat during the first 19 years of his career. Before serving his second big league drug suspension in 2012, Manny had hit 555 homers with a career batting average of .312.

Numbers don't explain Manuel Aristides Ramirez, the son of Aristides and Onelcida Ramirez. Manny was born in Santo Domingo in the Dominican Republic, but grew up in Washington Heights in New York (at 168th and Amsterdam). He graduated from George Washington High School in the Bronx in 1991; in his senior season he hit .650 with 14 homers in 21 games and was the New York City Public High School Player of the Year. The Indians made him the 13th overall pick of the 1991 amateur draft. Four years later, Manny batted .308 with 31 homers and 107 RBI for the American League Champion Cleveland Indians. The '95 Indians had one of the best lineups in baseball history, featuring Albert Belle, Eddie Murray, Jim Thome, Kenny Lofton, Carlos Baerga, Omar Vizquel, and 23-year-old Manny Ramirez. They lost the World Series in six games against Atlanta.

There was nothing accidental about Ramirez's batting prowess. Anytime he was asked about his philosophy on hitting, he'd say, "See the ball and hit the ball" (similar to what Tito Francona told Terry when he went off to professional baseball from the University of Arizona), but teammates and coaches saw work and dedication behind Manny's numbers. He worked diligently in the cage, watched a lot of video, and did extra conditioning to keep himself in shape. Still, a lot of players did those things, but no one else could hit like Manny. He had unusual balance and eyesight and tested off the charts in hand-eye coordination. He never looked off balance at home plate, and he never pressed. The science of hitting does not reward those who try harder. Thinking too much only makes things worse. One of the beautiful things about Ramirez seemed to be that he was incapable of carrying baggage to home plate. Every at-bat was a clean sheet of ice. He did not walk to the plate thinking, *I struck out last time,* or *This pitcher owns me.* It was always, *See the ball and hit the ball.* His preparation and physical gifts did the rest of the work.

"He was just better than everybody else," said Francona. "He had some natural gifts. Our strength and conditioning coach put together this contraption; it looked like a small hula hoop with about eight Wiffle balls connected to it. Manny would stand in his batting stance, and our guy would spin this thing toward him and call out a number or a color, and Manny would reach out and grab the ball he was supposed to grab. It wasn't easy for most people to catch the damn thing, let alone get the right-numbered ball or the right-colored ball. Manny could get it just about every time. It was incredible."

There may be profound depth to Ramirez, but anecdotal evidence suggests otherwise. Manny was the one who thought his Cleveland Indian teammates were talking about pitcher Chad Ogea (pronounced *oh-jay*) when they were watching the O. J. Simpson white Bronco chase in 1994. He forgot to cash checks, then asked Cleveland sportswriters if he could borrow $20,000. After he broke his finger sliding headfirst into home with the Sox in 2002, he rehabbed at Pawtucket and decided he loved McCoy Stadium and wanted to play there full-time. It was on the base path at Pawtucket that Manny lost a diamond-encrusted earring while making another ill-advised headfirst slide. And video highlight of Manny leaping and cutting off a throw from center fielder Johnny Damon — Manny was standing only a few feet in front of Damon — lives forever on blooper reels.

"It was the one time he really hustled," chuckled Francona.

There was a dark side too. Manny sometimes disrespected the game. With the Red Sox in 2002, he famously turned and walked to the dugout — never running out of the batter's box — after hitting a ground ball back to the mound during a game against Tampa Bay. Grady Little, who kept Ramirez in the game after the disrespectful stunt, later told reporter Tony Massarotti, "If someone gives you a dog and that dog has a habit of peeing on the floor, can you change them?" Manny hit .349 in '02, but had nothing to say to the media after winning the American League batting title. In '03 he bailed on the All-Star Game, claiming a hamstring injury, then changed his story and said he had to go to Miami to care for his mother who had allegedly fainted while working in her garden.

More obtuse than outrageous, Ramirez was rarely a disruption in the clubhouse. He had hugs for everybody. But he also had an unusual scorekeeping system regarding how his teammates were treated. If another starter asked for a day off, Manny suddenly needed a day off. In August 2003, Pedro hit the shelf with a diagnosis of pharyngitis. Two weeks later, Manny missed a critical three-game series with the Yankees, claiming to be suffering from the same rare throat illness that afflicted Pedro. There was a full-blown media storm when Manny — too sick to go to the ballpark — was spotted in Boston's Ritz Hotel lounge with Yankee infielder Enrique Wilson. Manny skipped a scheduled doctor's appointment the next day. Things got ugly in '03 when the Sox had to go to Philadelphia for a makeup game and Manny said he was too weak to pinch-hit. Sensing a clubhouse mutiny, Little benched Manny the next night in Chicago. A month later, all was well and Manny hit the game-winning home run in the clincher against the Oakland A's. He didn't speak to the media in 2003.

Naive about the true weirdness of "Manny being Manny," Francona thought he knew what to expect. He'd coached third base in Detroit in 1996 when Manny hit .309 with 33 homers and 112 RBI for the Indians. The Sox manager had worked in the Cleveland front office with John Hart and Mark Shapiro when Manny was negotiating with the Indians and Red Sox at the 2000 baseball winter meetings in New Orleans. After Manny signed with Boston, he tried to convince the Tribe's equipment manager, Frank Mancini, to quit his job and move to Boston to take care of Manny's needs.

Sitting on the Oakland bench next to Ken Macha in the 2003 playoffs, Francona watched Manny hit a three-run homer that sent the A's home for the winter.

"When you're on the other side, he's always friendly," said Francona. "He's got that smile, and he calls everybody 'Papi.' He was always friendly to me. That's his nature. When you don't have to be the guy that's telling him about the things he's not doing, it's a lot easier. But it's different when you are his manager. I didn't know what to think when he wouldn't shake my hand. I knew he had had it out with ownership earlier in the day, but he was walking around the clubhouse laughing, saying, 'The Red Sox put me on eBay.'"

On the day the Sox held their first full-squad workout in 2004, Manny saw on the chair in front of his stall what every other player saw on his chair: a sheet of white paper headlined "The Boston Red Sox 2004 Team Rules":

- The Red Sox travel as a team — you must receive permission from the manager to make your own arrangements.
- Beer and wine will be served on flights — please do not abuse this privilege.
- Headphones must be used at all times while traveling. (If music becomes an issue in the clubhouse, it will be dealt with by the manager.)
- Wives and children are permitted on any return flight to Boston. Please notify Jack McCormick respectfully in advance. (*Note:* in September, because of expanded rosters, there will be no room for excess traveling party.)
- Dress codes for road trips are posted in the clubhouse two days before trips begin.
- Dress presentably at all times on the road (no shorts or flip-flops; jeans are acceptable) — days off on the road are exceptions.
- Curfew:
    Night games — 2:00 AM
    Day games — 1:00 AM
- Beer will be available in both home and visiting clubhouses after the game only. No beer is to leave the clubhouse (this will be strictly enforced).
- No player is permitted to leave the clubhouse until the game has been completed and the manager has returned to the clubhouse.
- Players needing treatment for injuries must set up reporting times with trainers and not be late.
- Players who become ill must contact trainer prior to reporting time so appropriate measures can be taken.
- All medical appointments with physicians must be kept and on time.
- Everyone will be on the top step for the national anthem and will stay in the dugout during all games with the exception of commonsense situations.
- Miss a game or practice — loss of day's pay and subject to suspension and/or fine.
- Any fines will be at the manager's discretion.

- Players must adhere to club and MLB policy regarding club-house security and visitors. Please exercise discretion when bringing family into the clubhouse. It is not a play room for your children, and clubhouse personnel are not hired to police the activities of your children. If at any time children become a problem or a distraction, it will be dealt with appropriately.

ALWAYS BE ON TIME!
ALWAYS GIVE EVERYTHING YOU HAVE ON THE FIELD!
BE PROFESSIONAL ON AND OFF THE FIELD!
Theo Epstein (general manager)
Terry Francona (manager)

"I put those out there each year to protect myself," said Francona. "Every year, the sheet was a little shorter. I wasn't going to check cur-few. No manager does. But if somebody did something stupid at night, I could say, 'This is the rule.' It was all just basic commonsense stuff. It got a little shorter each year I was there. Those things at the end — be on time, be respectful, play your ass off — that's what all those other things meant."

On the first full-squad day, just before 9:00 AM, Francona got ready for his first annual "big" meeting in the lunchroom at the front of the minor league complex. The late February meeting was the only meeting of the year that featured remarks from ownership, the general manager, the manager, the traveling secretary, the public relations director, the equipment manager, and even a representative of the club's charity wing (the Red Sox Foundation). It was a chance to get everything out in the open at once. Henry usually said he wasn't going to speak, then would change his mind at the last moment. For Henry and Epstein, it was a chance to remind the players that the new manager had the backing of the front office. McCormick would talk about players' tickets and travel specifics. Veteran clubhouse manager Joe Cochran would put out his cigarette and tell the fellows not to leave too much laundry on the floor. Meg Vaillancourt, the director of the foundation, would take too much time talking about community responsibilities that nobody wanted to hear about in February. ("She'd ask for 20 minutes, and Jack would tell her, 'Meg, you got three minutes,'" remembered Francona.)

The meeting was supposed to start at 9:00 AM. As the wall clock reached the top of the hour, Francona nervously noticed that Manny wasn't there. This had the potential to be a big, bad story — something that could have blown up Francona's first day in the same way Roger Clemens sabotaged Butch Hobson back in the 1990s. Anxious about Manny's absence, Francona went over to David Ortiz, whom he did not know particularly well.

"David, Manny's not here," said Francona. "Can you go get him? I don't care if he listens, just get him in here."

Ortiz nodded. He appreciated that the new manager did not want to make a scene. He thought it was a smart move to get a player to bring another player. He went outside and came back into the lunchroom with Manny. It was "game on" for the meeting.

"It's always the longest one of the year," said Francona. "You try to satisfy everybody, which is not easy. I ended up speaking less than anybody. It wasn't a Knute Rockne 'Let's run through the wall' speech."

When Francona made no reference to the way things had ended in October, Tim Wakefield and a raft of other survivors of the Grady Little Game were no doubt relieved. In 1987 beleaguered Sox skipper John McNamara, still bleeding from the Bill Buckner Game 6, told his team in Winter Haven that no one was to speak of the heartbreak of October '86. McNamara is believed to be the first manager in history to start spring training by urging his team to erase the fact that they'd just made it to the seventh game of the World Series.

Manny said nothing during the lengthy session. When it was over, he hit in the batting cage, ignored a few media requests, then went home.

It was different the next day. With various workouts taking place on multiple fields — and Sox fans scattered throughout the complex — Francona was overseeing a baserunning drill involving several of the Sox veteran stars when Ramirez walked over to the new manager, draped his arm around the skipper, and said, "Papi, I'll hit third, I'll hit fourth, I don't care. I'll do whatever you want."

"I was glad, but it was an early glimpse," Francona said later. "With

Manny, you just never knew when the button would go off. You could see it coming most of the time, but by the time it got to that point, it was too late. You weren't going to get him back for a while."

The slow, early days of spring training allowed Francona an inside look at the new generation of baseball executives who were running the Red Sox. Francona grew up in a hardball world in which most ballplayers, coaches, managers, and executives shared space in coffee shops, at poolside, and in smoky lounges at a "team hotel" in Florida or Arizona. The spring scene in sprawling Fort Myers in 2004 was something else altogether. Twenty-nine-year-old Epstein saved the ball club $30,000 in spring hotel expenses by renting an eight-bedroom house on a canal in Cape Coral. Boston's hardball fraternity was stocked with potato chips, bottled water, and beer. ("It was usually Bud Light or Miller Lite," said Epstein. "But sometimes one guy would bring some more sophisticated beer and get mocked for it.") They ordered a lot of Chinese takeout, and if you stood on top of the empty pizza boxes at the end of a weekend, you could almost see the Sunshine Skyway Bridge spanning Tampa Bay. Assistant general manager Ben Cherington was the only one of the group who was married, and his wife, television reporter Wendi Nix, was on assignment back in Boston for most of the spring. The young Turks invited Francona to come over at night and play a little Texas Hold 'Em.

It was Francona's first chance to see the new-generation executives at play. He enjoyed them. He brought his high school senior son, Nick, a few times and once fell asleep on a couch while Nick sat at the poker table with Theo, Cherington, O'Halloran (nicknamed BOH), and Hoyer. These Sox officials wound up drafting Nick Francona with their 40th-round pick four months later, but the manager's son eschewed professional baseball in favor of the University of Pennsylvania.

Theo's frat house was the site of Francona's first encounter with the estimable Bill James.

It is impossible to overstate James's impact on 21st-century baseball. He is the godfather of sabermetrics — officially defined as the study and mathematical analysis of baseball statistics and records with the goal of discovering objective knowledge about the basic principles that underlie the game. James grew up in a tiny town in Kansas, studied economics and literature at the University of Kansas, and was a night watchman at a pork-and-beans factory when he self-published his first *Baseball Abstract* in 1977. In his 1985 *Abstract,* James wrote, "Baseball statistics, unlike the statistics in any other area, have acquired the power of language." Tall, bearded, and quiet, James rarely made eye contact and had limited interpersonal skills. He didn't believe in fielding percentage because the assignment of errors was completely unscientific. Errors were *scorer's decisions,* therefore subjective, unquantifiable, and unreliable. Ever-contrarian, James believed that standard numbers of value like batting average and RBI were overrated. He told his readers that "baseball statistics are not pure accomplishments of men against other men, which is what we are in the habit of seeing them as. They are accomplishments of men in combination with their circumstances."

James invented new measures of player accomplishment and value. He devised a formula for runs created: an estimate of the number of runs each hitter contributes to the team. He created the "win share," which compares players across positions, teams, and eras and measures in a single number the total sum of a player's contribution to his team. He created "secondary average" — a method of summarizing the things a player does to create runs, other than batting average. Cofounder of STATS Inc., James accurately described himself as "a mechanic with numbers."

He also gave Francona a headache, and the new manager was only too happy to break chops when he found himself opposite James at the poker table in the spring of 2004.

"I yelled at him," Francona said. "It was all in good fun, but I let go a little bit when I saw him that first night over there. I'd had a few beers, and I was telling him, 'You may know numbers, Bill, but you're not street-smart! I can read you!' The cards fell my way, which made it even better. I took him to the cleaners at the poker table. It was fun, but I don't know if he appreciated my humor as much as everybody else did."

"Tito's a really good card player," added Mills. "He feels he can match up with anybody. He's got that confidence. So those guys might have had the upper hand in the clubhouse, but at the card table it was like, 'Don't mess with me.' When they'd mess up at the card table, he'd say, 'Hey, I hope you're doing better with that hit-and-run stat you had, or the numbers on hitting the 3-1 fastball.'"

"There was a lot of smack-talking that night," said Epstein. "Bill James was doing a lot of math in his head. People were talking shit at one point, and Bill, who is pretty stoic and guarded, looked up from his cards and said, 'Up yours, Tito,' then, 'Fuck you, BOH.' We all cracked up. It was pretty funny."

James never visited the Red Sox clubhouse. He was aware of how baseball people viewed him. But he had his revenge over the next eight years of Francona's life as Theo and his baseball men bombarded the manager with data, much of it rooted in the philosophy of James and the generation of number-crunchers he'd inspired.

Early in the spring of '04, before the first exhibition game was played, Epstein and Hoyer met with Francona and Mills and presented the foundation of the Red Sox statistical approach. It was a skull-imploding three-hour meeting that wore down the manager and his aide-de-camp.

"It wasn't just numbers," said Francona. "There was a lot of stuff about 'types' of pitchers and how a certain type of pitcher would be good against a certain type of hitter. It was interesting. We had talked about it all during the interview. I think maybe Grady wasn't interested, but I wanted the information. It was stuff that I used. But there was a little panic that day after the long meeting. You could skew those numbers any way you want."

When the meeting was over and Epstein and Hoyer left the room, Mills turned to Francona and said, "Whoa. We'd better win."

"I looked at it a little different than he did," said Mills. "In my view, all that stuff was just going to prove that he was making the right decisions anyway. But our heads were spinning a little bit because it was the first time we'd seen anything like that."

From the day he got the job, Francona's computer skills were overstated by a media machine eager to demonstrate the Red Sox departure from old-school Grady Little. It was easy to see the computer on Francona's desk and paint him as a hardball Bill Gates, but Francona's wife, his four tech-savvy children (aged 11 to 18 when he got the job), and Mills knew better.

"I didn't know how to use a computer," Francona said. "When I was in Philadelphia, I didn't have one, and I did everything by hand. I had a computer in Oakland when I worked with Macha, and it was kind of a running joke. You have no idea how many printers I broke on the road. It was a fiasco. It was like the games couldn't start because I didn't have a printer working in my hotel room. I was horrendous. Unfortunately, the only thing worse than my computer skills was my penmanship, so it was better for me to put the lineup into the computer so everybody could read it. That's what I did in Boston. I'd have the computer on the desk, and I'd type the lineup into it. I liked having the information, and I could use the information if somebody made it easy and pointed me in the right direction. Our computer guy on the road was Billy Broadbent, and he was never very far away. He knew if I called, he was the only one who could help me. It was amusing to everybody. Theo and those guys knew I was trying, but if I veered off the path, it was death."

"He's better than my parents," said thirtysomething O'Halloran. "But not by too much. And he went through computer equipment faster than any human being alive. I was the one approving those expenses, and his computer was always broken. He went through four or five laptops a year. Sometimes it was just tobacco juice on the keyboard. I'd get a message that Tito needs a new laptop, and I'd go down to his office and say, 'What the fuck is wrong with you?' It was ridiculous. In fairness to Tito, he did seem to have a lot of legitimate issues. Sometimes our IT guys could not fix it. It was a mystery what he was doing to it."

No computer program could have prepared Francona for New England's fixation on the New York Yankees. He was walking into a firepit of frustration built on eight decades of suffering, compounded by the '03 playoffs and the winter of frustration. New York fans and the New York media were having a field day with the Red Sox in the spring of '04, and the Sox had no answer. The Yankees had a 26–0 lead in World Series championships since the Sox last won in 1918. The Yanks had pilfered Babe Ruth and a succession of Sox stars in subsequent seasons. Whether it was Lou Gehrig, Joe DiMaggio, Bucky Dent, or Aaron Boone, the Yankees stole the Sox lunch money every time. And New York fans taunted the empty-handed New England fans. It was not easy being a Red Sox fan and a college student in Boston or New

York. You had to hear it from the New Yorkers, and there was no good response. The Empire State kids, meanwhile, could afford to be dismissive of their pathetic Boston counterparts. It was a one-sided dynamic.

In 2004 John Henry was firing off emails to reporters, complaining about the Yankee spending and comparing George Steinbrenner to Don Rickles. George retaliated by comparing Henry to Ray Bolger's scarecrow character in *The Wizard of Oz*.

The Red Sox had a week of exhibition games under their belts when Francona got out of bed on the morning of March 7, grabbed his coffee from the Homewood lobby, and drove in the dark toward Edison Road in sleepy Fort Myers. The Red Sox were scheduled to play the Yankees that day. His jaw dropped when he neared the City of Palms Park parking lot and saw fans sleeping on the sidewalk at 5:30 AM.

Game 8.

"I thought to myself, *Holy shit. I better change the lineup and put somebody in there that the fans will know.*"

A-Rod and Jeter both made the two-hour-and-15-minute bus trip from Tampa, and more than 250 media members covered the spring training game, an 11–7 Yankee win. It remains the most overhyped spring training game in baseball history.

Francona took comfort knowing that Epstein would not be bothered if the team lost Grapefruit League games. The Sox were not worried about selling tickets for the '04 season, and Boston fans were sophisticated enough to understand that spring training scores and standings didn't matter. The Sox went 17-12-1 in Florida in 2004.

They traveled to Baltimore for the start of the season, and Francona got to Camden Yards at 12:30 for an 8:05 PM start.

"When I woke up this morning, I was right in the middle of an inning," he said in his late-afternoon press conference. "I know that sleep as I knew it is done. That's just the way it goes. You guys may think I'm not looking too good right now, but wait until the end of the season."

The writers didn't have to wait until September. Francona looked horrible by midnight after he was tested by his ace, the ever-petulant Pedro Martinez.

Pedro was not sharp in the opener. He'd had a bad spring, was peeved with the attention Schilling was getting, and believed his lack of a contract extension was disrespectful. In his mind, Henry and Lucchino were making him pitch for his supper. They were making him prove himself instead of rewarding him for winning 101 games in six spectacular seasons in Boston. He was lifted in the seventh inning of a 7–2 loss to the Orioles and left the ballpark before the game ended, walking past a posse of reporters as if to bring attention to his defiance.

A player leaving the ballpark before the conclusion of a game violated everything Francona had learned in his formative years, but it wouldn't do any good to publicly call out Pedro Martinez on your first night on the job with the Boston Red Sox. The new manager was furious when he learned that Pedro left, and he took note of Martinez's going out of his way to make sure everybody knew. But when he was asked about it, Francona said, "In fairness to him, and everybody else, that [rule about leaving early] wasn't conveyed correctly on my part, and I take responsibility for that."

"That was bad for me," said Francona. "Publicly, I took responsibility for it, but we had to have it out. Pedro was testing me. I didn't want to embarrass him, so I went to the park the next day looking for him. It was an off day, and we had an optional workout. Somebody told me that Pedro was with our trainer, Chris Correnti. Chris saw me and said, 'Careful, he's about to blow,' and I said, 'Fuck that. Where is he?' I got Pedro to come into my office, and we had it out. He got real quiet. That's what he would do when he was mad, he'd get all stone-faced. I just wanted him to know how things should be. I wanted him to understand that I'd take a bullet for somebody, but you got to do things right."

Over the next eight seasons, when Sox players would misbehave, Francona would address it internally, then go in front of the media and minimize the transgression. It fortified his image as a manager who was too nice, too much of a players' manager, but Francona cared more about what it did to help the team win. Ripping a player might feel good in the moment and satisfy fans and media, but it was not the way to get maximum production from a roster of 25 ballplayers. Not in 2004. Building trust with the players was more effective than playing tough guy for an ESPN sound bite.

Martinez took his sweet time getting to the ballpark before the second game of the season, strolling into Camden Yards an hour and a half before the first pitch, while all of his teammates were already on the field stretching.

"Overall, Pedro was tough on me," said Francona. "He was one of the best pitchers in the game, but he was used to doing things his way, and that was difficult for me. He had the contract thing going too, and he and Schilling didn't see eye to eye. He had so much pride, and it gets in the way sometimes. He pitched great in his second game for us, and I took him out right in the middle of the eighth when he got to 106 pitches. He was pissed. He was staring at me as I walked out there because he was on a roll. That was my way of saying, 'Okay, I can play this game.'"

It was also a nod to Bill James, who'd calculated that opponents hit .231 against Pedro before 105 pitches and .364 after the 105th pitch.

The Sox split four games in Baltimore, losing the series finale when journeyman Boston southpaw Bobby Jones walked four Orioles in the bottom of the 13th. Things got worse after the late-night loss when a defective wing flap kept the Sox team charter plane sitting on a runway at Baltimore-Washington International Airport for four hours. By the time they switched to another aircraft, landed in Boston, and bused to Fenway, it was 7:30 AM. Players were told that they needed to be back at the park by noon for a 3:00 PM home opening start, awash in pageantry.

Standing in the clubhouse at 7:30 AM, Francona had to make a decision. He could retreat to his temporary home at the Brookline Courtyard Marriott near Coolidge Corner ("They gave me a handicapped room, which seemed appropriate"), or he could sleep on the dirty old couch in his Fenway office.

He pulled a Red Sox fleece over his shoulders and lay down on the couch for an hour.

The home opener was worse than the finale in Baltimore. The Blue Jays routed five Boston pitchers for 14 hits in a 10–5 drubbing, and Francona was forced to use first baseman/outfielder Dave McCarty on the mound in the ninth. It was the first time the Sox had used a position player on the mound in seven years, and given that it was necessary on *opening day* at Fenway made Francona look a tad ill prepared. It was not his fault, but it didn't look good.

"That was a helpless feeling," he said.

Theo brushed it off, chalking the whole thing up to "unusual circumstances."

"Theo was always good about things like that, and I appreciated it," said Francona.

The Sox met the Yankees seven times in a ten-day stretch at the end of April, winning six of the seven, including a three-game sweep in New York. Selig called it "a playoff atmosphere in April." Francona wondered how any baseball team could sustain such intensity. What was it going to be like if they played the Yankees again in October?

His first in-season team meeting came after a fifth-straight loss in Cleveland in late May.

"Guys, I just want to let you know that we're good and we're going to be good," he told them. "We're in it for the long haul. I'm not panicked. Just go about your business and pay attention to detail and we'll be okay. I love every one of you guys. You're going to be fine, trust me. That's it."

The Sox won their next four games, and 15 of their next 23.

Meetings are risky. It's the Earl Weaver Theory. The Hall of Fame Orioles skipper was reluctant to call a team meeting for one simple reason:

"What happens if you lose the next game?" asked Weaver. "Then what do you do?"

"I didn't like team meetings where the manager ranted," said Francona. "When you get to that point, you've waited too long. All you are doing is relieving stress. It didn't help. I was in those meetings as a player. When I was with the Cubs we had one, and Dallas Green was the GM, and he wanted to yell at us. We had the meeting in the old clubhouse at Wrigley, and when I plopped myself down in the front, other players told me not to sit so close to the front, and they were right. Dallas Green was ranting and raving, and there was stuff flying everywhere, and I was scared to death. It didn't help us win, though. It never does. When I was managing the Phillies, I called them together one night in Houston and I was out of control. I was so mad. I was lurching. You get so mad you don't even know what you are saying. Scott Rolen told me he thought I was gonna fall down because my knee was so fucked up. I asked Millsie, 'What did I say?' He said, 'I don't know. Nobody knows!' And that's what happens in a meeting like that."

It was early in the 2004 season, and Francona was looking at everything for the long haul. March, April, and May are months in which a new manager is building relationships. Francona wasn't worried about Johnny Damon's shoulder-length hair or David Ortiz's jewelry. Francona had worn a necklace as a teen, and he knew it bothered his dad ("He thought that made me rebellious"), but those things didn't matter in 2004.

"We never had a rule about not wearing jewelry on the field," he said. "Still, not one player did it. I'm kind of proud of that. I didn't feel right making it a rule. I told them I didn't like it and I wouldn't appreciate them wearing jewelry, but it was not a rule. It was a personal thing, that I didn't think was appropriate to make a rule. When we'd trade for a guy, I'd tell him he could wear stuff, but I didn't like it. Nobody ever did. David [Ortiz] was the best. He'd forget sometimes. He'd have that diamond earring in, and then before he'd go up to hit he'd hand it to me, and I'd put it in my pocket and I'd be thinking, *Holy shit, that's $50,000 in my pocket.*"

Eliminating traditional boundaries was part of Francona's strategy. With Epstein's blessing (and participation), the manager and his coaches played cards with the Sox ballplayers.

Francona believed that a deck of cards could get a team through a 162-game season better than any training regimen or motivational program. Spades, hearts, clubs, and jacks kill downtime, engage minds, and teach strategy. Egos are stroked or dented, and money changes hands. It's *action* — the lifeblood of young men who make their living competing against one another. Poker. Pluck. Hearts. Texas Hold 'Em. Boo-Ray. Tonk. The names of the games change, but the concept is forever. It's an important part of managing and can be a foundation for team-building.

"We started playing on flights early in the season," said Francona. "I'd always play with the coaches. Millsie and Dale Sveum. I've always contended that no one could be a good big league manager if he's a shitty card player. It's got all the things you need to do to manage. It's the same mentality. You have to not be afraid, gamble a little bit, think quickly. I started off playing with the coaches, but then David came up front one day and called us a bunch of pussies. We were playing Boo-Ray, a guts game where you hold the cards. He got into it, and then Mark Bellhorn, who was quiet but really funny and loved to gamble. Johnny Damon was next. The coaches and me would be at the end of the first-class section, and the players would be right behind the bulkhead, and we didn't get in anybody's way. The games got really good. When Theo traveled with us, I asked him if he was okay with me playing poker with the players, and he didn't care. He joined right in. I started to really look forward to it. I'd get all my work done on the plane as quickly as possible so I could get in those card games. I felt the team coming together with it. There would be a group of us playing, and another group standing over us screaming and yelling. We'd have 12 or 13 guys standing around."

A manager could get a lot of good work done in subtle fashion during card games. It was something Francona had learned on those ten-hour bus rides when he was managing in the Southern League. In the big leagues, airspace over the heartland was a place where he could talk to a player and make a suggestion or correction without tension or fear of misunderstanding. In his rough years with the Phillies, Francona had invented what he called "the Mississippi River Rule": no sulking after the plane crossed the Mississippi; break out the playing cards.

Francona was a master of building relationships at 35,000 feet.

"One of Tito's greatest strengths was to empower players and put them at ease and let them be themselves and feel great about themselves," said Epstein. "In a collegial way, he created this atmosphere. He created this atmosphere where you show up to the clubhouse and the outside world ceases to exist. It was just a group of guys doing what guys always want to do, hang out, have fun, make fun of each other, be themselves, show off, and find some sort of conflict on which they could all be on the same side, which was that night's game. That 2004 team had great camaraderie. They could go out and stand shoulder to shoulder, play hard together, then come in and laugh about it at the end.

"To accomplish that, Tito needed freedom. We couldn't have pulled that off with a lot of rules, so there was a conscious effort to create space for that to happen. It meant putting up with some stuff that wasn't at times the most professional, but it was a means to the end. It was also an incredible amount of fun. Some managers use fear, power, and intimidation. Tito managed brilliantly by giving in to it, accepting it for what it was, and removing his ego from it, selflessly and brilliantly. It created this insulated environment where the only thing that mattered was competing and having each other's backs. So of course we had very few rules, and of course the players were allowed to play cards, and of course Tito got in the mix, and at times I got in the mix. We had to because of what we were trying to do. We were trying to let our hair down and get past the puritanical, anxiety-filled, rigid conflicts of the era that preceded us in the late 1990s and 2001. We just wanted to relax and go play, and I think Tito really pulled that off."

Francona noted, "It's amazing what you can tell someone at three in the morning on a plane when you're playing cards — something that you could never say in the dugout during the game. You're having a beer, and you're relaxing and talking about the game. I could say to Mark Bellhorn, 'Were you out far enough on that cutoff relay?' It was better than calling someone into your office. This was more like a conversation among peers, and maybe Bellhorn would say, 'Yeah, I could have been out farther.' I could say to David, 'When you hit that pop-up down the left-field line, if you had run harder, do you think you could have been at second?' It was a way to get them to see things. You'd get other players chiming in, and it ends up being good. I got a lot of shit accomplished on those plane rides around those card games.

"You don't know how it's going to end up. But I liked the way we were coming together that season. And I knew where we'd better be going."

---

# CHAPTER 6

# "This is not acceptable"

**F**RANCONA LOVED HIS CORNER office at Fenway, the same space where Joe Cronin closed his door and met with Ted Williams in the 1940s. It was remarkably unchanged through the decades. When the office door was open, anyone in the Sox clubhouse could see the manager sitting at his L-shaped desk. Francona's desk was outfitted with a landline telephone and a printer, computer, and monitor. There were three drawers on the right side of the desk, which was a tad inconvenient for the left-handed manager. The office walls were adorned with black-and-white "subway" tiling, and there was natural light from two back-wall opaque windows—protected by diamond-patterned metal grates. In the dismal Fenway years in the early 1960s, unsuspecting Sox fans walking down Van Ness Street toward Jersey Street could have tapped on those frosted windows and interrupted Pinky Higgins making out his lineup card or perhaps swilling some scotch.

The Sox manager's space was spartan. The new-millennium Red Sox were a $1 billion enterprise, and employees were equipped with state-of-the-art technology and the full force of John Henry's resources, but there were limits to what could be done with the old bricks and small spaces of a ballpark creeping toward 100 years old. The manager's office was a great example of these limitations. The office had its own toilet, encased in a small corner stall just a few feet to the left of the desk. Privacy was minimal. In army barracks fashion, the latrine featured a brown swinging door, offering maximum exposure and minimal privacy. You could see under the door. You could see over the door. And you could hear the manager turning the pages of *USA Today* as he sat on the throne.

Jack McCormick said, "I can't tell you how many times I'd knock on the office door and Tito would say, 'Come on in,' and I'd open the door and look across the room and see his feet underneath that stall door."

"We got a lot of work done that way," said Francona. "I used to mess with the players too. Ortiz would knock and I'd be on the toilet, and I'd say, 'Come on in,' and he'd go, 'Oh, no,' and I'd say, 'Let's go over the signs.' I loved that old saloon door."

"This is a subject that's unfortunately impossible to avoid," said Epstein. "This gets back to what really appeals to Tito. He loves baseball. He loves the game. He physically loves the clubhouse. Emotionally, I think he loves to let go of the outside world. Some people compartmentalize the job. Tito compartmentalizes the real world, throws himself into the clubhouse, loves every aspect of the clubhouse. He loves being down there and loves nakedness, vulgarity. Loves joking around, loves busting people's balls, loves playing cards. He loves everything about it. It's part of the fabric of who he is. So the social norms about going to the bathroom, those don't always translate to the clubhouse to begin with, and he took it to a whole other level because of how deeply he believed in the clubhouse ethos. He would find satisfaction in a way that wasn't always satisfactory to others. He would stimulate the senses, all of them, olfactory, auditory. It was a way to disarm people too. I think he felt like once you had a conversation with him where he was involved in a natural act like that, he felt like it brought you closer to him. You were sort of in. He did it to media, PR guys, front office. It was basically impossible to have a conversation with him without seeing things that only a toilet paper holder should see."

The office was equipped with its own shower stall, no small perk in a home clubhouse in which the coaches' room had no bathroom facilities.

"That office has the best shower head in baseball," said Francona. "It'll knock you right through the door."

The manager's office also featured a small closet and a pedestal sink next to two frontal exit doors. A 42-inch plasma screen covered the wall space between the two doors. Guests could sit in one of three roll-away chairs or on the faux-leather couch situated against the wall to the right of the manager — the couch that sometimes served as Francona's hotel bed.

"I swear that couch had Johnny Pesky stains on it from when Pesky was a player," said the manager. "People would come into my office, sometimes not in a very good mood, and it was adding insult to injury to have them sit on that couch. I slept on it a couple of times a year."

There was a thermostat on the wall behind the desk, and Francona hung a few photos there to make the stark space feel homey: framed images of Tito Francona, Nick Francona, and the Francona girls. In later years the family pictures were joined by a photo of Muhammad Ali, a poster of David Ortiz hoisting Dustin Pedroia, a shot of Francona with Ted Kennedy, a tiny photo of Johnny Pesky in baseball underwear, and a framed letter from an angry Red Sox fan "telling me to stick the lineup card up my ass."

The Pesky photo always generated commentary and laughter.

"I used to tease Johnny that you could see his balls hanging on the floor in that picture," said Francona. "It was taken one of my first years there. Johnny was in spring training with us, and he was coming out of the shower, and I was kidding with him, and I put my arm around him and told someone, 'Take a picture of me and Johnny.' They took the photo and gave it to us the next day, and I said, 'Sign this picture, Johnny.' He laughed like hell. I used to fuck with him when he'd come into my office and say, 'I've got the only framed Johnny Pesky picture with his balls hanging on the floor,' and he'd laugh and say, 'Don't let anybody see that!' One year, after they did some remodeling in the office, the picture disappeared, but it's out there somewhere!"

The old Fenway office had New England barn–like qualities. Sox clubhouse workers kept sticky-traps in the corners of the office, and Francona sometimes found dead mice in the traps when he returned from road trips. When he found mouse droppings in his sink, he suspected they were deposited by mice scurrying on the exposed pipes overhead.

"Once a week I saw a mouse running under the toilet," said the manager. "That was okay, but it bothered me when I saw the stuff on the sink. I asked them to fix the place up for eight years, but they didn't do anything until after I left." (The office was overhauled, including a makeshift partition shielding the toilet and shower area, in the months before Bobby Valentine took over for Francona.)

He loved his office. He loved his job. He knew this was an opportunity to win. More than that, he was expected to win. This was nothing like Philadelphia. The expectations were fair. He'd never had a team like this before.

The 2004 Red Sox were not a wire-to-wire champion. They spent the entire summer in second place, sometimes more than ten games behind the Yankees. Given their payroll, star power, and expectations, they could be wildly mediocre. From April 29 to August 6 — a stretch of 86 games, which is more than half of a 162-game major league season — they were a .500 ball club, an imperfectly average 43-43. This took its toll on everyone around the team, particularly the new manager.

Francona and Epstein worked to find a mutual level of comfort during their first season together, but the exhaustive November interview process and six weeks of spring training had done little to prepare the manager and general manager for the pressure and rigors of the regular season in Boston.

The pre- and postgame presence of the GM in the manager's office was a new wrinkle in baseball. Terry Francona's dad, 15-year big leaguer Tito Francona, would have been shocked to see the amount of time the modern-day Red Sox general manager spent in the manager's office. This was not the way it was when Tito played in the 1950s and 1960s. GMs stayed in the executive suites and let the managers manage.

The notion of separate orbits for the manager and GM is nicely demonstrated by a story told by onetime Twins manager Ray Miller about his interaction with the late Howard Fox, GM of the Twins in the 1980s. The Twins were on a road trip when Miller got a call in his hotel room in midmorning. It was Fox, Miller's boss.

"How'd we do last night?" the GM asked the manager.

Miller was shocked. The general manager of the team had somehow gone to bed without checking to see if the Twins won or lost. He was calling to get the news firsthand from his manager. Thirty years later, this detached dynamic was unthinkable. GMs were demanding accountability before players stopped sweating. Francona would come into his office after a tough loss, and Theo would be sitting there with Francona's son, Nick, asking why Mike Timlin got up to throw in the eighth, or why the Sox pitched to Alex Rodriguez with first base open. Epstein wanted to talk to his manager after every game. Immediately.

"Theo, you can ask anything you want," Francona would tell his young GM. "If I don't have a good answer, that's on me. But sometimes there's a way to maybe ask it so it's not the minute I walk in the door. We all got emotions going."

"At first we had a 24-hour rule," recalled Epstein. "We wouldn't talk about anything. If something fucked-up happened in the game, I promised not to talk about it. We could calm down and both get perspective. We lived up to it for a while. Then one time he asked me, 'Hey, would you have done this?' And the rule was out the window for a while. We got away from it, but we really had to get back to it. There was some real-time stuff that had to happen. I would get down there in the ninth inning, and I picked up on the fact that after a tough game he would benefit from having ten minutes to himself to spit out his chew and brush his teeth. It was always a delicate balance."

Postgame toothbrushing was part of Francona's daily ritual. Immediately after every game, he took a few minutes to purge the Lancaster chew from his mouth and gums. It was good for his teeth, and it kept Epstein at bay for an extra moment.

"Sometimes I'd take some extra time brushing that shit out because I was trying to get my thoughts in order," said Francona.

"It was like a marriage in that respect," Epstein acknowledged. "We did little shit to piss each other off."

During one of the difficult summer stretches, Francona was surprised to find a thoughtful, concerned email from owner John Henry. He didn't feel like he knew the reclusive billionaire very well and hadn't had much contact with the big boss, other than occasionally shaking hands and chatting when Henry would come into the manager's office before a game or say hello from the owner's box next to the dugout. (Henry usually watched from his box upstairs, where his computer was more handy.)

The late-night missive from Henry was warm and caring.

"We were scuffling, and he said he was worried about me," remembered Francona. "He wanted to know if I was feeling okay. I got it late at night, in my hotel room after another tough loss, and it really made me feel good. It was a nice gesture. It didn't come across as fake, more like, 'Hey, I hope you can sleep. I hope you're okay.'"

Henry had a nice ritual of visiting with any Fenway fan who was struck by a foul ball, but otherwise he rarely interacted face to face. Like stat guru Bill James, the Red Sox owner found optimum comfort in front of the keyboard. Inanimate objects never pushed back.

John Henry was born on September 13, 1949, in Quincy, Illinois, and his family moved to Forrest City, Arkansas, when he was a small child. By his own admission, he was a loner as a young boy. He grew up on a farm and his recreation was listening to Harry Caray's St. Louis Cardinals broadcasts on the radio. Henry went to his first Cardinals game at Sportsman's Park in St. Louis when he was ten, while his dad was being treated for a brain tumor in a St. Louis hospital. After graduating from high school, Henry enrolled in a succession of California colleges, but never graduated. He spent much of his youth writing music and playing bass in his band, Elysian Fields. When his dad died in 1975, Henry moved back to Arkansas and ran the family soybean farm. In his late twenties, he took to studying commodities, and he started John W. Henry & Co., offering managed futures funds, when he was 31. The company tracked prices and identified market trends. It was all about numbers. And it was wildly successful. Within ten years, Henry had enough money to consider buying his own baseball team. Instead of buying the Kansas City Royals (by this time Henry was a Californian, and he didn't think he could move back to middle America), he moved to Boca Raton, Florida, and bought into the West Palm Beach Tropics, who were part of the Senior Professional Baseball Association. The Tropics were managed by Dick Williams. After one season with the Tropics, Henry bought a 1 percent share in George Steinbrenner's New York Yankees. Then he bought his own big league team. From

1998 through 2001, Henry was chairman and sole owner of the Florida Marlins. He was frustrated that he was never able to get a stadium built in southern Florida. In 2001, hemorrhaging money and unable to get his new ballpark, Henry looked into moving the Marlins or buying the Angels of Southern California.

The Red Sox were for sale in the autumn of 2001. John Harrington, keeper of the Yawkey Trust, was mulling offers from as many as six groups when Henry, Tom Werner, and Larry Lucchino inadvertently aligned.

The unlikely trio came together because Henry wasn't able to buy the Angels, Werner (a former owner of the San Diego Padres) didn't have the means to buy the Red Sox, and Lucchino was looking for a new gig after a nasty split with John Moores, his partner in San Diego.

The 2001 purchase of the Red Sox got its unlikely start when Werner partnered with Les Otten, a charming businessman who had made and lost a fortune in New England ski resorts. Werner was an accomplished television producer with *The Cosby Show, Roseanne,* and *That 70s Show* on his résumé. Born to wealth and educated at Harvard, Werner experienced a failed tenure as owner of the San Diego Padres in the early '90s. He'd been the mastermind behind Roseanne Barr's crotch-grabbing national anthem, and while he owned the Padres he was cited by the *Dallas Morning News* as "the single-most hated man in Southern California." According to Seth Mnookin's *Feeding the Monster,* "Werner used to beep his horn at a heckler holding a 'Honk If You Hate Tom Werner' sign at the entrance to the Padres stadium in order to avoid detection." Baseball commissioner Bud Selig was a big fan of Tom Werner's.

Werner and Otten stepped forward first when Harrington put the Red Sox up for sale in the winter of 2000–2001, but it was quickly apparent that they didn't have enough money. With the likes of cable billionaire Charles Dolan and Boston business tycoons Joe O'Donnell and Steve Karp stepping forward as well, Werner and Otten needed help. Werner knew Lucchino from their days together with the Padres, and he asked Lucchino to join him with the Red Sox bid. Lucchino quickly sized up the situation. It was obvious there wasn't enough money. Ever the facilitator, Lucchino brought Henry to Boston, even though Henry and Werner had never met.

"I told Larry I was only interested if I could be the lead investor," said Henry. "That's how it happened."

Henry flew to California and met with Werner at Mr. Chow in Beverly Hills. Henry demanded, and got, total control. He was bringing the money.

The Sox were awarded to the Henry group for a purchase price of $700 million. The package included the Sox franchise, Fenway Park, and 80 percent of the lucrative New England Sports Network (NESN).

Selig, who knew and trusted Henry, Werner, and Lucchino, brokered the transaction. Rather than take a chance on outlier Dolan (the highest bidder) or local favorites O'Donnell and Karp (who wanted to build a new ballpark near Boston's waterfront), Selig awarded the team to the Henry group. The attorney general of Massachusetts was among many who termed the transaction a "bag job," but nobody tells Major League Baseball what to do.

A few months after the transaction was formalized, Selig denied all involvement, telling a *Boston Globe* reporter, "I had nothing to do with any of that," then adding, "but someday you'll thank me for it."

Within the walls of old Fenway, the roles of the new owners were defined immediately. Lucchino would be club president and CEO and run the team on a daily basis. Werner would serve as "chairman," overseeing the club's television operation and spending much of his time feeling "marginalized." Henry would be the principal owner, the only vote that counted.

"John was great to me," said Francona. "He'd stick his head in my office and say hello before games sometimes. He wasn't a real hands-on people person. When you own the team, you have a right to do whatever you want. He'd bring people into my office all the time. Tom liked to bring people in the clubhouse, and that got to be a point of contention a little bit because it was often women who would be with him. The players were a little uneasy. I didn't make a point of going out there to meet them all the time, probably because I was hoping they would leave. The only times I saw Larry was if we were having a weather issue. He would come down to talk about it. That's a tough time. When there's a weather issue, I've got the starting pitcher sticking his head in my door and he's pissed. Everybody's on point and everybody's ready to go. There's some anxiety going on. And that's generally when I'd see Larry, so it could get a little tense.

"John, Tom, Larry, me — we all wanted the same thing. I just wanted all of them to know that I was doing what I could do to help us win as many games as possible. I didn't want them to think I was being stubborn, and sometimes that was frustrating. Sometimes that was a hard message to send."

Everyone knew there was tension between mentor Lucchino and protégé Epstein. Epstein bristled at the notion that he was Lucchino's creation and was uncomfortable with the number of Sox employees who'd migrated from San Diego with Lucchino. Many of them had been with the Padres when Theo was handing out press notes and didn't yet have his driver's license.

Francona knew none of the Lucchino-Epstein history and didn't care. He didn't go looking for Henry or Werner either.

"I made a choice early on that I was answering to Theo," said Francona. "I think John was doing the same thing, going through Theo to communicate with me. When Theo would ask me about some specific decision or strategy, I sometimes sensed it was coming from John, and I'd tease Theo about that, and he'd say, 'Yeah.' Whenever John did send me an email and I wrote back, I'd always tell Theo. Everybody knew I was going through him on everything, and that made things simpler. There was a little bit of tension with Larry and Theo, and it was easier for me to stay out of it."

The night of Thursday, July 1, in Yankee Stadium was one of the most important nights of the 2004 season. Nearing the midpoint of their season, the Sox were seven and a half games behind the Yankees, trying to avoid a three-game sweep in the Bronx. They had suffered an excruciating defeat in the second game of the series when the Yankees rallied for two runs in the seventh and two more in the eighth. A Nomar throwing error was the key play in the eighth, and this came one night after Garciaparra made two errors in the Yankees' 11–3 win in the series opener.

The season was not going well for Nomar, who had developed a serious Achilles tendon problem. He was still angry at the front office for attempting to trade him and replace him with Alex Rodriguez. He was fretting about his next contract. And he was turning into a defensive liability, which did not go unnoticed by Theo Epstein and his young men in baseball ops.

The root of Nomar's Achilles tendon injury was shrouded in mystery. He'd slumped at the end of 2003, batting .170 in September. There was a rumor that he'd bruised his right heel playing soccer over the winter, but others believed that Garciaparra sustained the injury working out in Scottsdale. Nomar had a story about a batting practice ball that bounced off his heel. Wiseguy reporters thought it was his ego that was bruised.

When Garciaparra woke up on the morning of July 1—the first day of the last month of his Red Sox career—he decided he was not going to play the third and final game of the series. He was sitting out one of every three games during this stretch of his comeback, and he opted not to change the routine even though the Sox were scuffling. Committing three errors in two games and seeing the image of himself booting a ball under the headline "April Fools" on the back page of the *Daily News* did nothing to improve his mood.

Remembering that his shortstop would not be available for the series finale, Francona's mind flashed to his first impressions of 20-year-old Nomar a decade earlier in the Arizona Fall League:

"He was an off-the-charts good kid. Michael Jordan loved him. We all loved him. In those days, the Fall League went into December, and Nomar hand-delivered Christmas cards to everybody. Nobody does that. That's good manners. He was polite and asked the best questions of any kid I'd been around. He was hungry to be good. I remember getting a visit from Red Sox manager Kevin Kennedy and his coach Tim Johnson. They wanted an update on their first-round pick, so they stopped by and had a few beers in my office. They asked me if I thought Nomar could make a switch to second base. I told them, 'I don't know who the fuck your shortstop is, but move that guy to second! Just let this kid play short. He throws on the run sometimes, so you need to have a first baseman who can catch it, but you want Nomar to be your shortstop. This kid can play!'"

It was good advice. In 1997, still playing shortstop, Garciaparra won the American League Rookie of the Year Award, hitting .306. Two years later—a year in which the Sox played the Yankees in the American League Championship Series — Garciaparra won the AL batting title with a .357 average. In 2000 he hit an astonishing .372 and became the first right-handed batter to win back-to-back batting titles in the American League since Joe DiMaggio.

The DiMaggio comparison was not a stretch, not in the early years. Nomar wore number 5, just like Joe D. Ted Williams was fond of saying how much Nomar reminded him of DiMaggio. Even Joe's brother, Dominic DiMaggio, a former Sox center fielder, got on board with the comparison.

Those were heady days for Garciaparra in Boston. After Roger Clemens and Mo Vaughn . . . before Pedro Martinez and Manny Ramirez . . . Nomar was the face of the Sox franchise. Nationally, he was part of a "who is the best shortstop?" debate, posing for magazine covers with A-Rod and Jeter. Red Sox and Yankee fans enjoyed the Garciaparra-Jeter argument, just as they'd debated the merits of Teddy Ballgame versus Joe D, and Carlton Fisk versus Thurman Munson. Even Nomar's name was fun. A *Saturday Night Live* parody poked fun at New Englanders' pronunciation of "No-maaaaah."

In 2001 Garciaparra ruptured a tendon in his right wrist in the spring, underwent major surgery, and played in only 21 games. He was never the same player when he came back. The ball no longer jumped off his bat. He went from Hall of Fame–bound to being just a very good, injury-prone player. As his skills diminished, it became increasingly difficult for him to stay out of the trainer's room.

What didn't change was his maniacal series of routines. All baseball players are superstitious, but Garciaparra took it to a new level. When he stood in the batter's box, fans would see a series of toe taps and batting-glove tugs before every pitch. New England Little Leaguers easily mimed his motions. Kids paying close attention might also have noticed that Nomar put two feet on every dugout step every time he went on and off the field. It was like watching a two-year-old climb stairs.

There was much more. Garciaparra became a virtual prisoner of his routines, and it was sometimes hard for the rest of the people in the clubhouse to be around him.

Garciaparra loved the fans and they loved him back, but he didn't appreciate attention away from the park. Boston was far removed from his southern California home in every way. Nomar grew up in an environment where baseball players went unnoticed at restaurants and in crosswalks. It was never like that in Boston, and Garciaparra did not understand why baseball was so important. He particularly loathed the Boston media. It was Nomar who got the Sox to embed a red line in the clubhouse carpet — a line in front of players' lockers that could not be crossed by reporters. He had little use for the "new" Sox ownership, which had failed to extend his contract, then tried to trade him. He was also involved in a one-man licensing dispute with Major League Baseball. Nomar was the only player in the big leagues who objected to the MLB logo on the back of his helmet. He was fined for removing the logo. After he reluctantly agreed to allow the logo on his helmet, he smudged it with pine tar so that it was unrecognizable. Then he was fined again.

"He was just Bostoned-out," said Francona. "I told him that. Things that didn't used to bother him seemed to be bothering him at that point."

The final night in New York in July '04 was one of the better regular-season games of any season. It looked like a Red Sox victory when Manny Ramirez broke a 3–3 tie with a solo homer in the top of the 13th, but the Yankees rallied for a pair of runs after Sox reliever Curtis Leskanic retired the first two Yankees in the bottom of the inning. When the four-hour-and-20-minute epic ended, Boss Steinbrenner issued an official statement, declaring, "This was the most exciting game I have ever seen in all of sports."

It wasn't received with the same glee in New England, largely owing to the appearance of Nomar sitting on the pine while managers in both dugouts emptied their benches over the course of 13 innings. Francona used a five-man infield (lefty outfielder Dave McCarty playing second base) to get out of a jam in the 12th. Kevin Millar played third base, first base, and left field in the same inning. Torre used every player on his bench. The highlight came when Jeter dove into the stands behind third base, snaring a foul pop by Trot Nixon. Jeter walked back from the play holding the ball aloft, blood pouring from his chin. He was taken to Columbia-Presbyterian Hospital for stitches and X-rays. While all this was going on, Nomar sat. In one damning TV shot, Garciaparra could be seen in the background, sitting, while most of his teammates were on the top step of the dugout, supporting a Sox rally. The image was devastating.

"He wasn't supposed to be in the lineup," Francona recalled. "We had an arrangement. When he came back from the Achilles injury, it was real obvious he couldn't play every day. And I told the media that. He'd play two out of three, and it had to stay that way regardless of who we were playing. I checked with him that night because I knew he was going to get crushed by the media. So I went to him. He was in the whirlpool, and I wanted to get across to him that I wished he would play. The way we were playing, I thought people were going to start taking shots at him, and I didn't want them to. When you start looking at people's injuries, it's really difficult. It wasn't that I was mad at him. The weird part was, when he started getting loose in the 12th inning, I was stuck. I was confused. He was down in the tunnel kind of loosening up, and he said, 'I'm getting ready.' I was like, *What do I do now?* On a long night when a guy is unavailable for four hours, it puts the manager in a tough situation when he's suddenly available and it's the 12th inning.

"The next day in Atlanta I told him, 'Maybe it's time for you to move on.' I asked if he had thought about that, and he said he had. I wasn't telling him that I didn't think he could play or that I didn't like him, but I was telling him that maybe it was time for him to move on. He had a lot going on. Obviously there were feathers that were ruffled. Thankfully, I wasn't part of that. We didn't ever bump heads, but it was obvious he wasn't happy, and that puts a strain on your team."

The strain was all over the 2004 Red Sox. Epstein had not planned on going to Atlanta for the next series, but changed his mind and went south to check on his team. He wanted to talk to Nomar and would reluctantly answer questions from the media. Most of all, he wanted to talk to his manager.

"It's really uncomfortable to be the GM and have the team go off the rails when you're not there," said Epstein. "That was the meat of the three-month stretch where this great team was playing .500 baseball and was playing really soft and was letting bad defense and mental mistakes sabotage game after game after game. We weren't a good outfit at that time. We weren't playing the game the right way, and it was tough to watch. People were getting pretty upset, and we all thought we were wasting this great opportunity with this great team. So I flew to Atlanta in part to answer questions about Nomar and calm that whole thing down and pat Nomar on the butt and try to stabilize things, but I really wanted to talk to Tito. I felt like he needed some support and that he was on the verge of letting the outside storm — which he was usually good at insulating himself and the team from — he was on the verge of letting it get the better of him. He wasn't being himself."

It was their first formal closed-door meeting dealing with an uncomfortable topic.

"Tito, you're not yourself," started the GM. "Remember what we talked about when you had your interview? Remember everything you said you stood for in the interview? Everything that's important to you? The type of leadership you said you could pull off? How you said you would handle situations and deal with the media and interact? Well, that was only six months ago, and there's a lot going on now. There's a storm going on, and the wolves are circling. Let's make sure we do this our way. We only get one chance to do this. All the things you said? That's the guy we hired. Go be that guy. Don't worry about what happened in this one game, or what this one player is doing, or what somebody wrote. Be that guy you were in the interview. That's what this team needs, and that's who we hired, and we need to be able to look in the mirror and know that we did it our way."

Francona was a little surprised. He didn't sense any urgency.

"Theo, I don't even know what people are writing or care," he responded. "Fuck, I know we're not playing great, but I believe in all the things I said, and we're gonna fight through this. That's the only way I know how to do it. We're okay. I believe in myself. What I told you in that interview, that's what's happening here. Our clubhouse is fine. Guys aren't complaining. We're spinning our wheels a little bit, but we're going to be fine."

Years later, Francona said, "I know Larry was worried and sent Theo down. And I know John was worried about me at that time. But there was nothing any different really going on. I knew we weren't playing great, but I didn't think we were panicked. When you go to New York and have a 13-inning loss, it exaggerates things. Those games were on television, and sometimes people tend to think that the games they see are the only ones going on. I never really knew what Larry and Theo were worried about there. Maybe I was naive, but I didn't think it was as big a deal as they did. It probably flew over my head. I didn't feel any different before or after Theo came down. The reality for us is that we wake up the next day and we've got another game. That's baseball. You do the same things the same way every day, and I never thought we got away from that."

In the first game after the Yankee Stadium debacle, Nomar was back in the lineup and managed to get three hits in another Red Sox loss. Theo thought his manager looked fine. The GM was comfortable with what he heard and saw from his manager.

The Red Sox were 48-39, eight games behind the Yankees, at the All-Star break. On the morning of the final game before the break, Manny said he couldn't play. He said his left hamstring was sore. Everybody knew the real reason. Pedro had been allowed to go home to the Dominican Republic a couple of days before the break. Sitting out the final game gave Manny his own little vacation.

It was always the hamstrings with Manny. It was the perfect ruse. A sore hamstring doesn't show up on X-rays. There's no bruising. For an elite athlete looking for a day off, it's the equivalent of "Not tonight, honey, I have a headache." It's impossible to prove or disprove.

While Pedro kicked back under a mango tree in the DR, Manny went to Houston with Schilling and Ortiz for the All-Star Game. When the Sox gathered in Anaheim for the second half of their season, Francona had Manny in the lineup for the first game of the series, but Manny said he could only serve as designated hitter. Manny had homered in the All-Star Game. This was another Manny test for the manager. Francona stood up to his slugger when the team moved on to Seattle. He told Manny he'd only put him in the lineup if he was well enough to play left field.

"Spending time with stuff like that is draining," said the manager.

Then came the game that took on a life of its own. Red Sox 11, Yankees 10. Varitek 1, A-Rod 0.

It unfolded on the weekend of July 23–25 (Nomar's 31st birthday was

July 23), the eve of the Democratic National Convention, which was held in Boston.

The Yankees won the series opener Friday night, and it rained hard overnight, casting doubt about the field's readiness when everyone returned to the ballpark Saturday morning for a nationally televised game scheduled to start at 3:15. Four hours before the game, Henry, Lucchino, and Epstein met with Nomar and his agent, Arn Tellem. Garciaparra expressed his unhappiness with everything related to Boston and the Red Sox.

Francona was not at the meeting. He was worried about the field. He knew Sox officials and television officials would want the game to start on time, and he prepared for Lucchino's "Doppler weather" visit. Routinely, the decision to play is made by the home team, then becomes the domain of the umpires once the first pitch is thrown. Before the inevitable visit from Lucchino, Francona inspected the field with Torre, Sox chief operating officer Mike Dee, and Fenway groundskeeper Dave Mellor. The outfield was swampy, and there were puddles on the warning track. Dee, the managers, and the groundskeeper agreed that the field was unplayable. Torre went back to his clubhouse. Still standing in the outfield, Francona took a call from Epstein.

"You need to come back to your office now," said Epstein. "There's a mutiny here. The players want to play."

Francona and Dee went back to the Sox clubhouse and found Epstein, Henry, Lucchino, Varitek, and Millar waiting in the manager's office. The players wanted to play the game. Pedro even volunteered to come in as a relief specialist. It didn't make the field playable, but it was a good sign that the Sox wanted to get back out there. Francona agreed. The game would be played.

"Somebody better tell Joe," Dee told the Sox manager.

Sitting at his L-shaped desk, Francona picked up the landline phone and punched "0."

Making a call, any call, from the manager's desk of a major league ballpark can be an ordeal. After the Black Sox scandal of 1919 almost killed big league baseball in the early 1920s, strict rules were put into place regarding phone calls to and from major league clubhouses. Typically, Fenway was among the slowest to change. Sox owner Tom Yawkey had hired Helen Robinson to operate the Fenway switchboard in 1941, and she was the guardian at the gate for more than a half-century. Yawkey ordered all calls to go through the switchboard, and the order was never rescinded. Helen was gone by 2004, but the telephone situation was unchanged. It didn't seem to matter that managers and players had access to cell phones. Phone calls within the ballpark remained decidedly old-school. It was the glacial pace of change in baseball. Francona considered himself lucky that he wasn't using a rotary device.

"The whole phone situation throughout baseball is amazing," he said. "I lived through it in Cincinnati with Marge Schott. Somebody's wife was having a baby late one night, and the switchboard guy went home cuz it was after midnight and there were no calls coming in or going out. It's the Pete Rose thing. In most clubhouses, the manager can't make a call. It's the stupidest thing I've ever seen. I've got a cell phone sitting right next to me. If I wanted to place a bet, I could place a bet on my cell phone. I can't tell you how many times I'd be in my office in another park on the road and I would try to call Theo, and the lady at the switchboard would be like, 'Who are you calling?' and I'd be like, 'We're going to make a trade in about five minutes, put me through!' In most of the ballparks you have to tell the operator who you are calling and log it in. It's not worth it. It's silly. At some parks it's impossible. In Chicago, you cannot make a call. You have to sign your life away. It's silly."

Some Yankee players were already showered and dressed for a cushy day off in Boston by the time Francona got through to Torre.

"Joe was pissed, and I don't blame him," said Francona.

The phone call put things in motion for a pivotal day in Sox history. The fireworks started in the third inning when young Bronson Arroyo hit A-Rod with a pitch. Rodriguez stepped back, yelled at Arroyo, then noticed Varitek getting up in his face. It escalated quickly, and the photo of Varitek shoving his catcher's mitt into A-Rod's grill wound up on the cover of Stephen King and Stewart O'Nan's best-selling book *Faithful*. It stands as the iconic image of the 2004 season. It represents the day the Sox wanted to play a game that the Yankees did not want to play, the day the Boston captain got into the face of A-Rod, and a day when the Sox beat the Yankees, 11–10, on Bill Mueller's two-run homer off Mariano Rivera in the bottom of the ninth. The game featured 27 hits, nine relief pitchers, four Red Sox errors (another one by Nomar), and five ejections.

"It was going to be a tough day to play to begin with," said Epstein. "We had our seventh starter, Arroyo, going. We were playing bad. It was the type of game where, if there was a gray area, you just bang it. The players, I think, felt management was trying to bang the game because we couldn't win that day. That created a burr under their saddle. They needed something to rebel against. It was like, 'Fuck this, we're going to play today and we're going to win today.'"

"I saw most of that from the clubhouse because I got thrown out of that game pretty early," said Francona. "Everybody thinks we stand in the tunnel, but at Fenway in those days you were doing yourself a disservice to stand in the runway. You couldn't see the game. So I went to the clubhouse and watched with the clubhouse guys."

He kept his lineup card with him. It was a Francona tradition. When he got tossed, he took the lineup card off the wall and monitored from behind the scenes.

"I remember thinking, *If we turn it around, this will be a game we point to*. But we didn't really turn it around. We were still big and slow, and we didn't catch fire for a couple more weeks."

Until a week after Nomar was traded.

The last week of Nomar's Boston career was messy. When the Sox were in Baltimore, three days before the July 31 deadline, Nomar met with Francona and told his manager that he'd be needing more time off and might have to go on the disabled list. Francona picked up his phone, dialed Epstein, and gave him the news. Epstein contacted Henry and Lucchino, who were in their Fenway offices upstairs, and asked them to come to the baseball operations offices. When he gathered the bosses, he called Francona and had him repeat Nomar's request.

"He was struggling more than we thought," said Francona. "He was frustrated. The whole Boston thing. He was getting criticized a lot. It was something that wasn't going to get better, and I wanted Theo to hear what I heard."

Three days later, while the Sox were getting ready for a night game at the Metrodome against the Twins, Nomar was traded to the Cubs as part of a four-team deal. It was a long, uncomfortable day. There was no manager's office in the visitors' clubhouse of the Metrodome, only a small room that managers and coaches had to share, and cell reception was poor. Francona had to keep trudging up and down the formidable steps to the field to get reception to talk to Theo. It was difficult for the physically challenged manager.

"Imagine if we couldn't have made that trade because the manager has a blood clot," he joked.

When the deal finally went down, Francona asked his coaches to vacate the tiny office, and he called Garciaparra in.

"Nomar, we traded you to the Cubs," said Francona, speaking just loudly enough to be heard over the office air conditioner. "You need to call Theo. Jack [McCormick] has a ticket for you to get a plane out of here."

The manager and shortstop exchanged a quick hug, then Garciaparra exited the clubhouse to make his call. After the call was made, Garciaparra came back into the room, said his good-byes, and hugged everybody, including a few media members.

In exchange for Garciaparra, the Sox received flossy Twins first baseman Doug Mientkiewicz and Montreal shortstop Orlando Cabrera. In a separate deal, Epstein also acquired speedy Dave Roberts from the Dodgers.

The young GM was nervous after making the blockbuster deal. Fearful of what he might see in the Sunday morning Boston papers, he took an Ambien and went to bed.

There was no immediate bounce from the dramatic deals. The Red Sox lost four of their first six games after the trades, including the opener of a weekend set in Detroit.

"A lot of people thought we were off and running after the July 24 game, or after the trade," said Epstein. "No. We played under .500 for the next two weeks."

On Saturday, August 7, Francona and Mills were in a cab, en route to Detroit's Comerica Park, when the manager's cell phone rang. It was Lucchino, and the CEO was hot. He kept saying, "This is not acceptable." The Sox had dealt an iconic ballplayer, putting Theo and management on the hot seat, and Lucchino did not like the way the ball club was responding. They'd been playing .500 ball for more than three months. At least four times during the rant, Lucchino told Francona that things were "not acceptable."

When Lucchino was done, Francona clicked off his phone, turned to Mills, and sighed.

"What was that about?" asked Mills. "You didn't have much to say."

"I don't know, Millsie," said the manager. "All I know is that apparently what we're doing right now is not acceptable."

# CHAPTER 7

## "Just crazy enough to think we can do this"

THE RED SOX WERE the best team in baseball in August, September, and October. The acquisitions of Cabrera, Mientkiewicz, and Roberts did exactly what the Sox brass had hoped. Before the trades, the Sox were big and slow and led the league in surrendering unearned runs. The deadline deals made them more nimble and defensively tighter. Francona started to substitute at the end of games when the Sox were leading. Mientkiewicz would come in to play first base, and Pokey Reese would come in to play second. Players accepted their roles. The Sox went 40-15 after Lucchino's "this is not acceptable" rant — 21-7 in August. They led the majors in runs, slugging percentage, and on-base percentage.

"We got on a roll," said Francona. "There was a definite difference. We were able to use our strength, which was our offense. And when we'd get a lead, we could get guys out and get our defense in, and it worked great. Using Kapler, Pokey Reese, and Mientkiewicz late in games made us really good defensively. We ended up playing everybody, so everybody was involved and that gave us a little personality, and it all came together."

But there were still a few tough nights against the Yankees, including Friday, September 24, at Fenway Park.

It was the final home weekend of the regular season, one last chance to play the Yankees before an expected rematch in the American League Championship Series. Pedro Martinez got the Friday night start and had thrown 101 pitches when he took the mound with a 4–3 lead at the start of the eighth. When Hideki Matsui led off the inning with a game-tying homer, Francona never moved. Then came a Bernie Williams double and a Ruben Sierra single before the manager finally lifted his ace.

Fenway fans booed. It wasn't any better across the Charles River at Quincy House, where high school senior Alyssa Francona, being recruited by Harvard, was watching the game with a large group of Harvard student-athletes.

But when comments about the Sox turned nasty — and turned on her father — Alyssa was ushered out of the room quickly. She went on to play softball at the University of North Carolina.

"I just tip my hat and call the Yankees my daddy," Pedro said after the 6–4 loss.

Francona defended his non-move and deflected all reminders of Grady Little and the recent past.

Waiting for Francona in the manager's office, Epstein was furious.

"We fucked up," muttered the young general manager.

"I could understand the irony there," said Francona. "But for me it was completely different. It was a regular-season game and we were trying to keep our bullpen in order."

The Red Sox recovered nicely, winning four straight, and seven of their last nine games, as they prepared to play the Angels in the American League Division Series.

Even though he'd never managed in the postseason, Francona was relaxed.

"I didn't know how I was going to feel, but I was surprisingly at ease," Francona said later. "We had prepared so extensively, I felt good about things. That doesn't mean you're going to win, but I knew what I was going to do. I didn't think there were going to be any surprises."

There were no surprises in round one. The Sox swept the Angels, winning the Game 3 finale on a tenth-inning walk-off blast by Ortiz at Fenway.

"Those boys are winning the World Series," predicted Angel veteran Darin Erstad.

The big surprise came when the Sox lost the first three games of the ALCS against the Yankees, dropping Game 3, 19–8, on Saturday night at Fenway. The humiliating loss featured 22 Yankee hits, including four homers, two by Hideki Matsui who went 5-6. The Yankees led 11–6 after four innings, and 17–6 in the seventh. Francona used six pitchers, including Tim Wakefield, who gave up his Game 4 start by volunteering for three and a third innings of mop-up middle relief. In their private suite upstairs on the third-base side of home plate, Tom Werner and Larry Lucchino opened a bottle of Glenlivet and wondered what happened to the dominant team they'd built.

"We had been steamrolling people," said Francona. "And there we were wondering what the hell hit us. Everything was coming apart at the seams. We got embarrassed. But in a lot of ways that's when I thought we were doing our best work. We ran out of pitching in Game 3, and everybody thought we were giving up, but we were doing just the opposite. The object that night was to keep Mike Timlin and Keith Foulke fresh. We were scrambling, trying to figure out how to keep the series going. Halfway through Game 3, we were already thinking about Game 4. When Wake came up the steps with his glove and volunteered to pitch, I was like, 'Here we go. He's going to eat up some innings, and we're going to save our guys and be ready for Game 4.' It kept me going. At that point it was all about how we were going to move forward."

When the four-hour-and-20-minute debacle ended, Francona hopped into a golf cart to make his way across the field to the makeshift playoff interview room, which was accessible through the garage door along the wall down the left-field line. As the cart rolled along the Fenway warning track toward the left-field corner, the manager was pelted with beer cups, popcorn boxes, and insults tossed from the stands. It reminded him of a long-ago night in the Dominican Republic when fans banged on his dressing room door after a game in which he pinch-hit for local favorite Tony Pena.

Speaking to the media, Francona stayed the course. He said the Sox would not try to think about winning four straight games.

"We're going to show up tomorrow and try to play one pitch at a time, one inning at a time," he said.

When he returned to his clubhouse, Francona was asked to break up a fight in the family room, where Shonda Schilling and Michelle Mangum (Damon's fiancée) were in each other's face over some "lucky" scarves that Shonda had brought in for the players' wives. Mangum hadn't been wearing her scarf, made a sarcastic remark about the scarves not doing any good, and was rebuked by Ms. Schilling, who suggested that Damon might not be 1-13 in the series if Mangum had been wearing her scarf. That was it: go time.

All hope . . . all dignity . . . seemed lost.

Reeling from the hideous beating and the reality that the promising season was almost over, Francona drove Jacque and the kids back to the Brookline Courtyard Marriott. As the car maneuvered down Beacon Street in Brookline, 14-year-old Leah recounted her tale of the catfight in the family room. Leah wasn't allowed to repeat some of the language she'd heard. Everyone in the car giggled.

"It was hard not to laugh," said the manager. "It was like some Greek tragedy. It's not that you don't care, but man, we're trying to win a World Series, and this is what we end up talking about. It seemed like we couldn't even lose in a good way!"

Theo Epstein found another way to deal with the disaster. Slumped on the old couch in Francona's office after the manager vacated Fenway, Epstein was rescued by Jed Hoyer, Peter Woodfork, and Jonathan Gilula, three of his baseball ops clones. Concerned for his well-being, they took the 30-year-old GM back to a friend's apartment above the Baseball Tavern, which was situated one block from Fenway, at the corner of Boylston Street and Yawkey Way. They did not want Theo to be alone. The apartment TV offered little comfort. Somebody found a replay of the Grady Little Game from 2003, and desperate Theo forced himself to watch more bad history. After five vodka tonics, he passed out on the apartment couch and slept in his clothes.

Saturday night's thrashing was somehow cleansing. There was nothing more to lose. Abject embarrassment was all around them. No team had ever come back from 3-0 in a best-of-seven baseball series. Why not the 2004 Red Sox?

Hours before batting practice, the always optimistic Mills went around the clubhouse, stopping for private pep talks with individual players. Making his rounds, patting guys on the back, and telling them all was not lost, Mills was distracted by a screeching sound. In the middle of telling Mark Bellhorn, "We can do this," Francona's aide-de-camp was overwhelmed by the fingernails-on-a-blackboard sound of first-base coach Lynn Jones unrolling packing tape, sealing up cardboard boxes for transport to his off-season home.

Mills went to Jones and gently reminded him that he was trying to inspire the troops. The season was not yet over.

"Okay," said Jones, putting down his roll of packing tape. "But it doesn't hurt to have a few things ready."

Kevin Millar was optimistic. He told every teammate, coach, writer, and fan the same thing: "Don't let us win tonight. If we win, we've got Pedro going tomorrow, then Schilling, and then anything can happen in Game 7."

"You can feel sorry for yourself and pack it in, or you can give it a shot," said Francona. "Whatever they believed, I believed. Totally. Four in a row is awful daunting, but we just decided to play pitch to pitch, out to out, inning to inning. It wasn't like we had to dig deep. It was about being the same. It's how I've tried to be my whole career. The idea is to be consistent. Anything else, players see through it. I didn't give up. Whether it was 19–8 or 0–0 in the first inning. We're trying to figure out what we do next to put our best foot forward.

"It's easy now to look back and say that I thought we were going to win. I just knew we were good enough. We had played as good as you can play for seven weeks. We'd put ourselves in a position where if we make another mistake, we go home. But I did think we could do it. When I walked through the clubhouse before Game 4, and they had *Animal House* playing on the clubhouse TV, I thought, *These guys are okay. They're just crazy enough to think we can do this.*"

Sox players prepared for Game 4 the same way they had for every other game. Gabe Kapler went to the weight room, Bronson Arroyo strummed his guitar, Tim Wakefield sat at his corner stall doing a crossword puzzle, and Johnny Damon brushed his glorious hair. Jason Varitek sifted through a three-ring binder filled with details about the Yankee hitters, and Pedro Martinez arrived too late for the pre–batting practice team stretching exercise in front of the first-base dugout.

"To a man, their body language was identical to what it had been every day, and that's what I liked," said Francona. "You can tell when people are faking it. They were not faking it."

Francona made sure his players saw that he wasn't doing anything differently. His Game 4 lineup included the same nine batters who'd started the blowout in Game 3. As ever, he wanted to avoid any appearance of panic.

"When things are getting out of control, players look at you," said the manager. "They take their cue off of you. If I'm losing it, it gives them the right to lose it. There were a lot of times when my stomach would be churning, but I didn't want them to think that I was panicked. That's not going to help."

Derek Lowe was Francona's Game 4 starter. A free spirit with a great arm, Lowe won 70 games, saved 85, and pitched a no-hitter in his seven seasons with the Red Sox after he was acquired along with Varitek in a spectacular Dan Duquette trade that sent struggling reliever Heathcliff Slocumb to Seattle. Everybody liked Lowe (six-foot-five, blond, a free spirit, Lowe was a classic "himbo" — the male version of a bimbo), but he was frustrated and anxious because he'd underachieved in the 2004 season, the final year of his contract. His ERA in his walk year was 5.42. He had an unhittable sinker, but lacked maturity and partied too hard, and his performance sometimes belied his immense talent. Francona loved Lowe's raw talent, but wondered about his reliability.

"D-Lowe and Larry Lucchino got into it a couple of times that year, and he was upset with me when I told him he wasn't going to be in our rotation at the start of the playoffs," said Francona. "I told him he'd probably have something to say about the outcome, but I had no idea he'd do what he did for us. I still can't believe more isn't made of what he did for us in that month."

Despite Lowe's solid start, the Sox trailed 4–3 in the bottom of the ninth, and it looked like the '04 season of great expectations was going to end with an embarrassing sweep at the hands of the hated Yankees. As the Sox got ready to bat against the indomitable Mariano Rivera in the bottom of the ninth, Lucchino took out a yellow legal pad in his upstairs suite and started composing his concession speech. At the same time, Francona met in the dugout with Bill Mueller, due up second, and Dave Roberts, who had stolen 38 bases in 41 tries in 2004.

Millar was leading off, and Francona planned to pinch-run if Millar reached. He told Mueller he might have to bunt if Roberts was on first base with no outs. He also told Mueller he might have to hold off on swinging the bat if Roberts had an opportunity to steal.

"Hey, you're giving me like 12 things to do up here," said Mueller.

When Millar led off with a five-pitch walk, Francona nodded at Roberts, and the five-foot-ten outfielder started walking up the dugout steps. Before exiting the dugout, Roberts turned to his left to take one last look at the manager. Francona winked at him.

Francona was happy to see Rivera throw over to first base three times before delivering a pitch. It helps the baserunner when the pitcher throws to first. It makes it easier for him to get loose and get into the rhythm of the game.

When Rivera finally threw a pitch, ball one, Roberts broke for second. Jorge Posada, the Yankee catcher was ready. Posada's howitzer throw was slightly to the shortstop side of second base, and Roberts, sliding headfirst, got his hand under Jeter's tag. Second-base umpire "Cowboy Joe" West, a man who loved the spotlight, signaled "safe" — with gusto.

Sitting on the bench, to the right of Francona, Mills checked his stopwatch. The baseball term for the time it takes a catcher to catch the ball, transfer it to his right hand, and get it to second base is "pop time." Mills clocked Posada at 1.79 seconds. Pop time was not a Posada strength, but Mills had never clocked anyone faster.

After the steal, Mueller singled to center, scoring Roberts with the tying run and keeping the Sox alive for a run at history. Francona high-fived Roberts when he came into the dugout, then went back to work.

After Ortiz popped out with the bases loaded to end the inning and send everybody into the tenth, Mills went into the clubhouse to see if his stopwatch was working correctly. He ran the video a couple more times. There was no mistake. The slowest recording he could get was 1.81 seconds. Posada had made the quickest throw of his life, but Roberts still managed to beat the throw.

Mills went back to the dugout, sat next to Francona to watch the tenth, and said, "You have no idea how close that was."

Roberts lives forever in Red Sox lore.

"When we got him, we hoped that he would accept being an extra outfielder and help us win games off the bench," said Francona. "That's a little bit tricky, but he accepted it about as professionally as I've ever seen. You tell guys, 'Stay ready, you might be able to impact the game.' And he was the ultimate example of that. You can't do more than he did."

Three innings later, at 1:22 AM, Ortiz crushed a two-run homer off Paul Quantrill in the bottom of the 12th, and Fox's Joe Buck told America, "We'll see you later today."

Game 5 started less than 16 hours after the stirring conclusion of Game 4. It didn't leave much time for sleep or introspection. After a few hours at the Brookline Courtyard Marriott, Francona was back at Fenway at 10:30 AM, getting ready for Game 5, sitting in front of a chalky glass of Metamucil, joined by his still-hungover 30-year-old general manager. Solidarity? More like superstition. Anything to prolong the series.

"Theo would do anything for luck," said Francona. "If he thought we could win, he'd have drank that stuff until he wasted away."

Game 5 turned out to be the longest postseason game in history, a whopping five hours and 14 minutes over 14 innings. Wakefield pitched the final three innings for the Sox, holding the Yankees scoreless, but he was throwing to Varitek instead of Doug Mirabelli, and Varitek was not accustomed to catching the knuckler. Varitek allowed three passed balls, including a pair that moved Matsui to third in the 13th inning.

"I kept asking 'Tek, 'Are you okay?'" recalled Francona. "You're asking a guy to do something he hadn't done all year. You don't want to embarrass him, but you want his bat in the lineup because he was an offensive force that year. When your captain says, 'I can handle it,' you let him go. My heart was in my throat, but I trusted him so much. I figured he'd put his face in front of it if he had to. We all thought we were going to win that game. We weren't sure how, but we just believed we were going to win. That was the feeling that permeated the rest of the year."

Ortiz won it again, dumping a single into center in an epic ten-pitch, six-foul at-bat. Appropriately, Wakefield got the win.

They flew right to New York to get ready for Game 6 the next night. And there was hope that Curt Schilling might be able to start Game 6.

Schilling was a Francona favorite long before the infamous Bloody

Sock Game of 2004. As a high school pitcher and junior college pitcher in Arizona, he was tracked by Red Sox scout Ray Boone. Ray Boone was the father of big league catcher and manager Bob Boone and the grandfather of Aaron Boone — who ended the Grady Little Game in '03. Ray Boone finished his 13-year major league career with the Red Sox in 1960, then went to work as a scout for the Red Sox. He signed young Curt Schilling out of Yavapai Junior College in 1986, and the Sox wound up trading Schilling (along with Brady Anderson) to the Orioles for Mike Boddicker in the summer of 1988. Schilling was an underachieving, ordinary pitcher over the first half of his career, but blossomed with the Phillies in the early 1990s, and when Francona came on board to manage the Phils in 1997, Schilling was the Phillies' ace.

Schilling's ego and outsized personality never bothered the manager. "He says stuff once a week that he shouldn't and he takes it back," admitted Francona. "He knows he talks too much, but he's always prepared. The things that irritate other people — his face on TV and his politics — I didn't care. I respected him a lot and have as much affection for him as any player I've ever had. He always had respect for coaches. We'd have arguments. But I respected how he competed. He wanted the ball, and he wanted to be in the big games. I never worried about Schill. Guys would poke fun at him when he wasn't around, but he had the ability to laugh at himself. I don't think any of it was malicious. They liked him on the mound, and so did I. When he was surly, he was ready to go. He expected everybody else to be ready to go."

Schilling also protected his teammates. Francona never forgot a game in Philadelphia in which Schilling threw a 100-mile-per-hour fastball at Deion Sanders, then asked, "What is he going to do about it, arm-tackle me?" In another game, Schilling gave up a certain victory when he was ejected in the fifth inning of a game in which he had a comfortable lead. He gave up the win because he was defending Scott Rolen, who'd been hit by a pitch earlier in the game.

Only Francona could call Schilling a "fat-ass" in front of other players. He could make fun of the big righty in press conferences. But he could not shake the image that Schilling was running the team. The thick-skinned manager was stung by the words of Philadelphia radio shock jock Angelo Cataldi, who wrote, "The running joke in Philadelphia is that Curt Schilling was the first player-manager in baseball since Pete Rose."

"It was one of the few times I wanted to punch somebody," admitted Francona.

The Sox got everything they wanted from Schilling in 2004. He led the majors with 21 wins and finished with an ERA of 3.26. The Sox went 25-7 in his 32 starts. Everything was going according to plan until he hurt his right ankle in the first game of the playoffs at Anaheim. He was routed by the Yankees in Game 1 of the ALCS, surrendering six runs in three innings of a 10–7 loss. It was his shortest outing of the season, and Schilling was depressed after the game. He was pitching with a torn tendon sheath and could not push off the rubber. A pain-killer shot did not help.

Before Game 5 in Boston, Dr. William Morgan performed a 20-minute procedure on Schilling's ankle, stabilizing the flapping tendon by attaching the skin around it to the deep tissue, creating a makeshift sheath to hold the tendon in place. A day later Schilling arrived at the visitors' clubhouse in Yankee Stadium and found a pearly white game ball in his left shoe.

"Schill gets a lot of needless flak, but that night was as legit as they come," said Francona. "I expected him to pitch well, and that wasn't fair. He was a mess. He had wanted this moment so bad. He wanted to be on that stage and be the guy. We all talked, and I said, 'Schill, this will probably hurt your career.' He had no business pitching that game, let alone winning. You can't get enough superlatives. When he was warming up down in the bullpen, I kept waiting for the phone to ring because the stitches were popping. He had to break one on purpose because it was pinching him. I know some people poke fun and say the blood on his sock was ketchup, but it was legit."

After the implosion of 2003, Grady Little talked about some of his players not wanting the ball in the big moment, players thinking about the franchise's past failures. Schilling was the antidote to that. He wanted the ball in the big moment, and he'd pledged to end the 86-year-old curse.

"He wanted to stick it up their ass," said Francona.

The manager tweaked the Sox pregame routine before Game 6, granting players permission to skip on-field batting practice and hit in cages under the stands. It was Millar's idea. Millar didn't want his teammates to have to watch any more of the incessant Yankeeography videos celebrating Ruth, Gehrig, DiMaggio, Mantle, Munson, and the rest of the Pinstripes who'd tortured Boston for eight decades.

Millar had another idea, one involving paper cups and a bottle of Jack Daniels. Just a sip to get warm and take the edge off. It became a celebrated ritual when Millar talked about it in television interviews after the Red Sox won the World Series, and the irony was lost on no one seven years later when Francona's firing triggered a series of stories regarding starting pitchers drinking beer in the clubhouse during games in the disappointing season of 2011.

"I saw the Jack in New York," confessed Epstein. "Millar came up to me and offered me a cup, and I said, 'I don't see that shit,' and turned and walked away."

"I never knew about it until the World Series," said Francona.

Schilling didn't have any Jack in his veins when he took the ball on the Yankee Stadium mound. With blood seeping into his sanitary sock, he pitched seven innings, allowing one run on four hits with zero walks and four strikeouts. Bellhorn's three-run homer off Jon Lieber in the fourth inning gave Schilling all the runs he needed in a 4–2 Sox victory. The Yankees never attempted to bunt on one-legged Schilling.

After the Game 6 win, late at night, Ortiz went to a restaurant in Queens with his extended family. A year earlier, in New York before the infamous Pedro-Grady Game, Big Papi had stayed in his room all night, fretting about Game 7. He barely slept. In '04, Ortiz rejected the safe route and went out on the town before the final game of the greatest comeback in sports.

"I said, 'I'm not going to stress out this time,'" recalled Ortiz. "We went to my favorite place. My family and friends, we stayed until two in the morning, eating and drinking. There were Yankee fans by the exit, and when I walked by them they make jokes about how I wouldn't get any hits in the game the next day. I told them I was going to go deep in my first at-bat against Kevin Brown."

The Sox skipped batting practice and sipped Jack Daniels again before Game 7. Pedro Martinez ducked his head into Francona's office and offered to pitch in relief. The panicking Yankees summoned all their ghosts. Yogi Berra gave a pregame pep talk in the ancient locker room, the nervous hosts made space for Red Sox owners in the Babe Ruth Suite, and Bucky Dent was brought in for the ceremonial first pitch.

The Yanks would have been better off leaving Dent on the mound to face Damon at the start of the game because broken-down Kevin Brown had a bad back and nothing else. With a few drops of Jack burning in their bellies, the scalding Sox scored six runs in the first two innings, sucking the blood from Steinbrenner's face and removing all drama from the outcome. True to his word, Ortiz homered off Brown in his first at-bat. Damon hit two homers, including a grand slam. Lowe, wearing spikes purchased at a nearby sporting goods store because he'd forgotten to pack his game shoes, stopped the Yankees on one hit through six innings.

Leading 8–1 after six, Francona brought Pedro in to pitch the seventh. Fans scratched their heads. The Sox owners looked at one another and wondered.

"Pedro had come to me before the game and said he wanted to pitch out of the bullpen," said Francona. "I was thrilled that he was 'buying in' to what we were doing. I also wanted to keep Bronson [Arroyo] behind him because Bronson was one of the guys that was fresh. I knew what I wanted to do. I wasn't as concerned about the beginning of the inning as I was about the end of the inning. I knew Bronson could put the fire out if we needed him. When Millsie asked me what I was doing, I said I just wanted to get the crowd involved."

Martinez quickly gave up a pair of runs, and Yankee Stadium came to life ("Who's your daddy?") for the first time all night.

On the bench, Mills looked at Francona and said, "Way to go! I think you accomplished your goal."

The Yankees cut the lead to 8–3, but Pedro got through the inning and the Sox kept scoring in the eighth and ninth. At one minute after midnight, with Alan Embree on the mound for Boston, Ruben Sierra grounded to Sox sub second baseman Pokey Reese for the final out.

The Red Sox are still the only team in major league baseball history to win a seven-game series after trailing three games to zero.

While the Idiots of '04 were soaking the visitors' clubhouse with champagne, Francona sat at his desk and answered a call from Torre. Tito Francona's long-ago teammate congratulated Terry Francona,

Six-year-old Terry Francona stands outside the dugout with his dad, Tito Francona, in St. Louis, 1965. *Courtesy of the Francona Family*

Family day at Fulton County Stadium in Atlanta, 1967. The Francona family (left to right): Terry, Tito, Amy, and Birdie. *Courtesy of the Francona Family*

Draft day, June 1980. Arizona outfielder Terry Francona celebrates with his father, Tito, after being selected by Montreal in the first round.   *UPI / Bob Taylor*

Francona was a career .274 hitter with 16 home runs and 143 RBI in 708 games over 10 seasons, from 1981 to 1990.

*Getty Images / Richard Mackson*

Francona is carried off the field in Montreal after avoiding collision with Pirates pitcher John Tudor. A series of knee injuries derailed his once-promising career and forced him to retire as a major league player at the age of 31 in 1990.   *The Canadian Press / Denis Cyr*

Three generations of Franconas at spring training with the Indians in Tucson, 1988 (left to right): Terry, his son, Nick, and his dad, Tito. *Courtesy of the Francona Family*

Francona with his oldest daughter, Alyssa, in 1990, his final year in the majors.

*Courtesy of the Francona Family*

Managing the world's most famous athlete, Michael Jordan, with the Birmingham Barons, 1994.

*Getty Images / Patrick Murphy-Racey*

Francona built a relationship with Curt Schilling during his first managing stint. They were under .500 in Philadelphia but reunited in Boston to win two World Series.

*AP Photo / Chris Gardner*

Francona and young general manager Theo Epstein were side by side for eight seasons in Boston. *AP Photo / Winslow Townson*

Boston's young, innovative baseball operations department provided Francona with plentiful statistical data. *Brita Meng Outzen*

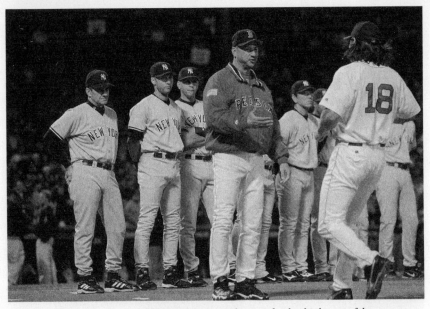

Francona greets Johnny Damon during pregame introductions for the third game of the 2004 ALCS, a game that ended disastrously for the Red Sox. *Getty Images / Jed Jacobsohn*

Even though they were adversaries at the height of the Yankees–Red Sox rivalry, Francona had great respect for Yankee manager Joe Torre (right), who had played with Terry's dad in the 1960s.

*AP Photo / Amy Sancetta*

With David Ortiz before the 2004 World Series at Fenway.
*Getty Images / Boston Globe / Stan Grossfeld*

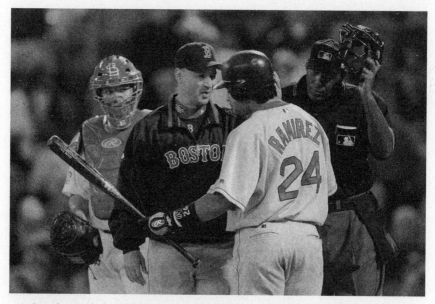

Consulting with slugger Manny Ramirez during Game 4 of the 2004 World Series in St. Louis after Cardinal catcher Yadier Molina accused Manny of stealing signs. Umpire is Chuck Meriwether. *Getty Images / Ezra Shaw*

Celebrating the 2004 championship on the field in St. Louis with Theo Epstein.

*AP Photo / Charles Krupa*

Francona wore his college baseball cap (University of Arizona) as a silent protest in the Duckboat parade for the 2004 World Champions in downtown Boston. He had had a small issue with Sox brass in the moments before the parade.

*Getty Images / Boston Globe*

then asked to speak with Wakefield, the man who'd given up the crushing home run to Boone one year earlier. Wakefield was a veteran who commanded respect in both dugouts, and Torre was among those who'd admired the way the knuckleballer handled himself in the 2003 ALCS, especially after giving up his start in Game 4. Francona waded into his chaotic clubhouse, found the veteran knuckleballer, and brought him into his office to speak with the manager of the vanquished Yankees.

"Time to play Finland now," said Theo Epstein.

The young GM was barely old enough to remember "The Miracle on Ice," but he had a great sense of history. America's 1980 hockey gold medal represents one of the great sports stories of all time (topped perhaps only by the '04 Red Sox), but it's often forgotten that the USA victory over the USSR was not the gold medal game. After Al Michaels asked America, "Do you believe in miracles?" the USA skaters had to beat Finland in order to secure the gold.

In 2004 the St. Louis Cardinals were Finland. They were the team standing in the way.

Boston fans were happy the Sox were playing St. Louis. The Cardinals are a signature National League franchise. St Louis beat the Red Sox in a seven-game World Series in 1946 and again in 1967. New England baseball fans were in no position to be choosy, but winning a World Series against the Cardinals had more meaning than beating the Houston Astros or the Colorado Rockies.

The Cards were not only traditional, they were beatable. St. Louis won 105 games in 2004 and had three players (Albert Pujols, Scott Rolen, and Jim Edmonds) who finished in the top five in the 2004 National League MVP vote. But the Sox felt they could beat the Cardinals. On Friday night, October 22, while most of the Sox front-office employees ate and drank at a lavish World Series gala at the John F. Kennedy Presidential Library, Francona gathered with his coaches, his catchers, Epstein, and several scouts and members of the baseball operations department in a conference room outside Theo's office. They went over a scouting report that had been prepared by a team of scouts and baseball ops executives, including Dave Jauss, who'd played baseball with Dan Duquette at Amherst and came to the Sox as part of the Duquette regime. Jauss and his men were confident that the Red Sox would be able to handle St. Louis pitching. The report was very specific regarding how to pitch to Pujols (curveballs breaking back over the inner half of the plate). The reports were spectacular.

Two months after the World Series, when Francona learned that Jauss hadn't received a full playoff share, he sent him a personal check for $20,000 for Christmas.

"Dave Jauss busted his ass for us, and I wanted him to know I appreciated it," said Francona. "Those reports were great."

Epstein agreed.

"When the World Series began, Theo was a completely different person," John Henry said. "He was so confident. I have never seen a man in the game of baseball that was as confident as Theo was."

The Sox won Game 1, 11–9, on a two-run homer off Pesky's Pole by Bellhorn. Sox fans were rough on Bellhorn throughout the 2004 season. He was a player who seemed to do only three things at the plate: walk, strike out, or hit a homer. He was the model Francona player, and the manager stuck with him when others would have abandoned him.

Game 2, the final Fenway game of the season, was a 6–2 victory, another spectacular outing by bloody-sock Schilling (six innings, four hits, no earned runs).

The Red Sox made four errors in the first two games, but it didn't matter. Boston did not trail for a single inning in the entire World Series.

Game 3 was played in St. Louis on Tuesday, October 26. It was the first game the Red Sox had played in Busch Stadium since the fifth game of the 1967 World Series, when Jim Lonborg beat the Cardinals, 3–1. It was a return to the scene of Francona's worst baseball moment, when he blew up his knee in 1982. The game was played on the 18th anniversary of Bill Buckner's fateful error in the 1986 World Series, and it turned out to be the final game that Pedro Martinez pitched in a Red Sox uniform.

It was in the visitors' clubhouse at Busch that Francona saw the Jack Daniels ritual for the first time.

"I was already in the dugout when they were doing that the first few times," recalled the manager. "But before Game 3 in St. Louis, I came back into the clubhouse because I forgot something, and I saw them all in the middle of the room, toasting. I saw what they were doing, and I was laughing, thinking it was funny, and they were like, 'Hey, you've got to have one.' Well, I didn't ever think it was real. So I took a drink and I was like, 'Goddamn! What the fuck.' I walked by and saw Theo in my office, and I said, 'Theo, I wasn't here.'"

Pedro finished his Sox career brilliantly, hurling seven innings of three-hit shutout baseball in a 4–1 victory. The key play was a base-running gaffe by Cardinals starting pitcher Jeff Suppan. Manny hit a homer and gunned down Larry Walker at home plate in the first inning.

Manny was on his way to World Series MVP honors, but not without providing another memorable "Manny being Manny" moment.

With the Red Sox leading, 3–0, in the fourth inning of the fourth and final game, Manny stepped to the plate with two out and nobody aboard and got into an argument with St. Louis rookie catcher Yadier Molina. Cardinal starter Jason Marquis stepped off the rubber, and play was halted as Ramirez and Molina got into one another's face. Spanish swears were flying. Home plate umpire Chuck Meriwether motioned for Francona to come out of the dugout. When the manager arrived at the scene, Meriwether explained that he didn't speak Spanish and couldn't understand the core of the argument.

"What do you want me to do about it?" Francona first asked Meriwether. "I don't speak Spanish either."

Turning to his slugger, Francona asked, "Manny, what's this about?"

Ramirez pointed toward Molina and said, "He thinks I'm stealing their signs."

Francona chuckled, looked at Meriwether, and said, "Chuck, Manny doesn't even know *our* signs!"

The manager looked at Manny for verification. "You don't know our signs, do you, Manny?"

"No," Ramirez said, grinning sheepishly.

Case closed. Play ball.

It was probably gamesmanship by the Cardinals, but it didn't work. At the end of October in 2004, nothing worked against the Boston Red Sox. It felt like they could play every day until Christmas and never trail.

Game 4 was a perfect example. Even when the Sox made a mistake, it worked to their advantage. Batting with the bases loaded and two out in the third, Trot Nixon worked the count to 3-0 and didn't notice when third-base coach Dale Sveum flashed the "take" sign. It was not a complex signal system. Francona's take sign was Sveum holding up his right index finger.

"With the group we had, I wanted to keep it as simple as possible," said the manager. "I always worry that you're going to outsmart your own self. Sometimes simple is better. If you're outsmarting the other team and yourself, it doesn't work."

Nixon, one of the few disciplined, short-haired members of the '04 Sox, missed the sign.

"Trot didn't even look down at Dale," said Francona. "He was all locked in, probably had 12 Red Bulls going. So when he swung right through I went, 'Holy shit.' He hit a bullet off the wall in center."

A two-run double.

"When he came into the dugout, I said, 'Nice swing. You know you swung through the take,' and he said, 'Yeah, I know.' I said, 'Well, you hit the ball off the wall, and that's great. Let's just have one story here. You had the 3-0 green light, right?'"

"Right."

An hour later, Nixon, unable to remember the story they'd concocted, told the world that he missed the take sign.

The final out was a one-hopper back to the mound by Edgar Renteria. Foulke gloved the bouncing ball, trotted a few steps toward first, then made an underhand toss to Doug Mientkiewicz and the party was on.

Francona did not charge the field with the rest of the Red Sox bench contingent. He raised both arms triumphantly, turned to his left, and embraced Mills. The manager avoided the chaos on the field. It was the players' moment.

Francona's family was sitting in the lower stands on the third-base side; his dad, his stepmom, Jacque, Nick, Leah, and Jamie. (Alyssa was on her college recruiting tour.) A Red Sox official rushed to their side and escorted them past Sox fans who'd gathered behind the dugout and onto the field. One of the Francona kids got ahold of a bottle of champagne and sprayed Jacque. They never got to Terry. They were led down the dugout steps and into the clubhouse, where Tito looked into his son's tear-filled eyes and said, "Your mother was looking down on you tonight."

The manager went into his office and closed the door and had a moment.

"I was tired," Francona said. "I enjoyed watching everybody celebrating, but I wasn't in much of it. I was a little sticky from the champagne, so I took a shower. When Schilling made that toast in the clubhouse, I wasn't even around. The first time I saw it was on TV. We had to make a decision about whether we were going to stay the night or go home. After the celebration went on for a little while, I said to Jack, 'Let's go home tonight. We'll let these guys blow off steam a little longer, then go back to the hotel and pack up.' But I had no idea what it was going to be like."

Busch Stadium was only a few blocks from the Adams Mark Hotel, where the team was staying. Papa Tito and the rest of the Francona family took a Sox charter bus back to the hotel. Terry grabbed 14-year-old Leah and walked out of the ballpark through a loading dock, avoiding hundreds of Sox fans who had made the pilgrimage to St. Louis.

St. Louis fans could not have been more accommodating to the Boston road-trippers. In the late innings of Game 4, Busch Stadium gate officers opened up the ballpark for ticketless Sox fans hovering on the streets. Given the green light, hundreds of New Englanders poured into the National League ballpark to be with their team for the long-awaited moment. It was an uncommon gesture of grace and sportsmanship, and the generosity of the hosts carried out into the streets of St. Louis when Terry and Leah Francona made their way back to the Sox hotel.

"I love St. Louis fans," said the manager. "They all look like they just showered. If they ask you for an autograph and you say, 'No,' they say, 'Thank you,' anyway."

The scene was more chaotic once the father and daughter walked into the lobby of the Adams Mark.

"The hotel was a mob scene," said the manager. "You couldn't get through the lobby to pay your incidentals. Everybody was running late. When I made it to the bus, I sat for a long time, and there were fans outside almost rocking the bus a little. Everybody was happy, but it was late. I finally got off the bus and signed a few autographs. But I was out of gas."

The Red Sox Delta charter lifted off the ground in St. Louis at 3:22 AM and landed at Logan Airport at 6:30 AM — "by dawn's early light," in the words of team publicist and choreographer Dr. Charles Steinberg.

Hundreds of airport employees and police officers greeted the champs at Logan. Players and their families boarded buses, and the caravan made tracks for the Ted Williams tunnel. When they emerged from underground, helicopters hovered above the caravan and tracked them through the streets of Boston. Francona looked out his window and saw a truck driver standing on his rig, removing his cap and repeatedly bending at the waist while the bus passed. The image stayed with him forever. Nobody in Philadelphia ever bowed to make way for the Phillies.

"It was when we landed that I realized what we had done and what it meant," said the manager. "We couldn't believe it. I got emotional. There were people coming out of everywhere. How the hell did they know we were coming? First, at the airport, there were the people who come out on the tarmac and pump the gas. Then we got out to the streets, and they stopped traffic for us, and I was feeling guilty, but the traffic wasn't going anywhere anyway. When we went under bridges, there were people on the overpasses holding signs. It really started to hit me then. It woke me up a little bit."

Mike Timlin kept his hand-held camera rolling for the whole ride. When Epstein saw Timlin's finished product, the GM said it reminded him of REM's video for "Everybody Hurts."

The parade — and the missing sweatshirts — came two days later.

"I was on the same Duckboat as Terry," recalled Sox COO Mike Dee. "As we were going from the street to the water, fans were congratulating him and telling him how great he was, and he told them, 'If not for one stolen base, I'd probably have been fired!'"

"It was cold when we were out there on the river in those boats," Francona said. "I remember turning to Jack and saying, 'Man, it's cold, and I've got to take a piss. I wish we'd lost.'"

That was a lie, of course. Francona and his players had just done something nobody'd done in Boston in 86 years.

Talk show bookers called as Idiot Nation shifted into full fury. Francona went on *Conan O'Brien,* seated alongside Eva Longoria. Damon went on *Saturday Night Live* with Eminem and was named one of *People*'s "Sexiest Men Alive." He told the magazine he did naked pull-ups at home. Ortiz appeared on *The Ellen DeGeneres Show.* Schilling campaigned with President Bush and helped deliver Ohio for the Republicans. Quincy held Derek Lowe Day. Massachusetts congressman Ed Markey read a Red Sox poem into the *Congressional Record.* Senator Ted Kennedy sponsored a resolution that "the Curse of the Bambino" was lifted. The 2004 Red Sox were named *Sports Illustrated*'s "Sportsmen of the Year," the first time the magazine made the award to an entire team.

Francona signed a $75,000 contract with Metamucil and appeared in a full-page newspaper ad under the headline, "Congratulations Boston on your World Championship. Let's hope it becomes a regular thing."

In the middle of the post-Series euphoria, Francona's phone rang, and a reporter asked about Kevin Millar saying the Sox had been drinking shots in the clubhouse during the playoffs.

"I got to enjoy our victory celebration for about a half-hour," said the manager. "All of a sudden my phone's blowing up, and I'm getting messages about Millar talking about guys drinking Jack Daniels. I called Millar and said, 'What the fuck's wrong with you? We win the World Series, and I can't even enjoy it. Clean it up and keep me the fuck out of it.' And he said, 'Well, that's where we have a little problem, because I told everybody you were right in the middle of it.' Then I hear from Johnny Damon, and he says, 'Hey, Tito, I'm going on *Letterman* tonight. I'll fix it.' I was thinking I'd be the only guy to win a World Series, then get fired during the off-season."

In one of the few quiet moments as he ran around the country doubling his salary with speaking engagements, Francona paused and thought about how things perhaps needed to change.

"There was a lot that went on that made me uncomfortable," he said. "You don't want a Boy Scout troop. When those guys got out on the field, they played and they picked each other up, but I was always making adjustments because we had so much going on all the time. I told Theo, 'We can do this better, and still win.'"

# CHAPTER 8

## · 2005 ·

## "Everywhere we went, people were bowing and shit"

T HE WINTER WENT BY too quickly. A once-in-a-lifetime season that almost stretched into November gave way to a short off-season peppered with celebrations and speaking engagements. Francona said yes to just about everyone and found himself flying from banquet to banquet, spreading the gospel of the great comeback against the Yankees and the first World Series win for Boston in eight and a half decades.

"When I spoke in Boston, all I had to do was speak about the parade part, and they loved it," he said. "It didn't matter what I said. They'd all laugh. In other parts of the country I had to explain more. Everybody wanted to know about the identity of the Idiots. 'What are they really like?' I told them that what you see is what you got.

"It wore me out a little. I was still living at our home in Pennsylvania. There was one day when I went to Las Vegas and spoke, then I flew to Boston the next day and spoke. After I spoke in Boston, I jumped on a private jet, flew to San Juan, spoke that night, then got on a private jet and flew back to Boston and spoke the next morning. It was four speeches in three days."

Francona got another taste of leftover champagne in the latter stages of his annual drive from Yardley, Pennsylvania, to the Gulf coast of Florida.

Steering south on I-95 in early February 2005, the manager of the World Champion Red Sox noticed a massive SUV with Pennsylvania plates, heading north. Green flags flapping, decals shining, the vehicle was festooned in Philadelphia Eagles logos. Francona nodded and kept driving. A few miles farther south, he saw another widebody Eagles chariot. Then another. Suddenly, it dawned on him. Eagles fans were driving home from Super Bowl XXXIX in Jacksonville — a 24–21 loss to Bill Belichick, Tom Brady, and the New England Patriots.

This made him smile. He had a World Series ring. He'd just wrapped up an off-season of handshakes and paychecks. He was working in a town of multiple winners, no longer answering to the angry cynics of Philadelphia. Seeing more Eagles fans coming toward him, he started flashing his middle finger and honking his horn as he passed. Somewhere in that caravan perhaps was the guy who'd slashed his tires outside the Vet.

So long, cheesesteak-eaters. The Philly mob couldn't hurt him anymore. It was a nice way to start the 2005 baseball season.

His roster was significantly altered when he reported to the Sox minor league complex for the arrival of pitchers and catchers. Roberts had asked to be traded (he wanted more playing time) and was dealt to San Diego. Free agent Kapler signed to play in Japan. Pokey Reese went to the Mariners. Boston traded Mientkiewicz to the Mets for Ian Bladergroen. The Sox made no effort to sign playoff hero Lowe, and the veteran sinkerballer signed a four-year, $36 million-a-year contract with the Dodgers. Similarly, the Sox let free agent Cabrera walk to the Angels without putting up any resistance.

Fans always wondered about the Sox reluctance to re-sign Cabrera. He'd done everything the club asked and came up big in the big moments. His departure was without ceremony. Cabrera never got into any trouble and always seemed to wind up playing for a team that was in the playoffs. After the Sox let him go, Cabrera played with eight teams over the next seven seasons.

With Cabrera, Roberts, and Mientkiewicz all gone by spring training, the Sox had no players left from Theo's flurry of bold moves at the

2004 trading deadline. Nomar was gone, and all the Sox had to show for it was a World Championship. To replace Cabrera, Theo signed free agent Cardinal shortstop Edgar Renteria to a four-year, $40 million contract. It seemed like a big commitment given that the Sox had a superstar shortstop down on the farm who was almost ready: Hanley Ramirez.

The loudest loss was Pedro Martinez. Pedro compiled a 117-37 won-loss record in seven seasons with the Red Sox. He changed the culture in the stands and on the streets outside Fenway. Martinez's starts brought Dominican flags, salsa music, and people of color to Fenway Park. Pedro also nicely bridged the star gap from Roger Clemens and Mo Vaughn to Curt Schilling and David Ortiz.

The Pedro negotiations were sticky. The proud Dominican righty wanted a contract that would pay him as much as Schilling (between $12.5 million and $13 million per year), and he wanted four years. The Sox thought they had Pedro locked up when they increased their initial offer to $40.5 million over three years, but Met GM Omar Minaya topped Boston with a four-year, $54 million package. Pedro, a man who equates money with respect, didn't have to think twice about his decision.

Francona loved Martinez's talent and understood his iconic position in Boston, but wondered about his durability.

"I worried about his staying healthy over the length of that contract," said Francona. "He was throwing lower and lower."

Through the decades in Boston, it was common to pay players for what they had done, not for what they were going to do. Signing Pedro for four years would have been a classic example of paying royally for past performance. Epstein agreed with his manager.

There was another back story that figured in the Martinez negotiations. Lucchino is a big-picture thinker and likes superstars. He was always okay with special treatment for special talent. When old-school thinkers would make a case that the special treatment was bad for team morale, Lucchino would say, "This isn't fucking high school baseball."

In December 2004, Lucchino and Henry flew to the Dominican Republic and met with Martinez. They came back from the island uncertain about Pedro's desire to remain in Boston. Lucchino wanted to go the extra mile to keep Pedro, but he did not push Theo. The mentor and the protégé were starting their third season as CEO and GM of the Red Sox, and the two were increasingly suspicious of one another. Lucchino knew Theo had the ear of Henry. He also knew that Martinez had made life difficult for his manager throughout the championship season.

Had Pedro won a couple of Cy Youngs and pitched the full four years in New York, it would have been rough on Epstein and Francona, but none of that happened. Martinez broke down and averaged eight wins per season over his four-year deal with the Mets. In the third year of the contract he went 3-1, and in his final Met season Pedro was 5-6 with a 5.61 ERA. He finished his career with the Phillies in 2009. The Mets won nothing and teetered toward financial ruin as Pedro went home to the Dominican Republic and waited for a call from Cooperstown. He would occasionally talk about making a comeback, but few took him seriously.

The remarkable stability and good health of the 2004 Red Sox starting rotation is one of the underplayed stories of that magical season. Martinez, Lowe, Schilling, Tim Wakefield, and Bronson Arroyo made 157 of 162 starts, a feat that is rarely duplicated. Starting pitchers get hurt and wind up on the disabled list. It is a baseball certainty. But the Sox had the same rotation for six full months — which was all the more shocking considering the fragility of aging star hurlers Schilling and Martinez.

Epstein and Francona would spend seven seasons together after 2004, but never again benefit from five healthy starters over a full 162-game season.

In the winter of 2004–2005, the Sox spent considerable time and energy pursuing pitchers Brad Radke (Twins) and Tim Hudson (A's), but came away with neither. They even flirted with bringing back Carl Pavano, a long-ago Sox prospect who was sent to Montreal in the Dan Duquette deal that brought Pedro to Boston in 1998.

None of the deals materialized and within a week of Pedro's departure to the Mets, they'd replaced Martinez and Lowe in the pitching rotation with David Wells (two years, $8 million) and Matt Clement (three years, $25.5 million).

"Boomer" Wells was an interesting character. A certified strike-machine with a perfect left-handed motion, he was a large man with a large belly and a larger personality. He was the guy who wrote a book about his life in baseball and bragged about being drunk on the day he pitched his perfect game. He'd alternately charmed and enraged fans, teammates, and management when he pitched for the Yankees. He went 68-28 with the Bronx Bombers, an astounding winning percentage of .708. Bud Selig had Wells's cell-phone number on speed-dial. Everyone loved Wells's talent, but he was 41 years old and the Red Sox were his eighth team. He could be a manager-killer. He ripped Joe Torre and Yankee pitching coach Mel Stottlemyre on his way out of New York, but was known to be a favorite of Steinbrenner. His free spirit seemed more in keeping with the Red Sox Idiot Culture of 2004 than with the new (buttoned-down, professional) direction Theo and Tito were trying to follow.

Francona was surprised when he heard that the Sox signed Clement. At the winter meetings in Dallas, it was generally agreed that the Sox would not pursue the tall free agent righty. Leaving the meetings, Francona encountered Cleveland GM Mark Shapiro, and the two talked about Clement.

"I heard you guys are interested," said Shapiro.

"I don't think he's on our radar," said Francona. "Good luck signing him."

When Francona's plane landed, he got the news that the Sox had signed Clement. He called Shapiro to apologize.

Neither signing worked out as well as hoped. The bald, bawdy Wells blended nicely with the returning Idiots from 2004 and never lost his ability to throw strikes, but he was near the end of his career and wound up nearly challenging Francona to a fistfight after he was lifted from a playoff game. Clement pitched well enough to make the 2005 All-Star team, but got hurt and never lived up to expectations.

The Red Sox scene in Fort Myers in 2005 was an absolute circus.

"Everywhere we went, people were bowing and shit," Francona recalled.

Millar showed up somewhat heavier and spent a lot of time talking about Jack Daniels and shocking the world. The Red Sox had become a third major league team for the New York market. Gotham media outlets planted a flag at the Sox minor league complex, and there were daily questions about Alex Rodriguez and the Yankees. Schilling, Arroyo, and others took turns piling on A-Rod. Nixon called Rodriguez a "clown." Damon, Millar, Mirabelli, Wakefield, and Varitek appeared in a club-sanctioned Bravo broadcast of *Queer Eye for the Straight Guy*. Stephen King made the drive down from his Sarasota spring home and basked in the glow of his Sox best-seller until bullpen coach Bill Haselman told him to stay away from the manager, who did not appreciate being called "Frank Coma" in King's tome. Autograph seekers swarmed the Sox training site, and it was hard not to notice the absurdity of it all when folks witnessed a man standing ten-deep in a roped-off crowd, holding a sign that read, SON WHO HAS CEREBRAL PALSY WANTS AUTOGRAPH, PLEASE.

Francona was not immune to distraction. Listeners of WEEI sports radio back in Boston were treated to the sound of the manager's car being struck from behind when Francona fulfilled his radio obligation while driving on the cluttered roads of Fort Myers.

One of the spring highlights was young Jonathan Papelbon's memorable performance on the east coast of Florida against the Orioles. A big goofy righty from Mississippi, Papelbon's first mistake was thinking the Sox team bus was going to swing by the Sox minor league complex to give him a ride to the Orioles spring site in Fort Lauderdale. He wound up traveling across the state in a van with several veterans, including Nixon, who spent most of the ride telling him that Francona hated rookies. When the van arrived late at the Orioles park, the veterans told Papelbon to report to the manager.

"Mr. Francona, do you need to see me?" Papelbon said as he poked his head in the door.

"No, dumb-ass," said Francona. "But I need to see you pitch. And how'd you miss the bus?"

All was forgiven a few hours later when Papelbon took it upon himself to buzz Sammy Sosa after one of the Oriole pitchers hit one of the Red Sox batters. The kid did it on his own, and that impressed the manager.

The Sox interrupted spring training in early March and made the obligatory trip to the White House. Manny skipped the trip, and Theo avoided the team photo with President Bush, opting to sit in the crowd with Larry Lucchino's wife Stacey. Ever-curious and mischievous, Francona was thrilled when one of the White House operatives showed him the door to the Monica Lewinsky pantry.

"I had a lot of concerns going into spring training," said the manager. "I was worried for good reason because I was tired. I'm the one supposed to keep track of these guys, and I'd been getting pulled in a million directions. I think I actually made more money that winter than I did managing, but I shouldn't have done that. It was shortsighted on my part. I was worn out, going all over the place. Dr. Charles Steinberg and all those guys were in heaven. They let everybody in. We had the TV show with the homosexual guys. Are you shitting me? I told our players to go ahead and have fun, but not to forget that they were baseball players. With some of the spring road games, I was asked to leave certain players behind so they could do television stuff. It got to the point where it was bothering me. *Don't forget you're baseball players. If you don't get it done on the field, it doesn't work.* I found myself so many times making excuses for players. Part of my job is to make it easy for guys to play here, but at some point I think we were sacrificing some of the things we believed in."

The manager didn't like the 6,000-mile side trip to Arizona at the end of spring training. Boston to Phoenix to New York seemed like a bad way to get ready for the Sunday night opener at Yankee Stadium. Ramirez, Damon, Renteria, Varitek, Millar, and Schilling were allowed to skip the trip, which made Arizona general manager Joe Garagiola Jr. as unhappy as Francona.

"We were exhausted," said Francona. "It was a fiasco. I thought the trip was made for the wrong reasons. It didn't put us in the best position to win."

It was not a big deal, but it was a red flag. Ownership is in the business of spreading the brand and making money, and sometimes that goal is not consistent with the goal of the people who are trying to win the baseball games.

"That was the beginning of us starting to butt heads and me asking, 'What are we here for?'" Francona said years later.

Francona did reap one benefit from the superfluous trip; a newfound appreciation for minor league infielder Dustin Pedroia.

In every way, Pedroia was the poster-boy Red Sox future. Listed as five-foot-nine (he is five-eight), Pedroia has spent his personal and professional life making fools of those who doubt him. To the naked eye, nothing about Pedroia makes sense. He's too short and scrawny to be a professional athlete. He's too loud and cocky to be an asset to any team. His swing is too violent, especially for a little guy who was never going to be a home run hitter. In high school, college, and professional baseball, Pedroia had to overcome preconceived notions.

Francona heard all about Pedroia from Theo Epstein and Sox director of scouting Jason McLeod. They made Pedroia their second-round pick (65th overall) in 2004, and when the kid showed up for work, no one could believe the Sox "wasted" a lofty selection on the Arizona State infielder who looked like a waterboy. The manager was skeptical when he first saw Pedroia in spring training in 2005. He put him into a couple of Grapefruit League games, and Pedroia was a logical candidate for the long, unnecessary trip to Arizona.

On the flight to Phoenix, Tim Wakefield engaged in some harmless, old-fashioned rookie hazing, sending Pedroia to the back of the plane to fetch beers for the veterans.

"I didn't know how to act," said Pedroia. "I was just out of college. I went back and got the beer and put it into two bags. But I had to go to the bathroom, so I set the bags down. When I came out of the bathroom, the beers had spilled out of the bags and were rolling around everywhere, and Wakefield started crushing me. I was young, so I started yelling back at him. I didn't know any better. He was all over me. The next day he pitched and gave up about eight runs in the first inning, and I was saying, 'That's what you get for making fun of me and crushing me in front of the whole team.'"

Francona loved Pedroia's reaction to the hazing.

"It made me think we may have something with this kid," said the manager.

Unfortunately, he had more pressing matters on his mind as the Sox played their exhibitions in Arizona. Francona knew his pitching staff was not ready for a title defense.

Wells and Clement started the first two games of the season in New York, and neither made it out of the fifth inning as the Sox lost both games. After the second game, Francona went to dinner at Del Frisco's with his coaches, then walked back to the team hotel in Manhattan.

The next morning, he boarded the team bus at 8:00 AM, bound for the series finale at Yankee Stadium. He felt cold and clammy. By the time the bus arrived at 161st Street in the Bronx, Francona was passed out in the front of the bus. Mills shook him awake, and they walked in through the press gate, hearing the usual catcalls from Yankee fans who had gathered outside the stadium entrance. Francona got into uniform and did his game-day radio interview with Joe Castiglione, but still felt sick. He walked across the clubhouse (the Yankees had installed new carpet after the Sox celebration in October '04) to see trainer Jim Rowe, and Rowe called for an ambulance. Francona walked out of the stadium, but en route to Weill Cornell Medical Center, a 20-minute ride from Yankee Stadium, a paramedic administered nitroglycerin.

"I thought I was having a heart attack," said the manager. "But it was a little chunk of blood clot that was going through me."

Mills took over managing duties, and Epstein joined Francona at Weill, where they listened to the game on the radio. The Sox rallied for five runs in the ninth off Mariano Rivera and broke into the win column for the first time in 2005.

"I was lying there in the hospital, thinking, *We've got to win today. We just won the World Series, but we've got to find a way to win a game. We can't start off 0-3.*"

"Once we knew Tito was okay, it was actually kind of a cool experience to be there with him, listening to the game on the radio," said Epstein. "We were able to get away from the storm. We got away from all the noise. We were just in the hospital together. Obviously personal considerations were more important. To just enjoy a win together, it kind of created a nice feeling that things were going to be okay."

After the game, Francona was medevaced to Massachusetts General Hospital in Boston. He stayed at Mass General when the Sox went from New York to Toronto for a three-game weekend series. While the Sox were at the old Skydome, Epstein and Francona watched from the manager's hospital room, eating takeout. When the team came home from Toronto with a 2-4 record ("Mills should have been fired," said the manager), Francona was cleared for the most anticipated Fenway opener since 1912.

It was not a day anyone wanted to miss.

"It felt like they spent 86 years planning it," said Francona.

The 2005 Fenway opener pregame ceremony was a large production. The Sox assembled the Boston Symphony Orchestra, the Boston Pops, James Taylor, American soldiers who'd been wounded fighting in Iraq, and a conga line of ex–Red Sox, including Bobby Doerr, Dominic DiMaggio, Jim Lonborg, Fred Lynn, Carl Yastrzemski, and "Oil Can" Boyd. Lowe and Roberts were given permission to leave their new teams for the day for a trip to Fenway. Johnny Pesky and Yaz raised the 2004 championship banner in center field (Yaz was in his car leaving Fenway before the first pitch was thrown), and the 2004 players passed through the line of ex–Red Sox before they were presented with their rings. Bill Russell, Bobby Orr, Richard Seymour, and Tedy Bruschi threw out ceremonial first pitches. Francona was anointed the recipient of Bruschi's toss — both starred at the University of Arizona. There was even a moment of silence for Pope John Paul II.

The presentation of the championship rings was a highlight. Hearing his name announced over the public-address system, Francona broke into a trot as he went to accept his hardware.

"Millsie gave me shit about that," said the manager. "He said, 'Don't ever jog again. You look like you're 80.'"

At the urging of Torre, most of the Yankees stood on the top step of the third-base dugout, swallowed hard, and watched the ceremony they'd spawned with their abject failure.

"I've been on the other end of a few of 'em," said Derek Jeter. "I was a little jealous, but they deserve it. You respect what they accomplished. You know how hard it is to do. They've waited a long time, so I'm sure a lot of thought and effort went into it."

Bud Selig was there — the first baseball commissioner to witness a Sox flag-raising.

In an afternoon of memorable moments, one that stood out was the thunderous ovation showered on Rivera when he was introduced. It was a good-natured mocking by the crowd, but Rivera turned it in his favor by smiling broadly and waving his cap in appreciation. It's easy to be gracious when you are the greatest closer in the history of baseball. His résumé would survive the blip against Boston in Game 4 of the 2004 ALCS.

"I thought what Mariano did put our game in a nutshell," said Francona. "The way our fans cheered him and the way he reacted. It was respectful — by the crowd and by him."

Francona enjoyed getting his ring from John Henry, catching the ball from Bruschi, seeing the tears in Pesky's eyes, and the Rivera moment. But he couldn't let go of the fact that his team was 2-4.

"I loved seeing so many people with so much joy," said the manager. "But the whole time I was thinking, *Goddamn, we've got to be 3-4 after today*. You can win a World Series, but you're 2-4, and that's not good."

Francona was not the only person in the park mindful of a bad start. Renteria came home to Fenway for the first time batting only .167 (4-24) and was booed when he was introduced. Fans didn't understand why the Sox hadn't tried harder to retain Cabrera and had little patience with the bad start by a career National Leaguer. The sensitive shortstop never recovered and was gone from Boston after one ordinary season.

"In Boston and Philly, there's always that guy — the guy they boo — and Edgar was that guy for us that year," said Francona. "I couldn't believe the way they let him have it opening day. It had been a week. He wound up being one of everybody's favorites in the clubhouse. He was in the middle of the poker games on the plane, and you could always get him to laugh. He just didn't talk much. He wasn't moving real well [30 errors], but it wasn't a horrible year. He scored 100 runs. I just don't think he was ever comfortable with the whole Boston thing. His lower back bothered him, he couldn't bend over, and the cold weather drove him crazy."

The Sox won the home opener, 8-1. Renteria went 1-4, boosting his average to .179. In his office after the game, Francona said, "To be honest, I'm glad it's over."

There was never a shortage of brushfires around Fenway, and the 2005 season was no exception.

Jay Payton, a spare outfielder acquired when Roberts was dealt to San Diego, tested Francona in early July. Payton was a 32-year-old veteran of five big league seasons who played in the 2000 World Series with the Mets. He'd agreed to his role as a spare outfielder but, like a lot of big leaguers, had trouble accepting the part-time position. It's hard to get out of a slump when you only get to play once every three or four days. Slights, real and imagined, are close to the surface. Payton had been a big deal in college (he played at Georgia Tech with Varitek and Garciaparra) and with the Mets. Francona sensed a problem coming, and his fears were realized on July 6 in Texas when the typically poised Payton erupted when he learned he'd been maneuvered deep down in the batting lineup as a late-inning replacement. Francona was merely trying to avoid a succession of right-handed batters, but Payton took it as an insult to his batting skills.

Hovering over Francona, Payton said, "You have that little confidence in me!"

Mills tried to intervene. But it was too late. Everyone in the dugout heard it.

"If you want to be out of here that fucking bad, you'll be out of here," said the manager.

After the game, a 7–4 Sox victory, Francona called Epstein. The Sox were packing for a flight to Baltimore.

"Theo, I really need some help here," said the manager, admitting he'd lost his temper. "If you let Payton get on this plane, I'm going to lose a lot of credibility."

"Tell him not to get on the plane," said Epstein.

Payton was designated for assignment the next day and traded to Oakland for righty pitcher Chad Bradford a week later.

"I felt horrible about the predicament I put Theo in, but it worked out for us," said the manager.

There was plenty of grumbling elsewhere in the clubhouse. Millar was upset with his reduced playing time after the Sox acquired sweet-swinging John Olerud. Ever-polite Bill Mueller balked when he was asked to play second base for a day. Not accidentally, Payton, Millar, and Mueller were all in the final years of their contract. Foulke developed a knee problem and went to the disabled list, but not before insulting fans with a dismissive remark about getting booed by "Johnny from Burger King." Schilling became the temporary closer. Wells was typically full of bluster and bombast, but nobody paid much attention.

Players had a chance to express gripes with management in round-table discussions that were held at Fenway three or four times per season. The clear-the-air meetings were the brainchild of Lucchino, a CEO always open to argument and discussion. Sandwiches were served and grievances aired. The manager always attended.

"We did them in San Diego," said Lucchino. "When we came here, we would have Jack McCormick or Terry invite players to a round-table. We were in the middle of a massive redevelopment, renovation of the ballpark, and we knew that player creature comforts had to be addressed in many ways. There were a ton of ideas for how and when to do that."

"I thought they were very productive," said Werner.

"Those were good at the beginning," said Francona. "About once a month, we'd get it together. Jack would set it up for early in the afternoon on the day of a night game. John, Tom, Larry, Theo, and me would all be there. It was a good idea. Pretty cool. Some good things came out of it. They would listen to players complain about things about the family room. We got a steam room out of it."

The 2005 Red Sox were mediocre over the first two months, sleep-walking into the second week of June with a 32-29 record. Then they got hot, winning 12 of 13 and vaulting ahead of everyone in the American League East. Damon, Ortiz, Ramirez, and Varitek (wearing the captain's "C" for the first time) were All-Star starters, but the midsummer classic was not a particularly happy experience for the manager of the defending World Champs. It's traditional for the previous season's pennant-winning managers to manage the All-Star Game. Francona was honored, but uncomfortable when Timlin and Clement thought he failed to promote their All-Star candidacies.

There was confusion because in previous seasons the All-Star manager had always had great influence over the selection of reserves. The system changed after Torre was accused of putting too many of his Yankees on the All-Star roster, but players still believed the manager had the final say, and that's why Francona was on the hot seat when several players who were having good first halves, including Timlin and Clement, were omitted from the All-Star roster.

"That was our first team meeting of the season," said Francona. "Clement and Timlin and some other guys weren't on the team when the list of pitchers and reserves were announced. They had the red ass because they didn't think I was sticking up for them enough. I had to have a meeting to explain how things worked, and I was pissed that I had to do that."

With a World Series ring in his pocket and the prospect of a lengthy run in the Sox dugout, Francona made the decision to buy a home in greater Boston and move his family from Yardley, Pennsylvania. A lot of thought went into the move. Nick and Alyssa, the two oldest children, were already in college, but it was going to be traumatic to uproot the younger girls, Leah and Jamie, while they were still in high school and junior high, respectively. All the Francona kids were stellar high school athletes, and transitioning teenagers is never simple. Compounding the decision, Terry and Jacque couldn't believe the cost of housing in greater Boston. They'd bought their house in Yardley for $350,000 and were able to sell it for $700,000, but $700,000 didn't buy much in the suburbs of Boston. When Terry put down a deposit on a three-bedroom $1.65 million home in tony Brookline, Jacque and the kids made plans to permanently move in August. After a year and a half of living out of a suitcase, the Red Sox manager would be ready to say good-bye to his room at the Courtyard Marriott near Coolidge Corner.

And then his Red Sox career almost ended before the moving vans pulled up to 915 Hunt Drive in Yardley, Pennsylvania.

It was all about Manny.

Ramirez hit 45 homers and knocked in a whopping 144 runs in 2005. He also played in 152 of 162 games. Given his reputation as occasionally lazy and undependable, Manny was decidedly durable. In three seasons from 2003 to 2005, he played in 154, 152, and 152 games, respectively. It was later learned that some of his strength and game-readiness was owed to performance-enhancing drugs, but it was indisputable that Ramirez was available to play almost every day, certainly more than most players.

The tricky part was the manner in which Manny would take his time off. Francona appreciated his left fielder's work ethic, and it was a luxury to be able to put Manny's name in the lineup every day. Careful to keep his slugger from burning out, the manager scheduled an occasional day off for Manny. In late July '05, with the Sox steamrolling most opposition, Francona told Manny he'd have a day off on the final day of the club's seven-game road trip through Chicago and Tampa. It would amount to two days of rest because the Sox had a scheduled off day after the trip.

The Red Sox beat Tampa, 10–9, on July 26, but lost a couple of players in the process. Clement was knocked out of the game when a line drive off the bat of Carl Crawford struck him on the side of his head and ricocheted all the way into left field. Clement struggled the rest of the year, going 3-4 with a 5.72 ERA in 14 starts after the break.

In the same inning of the same game in which Clement was struck, Trot Nixon strained his left oblique swinging at a pitch and had to come out of the game. The Sox knew immediately that Nixon was bound for the disabled list and that the team would be shorthanded in the final game of the Tampa series. After beating the Rays, Francona sent Mills to Manny's locker to ask the superstar if he could forgo his day off and play the next day. They promised to make it up to him later.

Manny said no. He still wanted his day off. And Francona had promised.

"When Millsie told me that, I was kind of pissed," said Francona. "I went to Manny and said, 'Hey, we've got no outfielders. We need you today.' But he was in one of those moods. He just said, 'You told me I had a day off.'"

So Manny sat while Adam Stern made his second big league start. After the game, when reporters asked Francona if Ramirez had been asked to play because of an unexpected situation, Francona answered, "We told Manny he could have a day off, and he took it."

"I didn't bury him, but I didn't come to his rescue," said Francona.

When the Sox came home, it was a full-blown storm. Thursday was a scheduled day off, but the Sox were still headline news. Sox fans were reeling from a *Sports Illustrated* item in which the estimable Tom Verducci wrote, "Manny Ramirez wants out of Boston." Lucchino went on the franchise flagship station WEEI and confirmed that Ramirez had asked for a trade.

The Boston brass called Francona to Fenway for a meeting. Henry didn't like the idea that Francona would have his bench coach first approach Ramirez. Totally removed from the clubhouse, oblivious to the dynamics of dealing with Ramirez, the owner thought the manager should have gone directly to Manny first.

The meeting at Fenway did not go well.

Henry spoke first.

"Terry, we want you to make a public apology to Manny," said the owner.

"No, John," said the manager. "Fuck that. I thought this was going to be the other way around. Are you shitting me?"

Henry and Lucchino left the room.

Francona turned to Epstein.

"I'm going home, Theo," said the manager. "And I don't mean back to my house. I mean back to Pennsylvania."

"I think that meeting was about me," Francona said later. "John had a blind spot for Manny. Manny was the perfect player because of his numbers. But I was livid that day. I couldn't believe it. I thought we were going to try to fix things, but not by me apologizing. That would have buried me with the other 24 guys. I couldn't do that. I had had enough."

Epstein brokered a truce. He told the deflated Francona to go home and show up to work at Fenway for the Twins game the next day. He would deal with the owners. And Manny. There would be no apology from the manager.

"Theo cleaned it up," said Francona. "That's when Theo was at his best. He was good. I wasn't going to apologize, because I didn't do anything wrong. Let's face it: with a talent like Manny, at some point you've got to balance how good he is with the rest of these guys in that clubhouse who were killing themselves. I came back the next day and started trying to put it back together."

"I had to try to explain the realities of managing a team to the group," said Epstein. "I reminded John, Tom, and Larry about how difficult the manager's job is and how he has to make difficult decisions on handling players on a daily basis. One thing that truly sinks a team is if the manager loses the respect of the players. That's the quickest way for a team to completely go off the rails. I told them that Tito couldn't do that. He needed to have some autonomy on how he handled these situations. I also had to give Tito some perspective of the realities of owning the team and watching it from a removed position and how different perspectives were natural, but didn't necessarily have to lead to dire conflict."

One of the happier developments of the turbulent late summer was the return of Gabe Kapler. The thoughtful outfielder had signed with the Yomiuri Giants after winning the World Series, but rejoined the Sox at the end of July. Francona cherished veteran role players who sometimes served as assistant coaches. Kapler was one of the best. He was smart, sincere, and he understood his role, never complaining about playing time. Everyone could see he was a manager-in-the-making. Late in the 2005 season, Kapler spent some time counseling Francona in the weight room up the stairs from the home team clubhouse. (The Fenway weight room is shared by the Sox and their opponents because there's not enough room for a visitors' weight room — imagine A-Rod and Varitek pumping iron, side by side, the day after they slugged it out in July '04.)

"It was complete role reversal," said the manager. "I was at a loss, just going through a tough time. Kap walked by me and stopped by and asked how I was doing. We must have talked for 45 minutes. It was kind of strange. I'm the manager and he's a bench player. But it was such a caring conversation. He kept telling me that we were going to be okay. I was telling him things that were bothering me at the time. Right after that, I saw him going to people, pushing them in the right direction. That was special. It stayed with me. When you're going through a tough time and you're with people like that, it makes it a lot better."

In mid-August the Sox were 68-47 and had a five-game first-place lead when a Sunday home game against the White Sox was postponed due to rain. Over the objection of the manager, the game was rescheduled for a September day that was supposed to be an off day in between home series against the Orioles and Angels. Francona wanted to play the game, if necessary, after the conclusion of the season. His owners wanted to wedge it into the bridge between the series against Baltimore and Los Angeles.

"We had a meeting on that too," said Francona. "I tried to explain that we were on fumes and if we didn't play that game on that day, it was a win for us."

"Just play the backup guys," suggested Werner.

Francona gripped his chair and reminded himself to hold back after hearing the insulting suggestion.

The game was played on the busy Monday, and the Sox lost, 5-3.

Four weeks later, on the final weekend of the season, the Sox were in a steel cage match to make the playoffs and needed a victory on the final day of the season to guarantee admission into the tournament. Francona had to pitch Schilling against the Yankees in the regular-season finale. Boston's 10–1 win put the Sox in the playoffs for a third straight season, but the pitching rotation was mangled by the late-season urgency, and the Sox were not in a position to succeed when the playoffs started in Chicago two days later. The Sox played 30 games on 30 consecutive days in September.

"Our worst fears came true," said the manager. "We had to pitch Schill on the last day of the season."

The Sox playoff spot was clinched early during the Fenway finale when the White Sox beat the Indians in Chicago. Francona lifted many of his starters once the playoff bid was sealed, but it was too late to save Schilling for any of the first three games of the ALDS against Chicago. The manager felt a little resentment when he saw the Red Sox owners and their friends gathering in the dugout tunnel in the final inning of the 10–1 victory. Everybody wanted to get in on the wild-card clinch celebration, but Francona hadn't felt supported earlier in the month when his team needed a day off.

While the Sox celebrated their playoff bid on the field, Francona and Epstein stood off to the side and started talking about their playoff roster and rotation. It was an animated conversation that looked like an argument when it was aired on the evening news.

"The exuberance of making the playoffs went away pretty quickly that day," said Francona. "Obviously, we were glad we won, but we were not lined up to have a good showing in that first round."

Two days after the cheesy "wild-card champs" celebration on the Fenway lawn, the Sox opened their playoff run with a 14–2 loss at US Cellular Field in Chicago.

Clement was awful. He hit two of the first three batters he faced. He gave up five runs in the first inning and wound up surrendering eight runs before Francona lifted him with one out in the fourth. The White Sox hit five homers, three off Clement. The Red Sox were smothered by Chicago's Cuban righty, Jose Contreras. There was some irony there. Contreras was the object of desire in the winter of 2002–2003 when new Sox GM Theo Epstein was rumored to have trashed his Nicaragua hotel room after losing a bidding war with the Yankees. That was the signing that prompted Lucchino's infamous remark: "The Evil Empire extends its tentacles even into Latin America."

Three years later, Contreras was sticking it to the Red Sox again, this time in a White Sox uniform.

The Sox were still reeling from their regular-season weekend sprint against the Yankees, and Francona started thinking about Game 2 long before Game 1 was over. Saving Papelbon and Timlin, he used Chad Bradford, Geremi Gonzalez, and Bronson Arroyo for playoff mop-up duty. The goal was to have Wells win Game 2, then bring the series back to Boston, where Schilling would be waiting to pitch Game 4.

Wells looked like a good bet. The big southpaw was 10-3 lifetime in the playoffs and World Series.

The plan looked pretty good when the Sox burst to a 4–0 lead in Game 2. Wells cruised into the fifth with a two-hit shutout, then got into trouble. He gave up three hits and a pair of runs, but looked set to get out of the inning when Juan Uribe's potential inning-ending double-play grounder skipped between the legs of Sox second baseman Tony Graffanino. Buckner style.

Francona sent pitching coach Dave Wallace to the mound to talk to Wells. After Wells got the second out of the inning, Chicago rookie Tadahito Iguchi drove a 1-1 pitch over the wall in left for a 5–4 White Sox lead. Fireworks. Ball game.

There were more fireworks when Francona came to the mound to lift Wells with two outs in the seventh.

"He swore at me, but I wasn't paying attention," said Francona. "At that point, it's just 'Give me the ball, David.' When I got back to the dugout, I started getting reports that he was breaking stuff in the clubhouse. A couple of players said, 'David's back there having a fit.' I just said, 'Whatever. I'll get to him later.' Things like that didn't bother me."

The Red Sox went down feebly in the ninth. Later that night, after all the interviews were done, Francona was packing in the manager's office when he noticed the looming presence of Wells in the doorway.

"You know I really wanted to kick your ass when you came to get me," said Wells. "I really wanted to punch you out."

"Fuck, Boomer, give me a break," said Francona. "Are you kidding me? I'm tired. Do you want to have a fistfight? Fine. Let's go. But we've got to hurry because I want to get to the airport before the White Sox so we can get through security and get out of Chicago before them."

Wells was disarmed. Soon the two were sharing a beer.

"I knew how David could be," said Francona later. "He was like a little kid. He was fine. I took him out because I was thinking I might be able to bring him back later in the series. We were scuffling there just to win one game. It wasn't something I could explain to him out there on the mound that day, but when I told him in my office, he understood. That's just how Boomer was."

Through the magic of traveling secretary McCormick, the Red Sox beat the White Sox to O'Hare and took off first.

The airplane race was Boston's only victory over Chicago in that series. On Friday afternoon, October 7, Chicago won the clincher at Fenway, 5–3, officially ending the 2005 title defense. The Sox lost the finale, with Renteria making the final out on a feeble grounder.

It was the final Red Sox game for Millar, Mueller, and Damon.

"These were the best three years of my life," Millar said.

"I think it's been awesome," added Damon. "I'll remember how our players loved the city and how the city loved us and how they enjoyed us."

After all the glory and heroics of 2004, Schilling never got the ball in the 2005 playoffs.

"I certainly didn't expect to be on a postseason team and not get the ball, but it was my own fault," said Schilling. "I struggled all year, and then we went down to the last game of the season and I had to pitch."

"The White Sox that year were really good," said Francona. "They were the team that we had been the year before. Having Schilling would have been good, and I knew Wells could come back for Game 5, but we just couldn't get there. That whole experience taught me how important it was to have your pitching set up — more important than winning the division. We were on fumes."

Playing to the tune of Journey's "Don't Stop Believin'," the White Sox went on to beat the Angels and Astros (Chicago was 11-1 in the postseason), delivering a World Series winner to the Windy City for the first time since 1917.

There was no joy in Boston. The Red Sox simply had not been ready when the playoffs came around.

The day after it ended, a sleepy Saturday in Boston, players and coaches came to clean out their lockers. Francona didn't have much packing to do. Jacque and the kids were all moved into their new home in Brookline. It would be nice to be close to his office during the winter for a change. He was planning on watching some of his girls' volleyball games and scheduled a surgery for his right knee at Mass General.

He went to Fenway to fulfill media duties and say good-bye to a few of his players on the day after the final game. He was touched when Damon came into his office, closed the door, and presented him with a signed Red Sox jersey.

"I just think he was a special manager," Damon said in 2012. "I didn't know 100 percent that I wasn't going to be back, but I had a good sense that it was the end, and I wanted to give him a pretty special thing."

"That meant a lot," said Francona. "I don't save a lot of stuff, but I saved that. Johnny was not only tough, he was respectful. There were nights when he'd go home where he was beat to shit and I'd alert his backup. But he was so respectful, he'd come into my office when I was telling whomever, and he'd say that he'd call me in the morning and then I could call the player I had ready to play for him. And every night he would end up playing. I got a little nervous when he became a team spokesman, but he was so dependable. I loved him. He was good-hearted, and he grasped things better than people gave him credit for. He was the kind of guy who'd come back and check on me after I got thrown out of a game. He'd be like, 'Hey, you okay?' He always tried hard to do the right thing."

Officially, Damon was still a member of the Red Sox. He was due to be a free agent at the conclusion of the World Series, but his gift to Francona on October 8 demonstrated his slight chances of re-signing with the Red Sox.

"I was really happy and honored that he gave me the shirt," said Francona. "But I was also thinking, *This doesn't bode well for us having him stay here.*"

Damon wound up signing with the Yankees in December. His "defection" brought the wrath of Red Sox Nation down on Damon, and it would have been the biggest story of Boston's baseball winter had it not happened seven weeks after the Halloween night resignation of Theo Epstein.

# CHAPTER 9

## · 2006 ·

## "We will take care of your son"

TERRY FRANCONA WAS HAVING trouble walking by the end of the 2005 season. His right knee — the one that exploded on the warning track in St. Louis in 1982 — was constantly swollen, causing him considerable pain. Attending parent-teacher conferences at Brookline High School with Jacque in early October, he had trouble making the rounds from classroom to classroom. He scheduled knee replacement surgery at Massachusetts General Hospital.

There had been a funny moment at the end of the season. On the day players were packing their belongings and getting plane tickets home — the day Johnny Damon walked into his office with a parting gift — Francona got a check for $50 from Red Sox traveling secretary Jack McCormick. Teams are obligated to pay players' and managers' transportation costs back home after the season, and this was the first time Francona was not driving or flying some distance after the season. His family's new house was less than five miles from Fenway Park.

"I guess they figured $50 would cover the flight to Brookline," he joked.

Living near Fenway had its advantages during the off-season. In order to stay on top of the ball club's off-season activities, Francona no longer had to fly into Boston and check into the Brookline Courtyard Marriott. In the fall of 2005, he was at the ballpark enough to sense mounting tension between baseball operations (which by this time had been moved to the Fenway basement) and the CEO's office on the third floor. The Theo Epstein–Larry Lucchino situation was at a critical mass.

The story of Larry Lucchino and Theo Epstein is as old as the Bible, and by October 2005 the protégé was ready to break free of the mentor. Theo wanted to build a farm system that would make the Red Sox perennial contenders with homegrown talent. Epstein wanted "a scouting and player development machine." He was okay if the Sox regressed for a year in order to make things better in the long run. Lucchino, meanwhile, wanted to win every year. As a franchise, the Red Sox had no tolerance for rebuilding, even for a half-season. The Sox CEO wanted to stand at the podium with the World Series trophy every year.

Some of the Epstein-Lucchino disagreements were rooted in the core of their job descriptions. The GM acted in the interests of baseball operations. The CEO had his eye on the Red Sox brand and the bottom line. Sometimes decisions had to be made that took away from the effort on the field—like playing games when it would have been better to wait and reschedule, or making an out-of-the-way spring trip to Arizona when it would have been better to send the team directly from spring training to New York for the opener.

In the fall of 2005, Epstein's contract expired, and his salary was a point of contention. He was making $350,000, but everybody knew the Sox had offered Billy Beane $2.5 million per year back in 2002.

Trust had become a major issue between the two men. During the summer of '05, John Henry killed a deal for Colorado outfielder Larry Bigbie. The deal was squashed by Sox ownership at the request of Epstein. He thought he had a better deal to make with Arizona. When national baseball reporters got wind of the broken trade with Colorado, Lucchino was colored as a bully CEO who overruled deals made by his general manager. When it was reported in the October 30 *Sunday Boston Globe* that killing the Bigbie deal had been Theo's idea all along, some would have seen this as a demonstration of a front office working in harmony, but the young GM saw it as a violation of trust. It was the tipping point for Epstein. After 14 years together, he believed Lucchino was using friends in the press to make himself look good . . . at Theo's expense. Epstein decided to quit.

"Basically, I thought Larry was trying to mess with me, and Larry thought I was trying to mess with him," Epstein said later.

Epstein resigned October 31, walking out of Fenway Park wearing a gorilla suit.

Lucchino was conspicuously absent at a press conference two days later when a befuddled John Henry said, "Maybe I'm not fit to be the owner of the Boston Red Sox. . . . Did I blow it? Yeah, I feel that way."

Francona was not there either. He was working out his own strained relationship with Lucchino.

"It was a weird time," remembered Francona. "Theo had resigned, and Larry called me up to his office. When I got there, our scout Dave Jauss was just leaving. I said hello to Jauss and walked into Larry's office thinking this was going to be some kind of 'circle the wagons' talk. But Larry didn't even look up from his desk. He just said, 'I've been around a lot of baseball managers, but you, by far, make me the most uncomfortable.' And I was thinking, *What the fuck?* So I just said, 'I'll try to do better.' But I was stunned. That's just Larry. If he's got something on his mind, it just comes out. I walked back downstairs and saw Brian O'Halloran and said, 'I don't know if I've just been fired, or what. It was the weirdest meeting I've ever had.' It was morbid humor. Theo leaves in the gorilla suit, and then I have this weird meeting with Larry. We were actually kind of laughing about it.

"As much of a shock as Theo leaving was, that time is a little blurry for me because I was getting ready for the knee surgery, and nobody really knew what was going on with the ball club."

While Epstein toured the world with Pearl Jam, never losing touch with the Sox front office, the Sox were run by a committee of general managers. Jed Hoyer and Ben Cherington were named co-general managers, but there was no attempt to hide the considerable input of veteran Bill Lajoie. Another voice in the chorus was Jeremy Kapstein, a "Lucchino guy" with a reservoir of baseball experience. Sox fans know Kapstein as the man who sits directly behind home plate at every Red Sox home game. Wearing large yellow headphones and a blue windbreaker (no matter how hot it gets), Kapstein gets more television time than Katie Couric. What most fans don't know is that he once ran the San Diego Padres, was married to the daughter of McDonald's visionary Ray Kroc, and served as baseball's most powerful player agent during the seismically shifting 1970s. Jeremy Kapstein, then known as Jerry Kapstein, was the man who advised Sox stars Carlton Fisk, Fred Lynn, and Rick Burleson to play hardball with Tom Yawkey after the magical season of 1975. Kapstein was the Scott Boras of his day. He has served as an adviser to Lucchino since the "new owners" bought the Red Sox in 2001.

When Francona had knee replacement surgery at Massachusetts General Hospital on November 21, 2005, one of his first visitors was Jeremy Kapstein.

"It was the strangest thing," said Francona. "Jeremy was under the impression that he was the next general manager. I was just a day out of surgery, coming out of anesthesia, with the pain pump going, and Jeremy is sitting by my bed telling me how it's going to be. He was telling me I'd have a special cell phone to get in touch with him. I was in kind of a haze, and the next thing I know he's handing me a cell phone, and the guy on the other end says, 'Hey, Terry, it's Keith Jackson!' It was Keith Jackson, the old football announcer. I didn't know what that was about, but after Jeremy left, I called Jed Hoyer and said, 'Get your ass over here! I just had Jeremy Kapstein in my room, and he thinks he's going to be the next general manager.'"

O'Halloran and Hoyer went directly to Mass General and told Francona that, to the best of their knowledge, Kapstein was not going to take over as general manager.

The next day, on November 24, in a deal that Kapstein and Lajoie pushed over the objections of some of Epstein's operatives, the Red Sox acquired Josh Beckett, Mike Lowell, and reliever Guillermo Mota for hotshot shortstop prospect Hanley Ramirez, pitcher Anibal Sanchez, Jesus Delgado, and Harvey Garcia. The trade was announced while Theo was en route to South America to join Pearl Jam.

Theo breathed a sigh of relief when he learned that Jon Lester was not in the deal, jokingly telling a reporter that he would have "resigned" if the Sox traded Lester. Ramirez, Sanchez, and Delgado were all holdovers from the Dan Duquette regime.

"I was still in the hospital in a complete haze," said Francona. "We knew Hanley was going to be really good, but I don't think any of us thought he was ready to be that player yet. He got much better, quicker than we anticipated. Hanley was a little temperamental. I didn't see him play much. He'd played in a couple of split-squad games in spring training, but I'd traveled those days and didn't see him. I would come back and Millsie would say, 'You got to see this guy. This guy is a man.' I knew I was thrilled to get Beckett. I was thrilled anytime we got pitching. In my opinion, that's how you win."

Lowell was the supposed throw-in. In order to give up Beckett, the Marlins insisted that the Sox take Lowell and the remaining two years of his (four-year, $32 million) contract. The acquisition of Lowell meant switching Kevin Youkilis from third base to first base.

Damon signed a four-year deal ($52 million) with the Yankees the week before Christmas. To replace Damon, the Sox traded for Cleveland center fielder Coco Crisp. Admitting they failed on Edgar Renteria, they traded the shortstop to the Braves (eating a big chunk of Renteria's salary) and acquired slick-fielding Alex Gonzalez.

On January 20, 2006, Theo officially returned as general manager. The degree to which he actually left has always been in doubt. He never lost touch with temporary co-general managers Cherington and Hoyer, and there was a sense that he never really went away.

Henry lobbied hard for Epstein's return, and future Sox problems would be rooted in concessions made at that time. When Theo returned from his self-imposed sabbatical, he had more power and independence. He isolated baseball operations from the rest of the organization. Moving forward, there would be far less "interference" from Lucchino. Baseball operations became its own country in the basement offices of Fenway.

During spring training, Epstein sent Bronson Arroyo to the Reds for Wily Mo Pena, a deal that would fail badly for Boston. Arroyo went on to pitch the next seven seasons in Cincinnati without missing a single turn in the rotation, a streak of more than 320 consecutive starts. Pena clubbed 16 homers for the Red Sox over parts of two seasons and was playing for the independent league Bridgeport Bluefish by 2010.

"Theo really struggled with that on a lot of different fronts," said the manager. "Bronson was everybody's favorite guy, and he had signed his new contract when Theo wasn't here. It wasn't a talent-for-talent deal. He thought Wily Mo could be the guy. I knew how hard this one was for him. It was really tough. I just didn't know if Wily Mo could play or not. I don't think anyone knew if he could do it. He'll tease you because of what he can do, but there's a lot of swing-and-miss in there. And Bronson, for me, was a perfect guy to have. Like when the Yankees had Ramiro Mendoza. A guy who can throw 90 to 100 innings out of the bullpen. I thought that's what Bronson was for us. But he was making too much money to do that anymore.

"I think we both knew it might end up how it did, but I knew Theo really wanted to do it. I just told him, 'Theo, if you make the deal, I'll back you.'"

In mid-March the Sox announced a new three-year contract for their manager. Francona was bumped to $1.5 million per season, still a long way from the $5.5 million Torre was making in New York.

Toward the end of spring training, Francona called 22-year-old Jon Lester into his office and told the kid he'd be starting the season at Triple A Pawtucket. A standout southpaw at Bellarmine Prep in Tacoma, Washington, Lester was the second-round Sox pick in the 2002 draft, and he moved steadily, if not that quickly, through the system. The organization's top concern was that Lester tended to be a five-inning starter because he threw too many pitches in the early innings. He made 26 starts at Double A in 2005.

"Like a lot of young kids, I had those unreal expectations that I was going to make the team," said Lester. "Tito sat me down and told me to soak it in and be realistic with myself and make the most of my time in the big league camp. It's not something you want to hear, but I think it helped me. It relaxed the tensions I had going in. He made sure I knew there was nothing to prove. I wanted him to know the type of person I am and when the time comes, you can call me up."

There was a big decision to make at the start of the season. It was clear to everyone, including the manager, that it was time to pass the closer torch to Papelbon. Francona knew he had to talk with Foulke.

Foulke was one of the more unusual personalities in Boston sports. He had as much to do with the championship of 2004 as any player. Schilling went into the books as the personification of guts and glory. Damon, Millar, Ortiz, and Ramirez served well in their role as Idiots.

But the presence of Foulke was the biggest difference between the 2003 near-miss team and the 2004 champs. Foulke was the difference between heartbreak and history. He saved 32 games in the 2004 season, then made 11 postseason appearances, compiling an ERA of 0.64 and allowing a single run over 14 innings. He pitched in all four World Series games. He easily could have been named World Series MVP instead of Manny.

For all of that, he was never embraced. Foulke was a quiet man and not much fun. Francona had seen him up close when the two were in Oakland in 2003. He knew what the Sox were getting. When Francona was in managerial limbo in the autumn of 2003, working in the Sox offices though he had not yet been named skipper, the Sox asked him about acquiring Foulke, and he said, "If I'm going to be manager, I want you to sign him. If I'm not, I don't want you to sign him."

Knee trouble diminished Foulke's effectiveness in 2005. He saved only 15 games, spent 50 games on the disabled list, and went through a second arthroscopic procedure after the Sox were swept out of the playoffs.

"He was quiet, and he wouldn't let you get close to him, but he loved to pitch," said Francona. "He was unflappable. He had that changeup, so even if he didn't have his great stuff, he could get the job done, and that's what he did in '04. What happened after that was disappointing. He got bitter, and he didn't handle things real well. He got a little stubborn. He was kind of a smart-ass. That was his way of dealing with things, but he'd take the ball every day, and as a manager, you love that. He'd never turn the ball down."

Francona had fallen in love with rookie Papelbon's talent and attitude in the spring of 2005 when the big kid knocked down Sammy Sosa in a spring training game after Jay Payton had been hit by an Oriole pitcher. Papelbon was obviously the next closer. Forced into the bullpen in September 2005, he pitched 17 and a third innings, allowing just two earned runs, striking out 15 and walking only four.

Papelbon was born in Baton Rouge, Louisiana, and drafted in the fourth round in 2003. He pitched only a few games at Bishop Kenny High School in Jacksonville, Florida, and was playing first base at Mississippi State when coaches noticed that his throws across the infield to third were pretty impressive. Epstein liked him as a starter. Francona saw the perfect closer. The kid could throw 100 miles per hour and certainly had a closer's mentality.

"When I got drafted by this team, I was part of a Nation," Papelbon said in the spring of 2006. "I want to be part of something special here, and I want to be part of a team in a city where when you go out there, you're expected to win. That's the only way I know how to play. That's how I compete. If I don't do good, go ahead and boo me. It doesn't really get to me."

Papelbon worked a perfect one-two-three eighth inning in the Red Sox 2006 opening day win in Texas. After Papelbon was lifted, Foulke was hit hard in a non-save situation in the bottom of the ninth. Francona made up his mind that it was time to award the closer's role to the big kid from Mississippi State.

Francona believed that the casual hour of late-afternoon batting practice was one of the best times to deliver important messages to players. The manager's physical limitations made the batting practice counseling personally challenging, but it was worth the discomfort.

"It took me forever to get dressed, but I found the right mix of layers," said Francona. "It was important to me to be out there. I needed to work, and I didn't want to just sit around. I wanted to hit fungoes, and I wanted to try to throw batting practice. I wanted to be a working coach. I found that it was my time to visit with players when it was quiet — especially in Boston where there was so much media around."

While the Sox were taking batting practice before the third game of the season in Texas, Francona sidled up to Foulke near a protective screen behind second base.

"Hey, Foulkie," said the manager. "I just want you to know that if we get a save situation tonight, I'm going with Pap."

"Okay," said Foulke, always a man of few words.

A few hours later, with the Sox leading 2–1 in the bottom of the ninth, Francona summoned Papelbon from the bullpen. Papelbon needed just 11 pitches (two strikeouts) to retire the Rangers in order for the save. A closer was born.

When Francona met with the frothing press, they wanted to know if this meant that Papelbon was the new closer. Bumping Foulke, a World Series hero making $7.5 million, was a big story.

"This is by no means an indictment of Foulke," said Francona. "I think he's gonna be brilliant."

The manager was reminded that his move made it look like Papelbon was the new closer.

"I don't care what it looks like," he snapped. "I just told you the truth and how I feel. We won, and that's what we set out to do. It's a long year. I don't think Foulke is the guy we need yet, and I think he's going to get there."

The episode was a perfect demonstration of Francona's managerial nuance and accountability. Francona, the ex-player and the son of a major leaguer, never wanted to rip or embarrass a player. There was no getting called to the woodshed when Terry Francona was in charge.

"I wasn't a big fan of bringing guys into the office, especially a guy like Foulke," said Francona. "He would view it as being brought into the principal's office. I'd always stand behind the screen during batting practice and talk to guys. I got a lot done out there. This wasn't something I'd decided in one day. Foulke deserved respect, and I owed it to him to let him know ahead of time. Of course, word travels fast. About ten minutes after I talked to Foulkie, I had Schilling in my face saying, 'Are you crazy?' I just felt that telling the guy was the right thing to do."

Why say something contrary to the media?

"I think that's the right way to do it," said Francona. "I don't think you need to anoint the next guy and say he's going to the Hall of Fame. It was pretty obvious to me how good Pap was, but I had a lot of respect for what Foulke did for us. I don't care what it looks like — I'm going to do what I'm going to do. I'd rather have a person from the media think I'm dumb or arrogant or argumentative. This happened a lot. I would say I didn't see something. I'd rather they think I'm stupid than put an indictment out on one of our players. In the Pap situation, I had nowhere to go but to kill Foulke, and I didn't want to do that. I can't tell you how many times we'd leave the interview room and Pam [Pam Ganley, then a public relations assistant] would say, 'You just watched that,' and I'd say, 'What do you want me to do, kill our player?' I can live with someone thinking I'm a little slow if it helps us. It was better. I felt like it worked."

He was good at the media game, and the Red Sox knew it. Still, there were times when he could be testy, particularly with non-baseball reporters. Francona had a couple of media pet peeves. He hated it when someone's cell phone went off during an interview session. And he would get almost violent when he was addressed as "coach."

"Coach?"

This is a baseball thing. In every other sport the head coach is "coach." In Little League, high school, and college baseball, the head coach is the coach. But in professional baseball the top guy is the "manager," and the coaches are the guys who work below the manager. This is why most major league managers would rather be addressed as "moron" than as "coach." The salutation of "coach" strips them of their hard-earned stripes and also indicates that the person asking the question doesn't know anything about baseball.

The legendary Peter Gammons, a Hall of Fame sportswriter who has covered the Red Sox since the late 1960s, would never address a manager as "coach." In June 2006, Gammons was stricken by a brain aneurysm while driving near his home on Cape Cod. While Gammons lay unconscious after surgery at Brigham and Women's Hospital, his wife Gloria was surprised to see the manager of the Red Sox standing in the corridor outside his room after a home victory over the Mets. It was just a small measure of respect for someone who'd done so much for baseball and the Red Sox.

The Sox infield was in good shape for most of 2006. Youkilis proved to be a natural at first base. Lowell, coming off a horrible year at the plate in 2005, found his Fenway stroke and proved that he was more than a throw-in from the Beckett deal. Mark Loretta held down second base, and Gonzalez was sensational at short.

"Gonzo in 2006 had the best defensive shortstop year I've ever seen," said the manager. "Mikey Lowell told me in spring training that I was in for a treat, and he was right. Every week we saw something we'd never seen before. That's a nice feeling when the ball is hit to shortstop and you know it's an out. There was no play he couldn't make. He had a reputation as moody, but he wasn't moody. He just didn't talk to anybody."

Josh Beckett was another matter. Beckett spent much of his first year in the American League trying to throw fastballs past hitters. Too often the talented sluggers of the AL East were able to track his fastball and hit it out of the ballpark. Beckett gave up 36 homers and compiled an ERA of 5.01 ("Embarrassing," said Beckett) in his first year in the Junior Circuit.

"He was pretty much what I expected," said Francona. "He's stubborn, and I knew that someday we were going to like that about him. He was throwing 95, but it would be on a straight plane level. There's things you can get away with that you can't when you're pitching against the Yankees. Some of those lineups are difficult to navigate through. Beckett had that personality. I always thought he was at his best when he was in the clubhouse and being a smart aleck and arrogant. When he'd get quiet, he wasn't as good. I always liked it when he'd come into the dugout from a one-two-three first inning and start yelling. He did that just to get himself going and to get the rest of us going. He was a guy that kind of took charge of the staff, which I liked."

Despite Beckett's subpar performance, the Sox rolled into the All-Star break with a 53-33 record, a 100-win pace. Ortiz had 31 homers, and Papelbon had 26 saves and a 0.59 ERA at the break. They were 63-41, holding on to first place with a one-game lead over the Yankees, when Epstein stood still at the July 31 trading deadline. While the Yankees acquired slugger Bobby Abreu, the Sox did nothing and left themselves without a reliable lefty reliever for some big second-half games with the Yankees. In the final two months of the season, Abreu would bat .330 with seven homers and 42 RBI.

"I felt bad for Theo at the trading deadline," said Francona. "He said that if we had the ability to start over, we'd be really good a year from now, but you can't do that in Boston. I understood that. He understood it more than anybody. We were getting old in a hurry, trying to hang on. I think he wanted to go young, and it was probably time. We would have been better doing that, but he just couldn't. We tried to string it together, but then everybody got hurt. We managed to win 87 games, but we were done."

"It's a long-standing impediment for the Red Sox," said Epstein. "With the Red Sox, there's been so much focus on winning and building an uber team this year, so much focus on tomorrow's paper, so much focus on the Yankees. Some of that had to do with the timing of the end of the Yawkey regime. There's no doubt that we feel the only way to sustain success over a long period of time is to have a successful farm system. . . . Two years ago I said we were two years away. Finally, we're at a point where the farm system is going to start to pay dividends at the big league level."

Not making a deal at the deadline was another indication that Epstein was having his way with ownership. There was no more interference from Lucchino, the man who always wanted to win now, even if it meant sacrificing some people in the farm system.

The manager had some old-school techniques. Before each game, Francona assembled his own stat card. With help from advance scouts' reports and numbers supplied by baseball operations — data that was more inventive than ever — he marked the good matchups in green and the dangerous matchups in red, keeping the standard information in black. He put stolen base numbers on the left side of the paper and scribbled miscellaneous notes on the right margin. He never shared the card with his players.

Francona valued reports from advance scouts and found matchup information particularly useful, but the mountain of information and suggestion started to overwhelm him in the summer of 2006.

"Prior to every series, Jed would have a conversation with Tito about lineups," said O'Halloran. "It was just to talk about matchups, and Tito would eventually get to a decision on what he wanted to do. We had resources to come up with what we thought was best."

"I don't think they ever felt they had to push it on me," Francona said. "They knew what I liked. I liked to get on the plane for the next series and look at the advance book on the next team and see right away how we wanted to pitch their guys. As the years went on, they started to personalize more of the stuff, putting together individual stuff for each coach. There were days when I'd see Theo and he'd be so proud of it. We all knew these guys were busting their ass putting this stuff together, and I never took it for granted. I welcomed the help. There were things that meant more to me than others, and I told them that. They tried to go deeper than Mike Lowell being four-for-six against a certain kind of pitcher and that did help. They broke it down more than just the number. But the problem was that the number would change as we played. On Thursday Mike Lowell could be a good fit to play, but on Friday he could be a bad fit because of what happened to the numbers Thursday. It was a little too fluid for me.

"I guess the only time I objected were the times they were telling me how to do something. There was a difference. Just give me the information, don't tell me how to manage."

Epstein's research team extended their analyses to some outer limits.

"In those first years they had a guy who would send me lineups," said Francona. "This guy would tell me not to hit David Ortiz against Scott Kazmir because chances are, David's going to have a rough night. Well, I'm not sitting David. He's got a chance to be MVP, and you want me to start Doug Mirabelli at DH because Doug has better numbers against this guy? I was like, 'Fuck, I'm not going to do that.' You might win a game somewhere along the way, but it's not worth what you might lose from David overall. I told Theo, 'I want to meet this lineup guy,' and they were like, 'No, you don't.' It was kind of a running joke. I never found out who the guy was. He was smart and had some numbers, but come on, man. I didn't mind getting information, but it was strange not to know where it was coming from."

"He was on the payroll as an outside consultant," said O'Halloran. "I never met the guy. Jed met with him. I don't know exactly how he came into the organization. He was a bit of a mad scientist, but he was one of the sources that the front office would use. He was a whacko sabermetrician type. We stopped using him."

"There were actually two of them," Epstein confessed in 2012, after leaving the Red Sox for the Cubs. "Eric Van and 'Vörös' McCracken. We were always looking for little breakthroughs that would help us in the draft, in player projections. John Henry discovered Eric on the Sons of Sam Horn website and asked us to talk to him, so he was a consultant for us for a couple of years. He had some interesting things to say and some other things that were kind of off the reservation. The other one, McCracken, I think used Vörös as a pen name. We would joke with Tito about those guys, but they were not making out the lineup card or anything of that nature. These guys were literally in the basement, on the computer. They were stats-only consultants. They would occasionally chime in with these harebrained lineup ideas. Because I had a good relationship with Tito, I would throw them at him. He would get frustrated. It was the antithesis of being in the trenches. These guys were so far removed that Tito never even saw them."

(In 2011 Vörös McCracken was an *ESPN Insider* blogger. In a 2007 blogpost on vorosmccracken.com, he wrote, "I won a World Series while working with the Boston Red Sox. And now live in abject poverty and anonymity somewhere in the Phoenix metropolitan area." According to a *Boston Globe* feature story written by Mark Shanahan, Eric Van was a Harvard graduate with an IQ of 143.)

Tom Tippett and Zack Scott were nothing like Van and McCracken. Tippett and Scott were stat men who wound up with offices at Fenway Park.

Scott was a math major from the University of Vermont who worked for Tippett at Diamond Mind from 2000 to 2003. Scott introduced himself to Epstein backstage at a "Hot Stove Cool Music" fund-raiser (where Epstein played guitar alongside Gammons to raise funds for the Foundation to Be Named Later). Epstein hired Scott as an intern with baseball operations in 2004.

Toronto native Tippett was a Bill James lieutenant who'd been the owner and creator of Diamond Mind Baseball. John Henry loved Tippett's computer simulation baseball game, which allowed Henry to play out entire long-forgotten seasons on his computer. In the summer of 2006, Henry put Babe Ruth on the 1929 Red Sox roster to see how the moribund Sox of '29 (58-96 in real life) would have done if the great Bambino had still been on the Boston roster. Francona had played Diamond Mind games against the best and brightest of Theo's subordinates when he was auditioning for the Red Sox job in 2003. Tippettt was hired by Epstein in 2003 and assisted on a variety of technology and baseball research projects. Francona enjoyed Tippett's company and called upon the young software savant whenever he needed help with his NCAA pool picks.

In 2006 Epstein commissioned Tippett to create a proprietary software program that would provide Boston's baseball operations staff with easy access to a mountain of baseball data, analysis, and history. Eschewing unreliable one-year samples, the software made valuations combining offense, defense, and baserunning metrics. There were projections for every player in professional baseball, and the software stored interviews from each player's high school and college coaches and trainers, plus observations from opposing coaches. The numbers were updated daily.

"We had a contest to name it," said Epstein. "I nominated 'Carmine' because it was a play on Carmine Hose (an ancient Red Sox nickname) and it sounded like a tough guy and a hot woman, just what we want from our software. 'Carmine' won not only because it was the best nomination, but also because I was the judge."

Henry, who had made hundreds of millions of dollars through the application of proprietary formulas, loved Carmine and all of its powers.

"Carmine took on a life of its own with the media," said Francona. "It made it easy for the radio pundits to make smart-aleck remarks, but it was a tool in our system that dumb-asses like me could go to, to get information right now. And it was good. It had everything in one place. It's a really concise way of looking up information. It had eight different clicks. One would be a guy's birthday, weight, and height. One would be all the reports on him. One would be his stats. It just made it so easy. I always went to it at the beginning of camp, just to look at who was coming into camp. I'd go on Carmine and I'd know everybody. I'd know what their minor league manager thought of them. But Carmine didn't tell you when to play a guy."

"Carmine did not make all our decisions," said Epstein. "It was just one club in our bag."

In the hours before the first game of a five-game series against the Yankees — the day game of a day-night doubleheader at Fenway on August 18 — young Zack Scott sent a report to the manager explaining that Eric Hinske should start at third base instead of Lowell in the first game of the series. Hinske, Scott reasoned, had better numbers against Yankee starter Chien-Ming Wang, who was 13-4 going into the game.

This wasn't what the manager needed on the cusp of the big series. He called Scott to his office.

"Zack was a nice, smart guy, but he didn't have a good feel for the clubhouse," said Francona. "They were trying to get him more on the baseball side of things, but everything he did was by the numbers. It was information when he started out, but then it got to the point where it was starting to get a little too assertive. When he told me we should play Hinske in that game instead of Lowell, I said, 'Why don't you go up to Mikey Lowell and tell him he's not playing in this game against the Yankees?' We got a five-game series with them, and I've got better things to do than call Mikey Lowell in here and tell him he's not playing.

"That's when I said, 'Enough, Theo. We're into August and we're wearing thin, and all this is doing is aggravating me.' So we stopped doing that. It's well intentioned, but there is more that goes into lineups than numbers. There's personalities and egos. You're trying to instill some confidence. That's why you have managers. That's what I would tell John Henry. I'd say, 'John, we all know what the numbers are. So you could do this. But you need somebody with strong opinions, or you don't need a manager.'"

Torre said the same thing in his autobiography with Tom Verducci (*The Yankee Years*) when he told Yankee GM Brian Cashman, "The numbers are good. But don't you ever forget the heartbeat."

"It's not that these people were not smart," said Francona. "But when you have somebody watching video and telling me how they are attacking the hitter, you'd better have a degree in baseball. Tell me something that's going to help me, but don't manage the team."

After dealing with Scott, Francona put Lowell at third base and Hinske in right field in the first game of the day-night doubleheader. Lowell went 0-3 against Wang. Hinske hit three doubles in three at-bats against Wang.

Somewhere near a silo in Lawrence, Kansas, poker-faced Bill James was probably thinking, *Up yours, Tito!*

But the Red Sox lost, 12-4.

"We kind of stopped the lineup program at that time," said O'Halloran. "It became clear that Tito was getting annoyed by it and it was kind of counterproductive."

"The lineup was always the manager's decision," said Epstein. "I always said, 'We always want to give you as much information as we can. You don't have to follow it.' In this case, we were doing a lot of deeper-level lineup stuff, swing path and all that. Hinske could really hit right-handed sinkerballers. He had a swing path that was down in the zone and away from him. Guys like Wang he owned. Mikey Lowell at that point wasn't handling those types of guys. This was an area where Tito and I butted heads at times. Tito usually erred on the side of showing players he had faith in them. I felt like at times he went too far. If you're not going to play Mike Lowell from time to time, do it against the guys he can't handle."

Jon Lester, who'd been called to the big leagues on June 10, was involved in a minor car accident on Storrow Drive en route to Fenway while Scott was sitting in Francona's office. Lester shook off his back pain and started the nightcap. He gave up seven runs on eight hits and was pulled with one out in the fourth. The Yankees beat the Red Sox, 14–11. Lester could barely walk when he woke up the next morning.

*That car going under my truck must have messed me up,* Lester thought to himself as he struggled to get to the ballpark.

It turned out to be something far more serious.

The Yankees beat the Red Sox for a third straight time on Saturday, this time by a score of 13–5. After the matinee beat-down, friends of Theo Epstein's gathered at Rowes Wharf behind the Boston Harbor Hotel for a cruise aboard John Henry's 164-foot yacht, the *Iroquois,* to celebrate Theo's engagement. Epstein was Boston's most eligible bachelor, and news of his engagement to Marie Whitney had dominated the "Inside Track" gossip page of the *Boston Herald.* No one could have guessed there were storm clouds over Fenway on the night of their engagement celebration.

"The timing sucked," said Epstein. "The baseball gods always get you. Anytime you schedule anything, even if it's giving a speech to a charity function, or heaven forbid you have a dinner party, it's a guaranteed four-game losing streak. The baseball gods would have it no other way. I would have given anything to not have to go. It completely ruined the whole thing. I don't think I was medicated, but I should have been. We were all in a morbid trance at that time. It was one of the nicest things anyone ever did for me, but it did feel like, John, Tom, and Larry, with their looks, were throwing us overboard. We felt like they were going to make us walk the plank. And I was looking for ways to throw myself overboard. I was looking for ways to knock myself unconscious."

"I was tired, we were going down, it was all I could do to go to that thing," said Francona. "We were all teasing Theo about it. I had never been on John's boat before, so I didn't know I'd have to take my shoes off until I was walking on board. I looked up and saw Theo and said, 'I hope nobody is here taking my picture in the middle of all this,' and he said, 'How do you think I feel?' Anyway, we got out there on the water, and I told my wife I was getting seasick, and she said, 'We're in the harbor!' I guarantee you, I was the first one to leave when we got back. It was actually very nice, but that was a tough weekend, and the timing of the cruise was everybody's worst nightmare, including Theo."

Epstein stood before the media the next day and said, "We're not going to change our approach and all of a sudden try to build an uber team and all of a sudden win now at the expense of the future."

The Sox lost again Sunday, 8–5, with the bullpen blowing another lead for Schilling.

Boston lost the Monday finale, 2–1 ("Both teams were too tired to hit anymore," said Francona). Wells pitched seven stellar innings, then watched Foulke allow the winning run to score on a wild pitch. Naturally, Wells gestured madly from the dugout.

The image of Wells showing up his teammate wasn't the only sign that the Sox were unraveling on and off the field. Timlin had offended teammates by telling reporters that the pitching staff was not to blame for Boston's woes. Youkilis joined the chorus of players ripping Fenway's official scorers, saying: "We have no home-field advantage." There was disgruntlement in the clubhouse over Epstein's non-moves at the trading deadline, a strategy that looked ridiculous when contrasted with Abreu's impact on the Yankees.

"Who fucking loses five in a row in a series?" Epstein said. "There are no five-game series. The odds of that happening. The wheels were completely falling off the wagon."

And then there was Manny. Ramirez tore into the Yankees during the lost weekend. He hit homers Friday and Saturday and went 8-11 with nine walks over the five-game disaster. He became only the fifth major leaguer in history to record nine consecutive 30-homer, 100-RBI seasons. But he was upset with a scoring decision regarding a ball he hit to Jeter Friday night and had to be coaxed into the lineup Saturday. He came out of the games early Friday and again Saturday. He went the distance Sunday. On Monday, Manny decided he'd had enough . . . for the season. Noticing that Torre had given Jorge Posada, Jason Giambi, and Johnny Damon a day off (Jeter was allowed to DH), Manny took himself out of the game after getting forced at second base in the bottom of the fourth inning.

"I'll never forget that," said Francona. "He came off the field, walked down the dugout steps, yelled over and said, 'Hamstring!' and I said, 'Manny, which one?' and he pointed with both hands to both hamstrings. It was like, 'You pick. Fuck, I'm coming out.' It was funny later, but it wasn't funny at the time.

"I had had it with Manny at that point. David came to me after that night and said, 'Give me a chance. I'm going to see Manny tonight.' When I saw David the next day, he said, 'Fuck it. Whatever you want to do.'"

That was it. In essence, Ramirez had quit for the season.

The Sox had been two games out of first place when the series started. They were unofficially out of the race when it was over. Over the five games, New York pounded Boston's pitching for 49 runs, an average of almost 10 per game. Abreu went 10-20 in the series with seven walks. Yankee lefty batters Jason Giambi, Melky Cabrera, and Johnny Damon feasted on Boston's right-handed pitching. Damon was booed every time he came to the plate. Sox fans wore T-shirts with Damon's image and the words: LOOKS LIKE JESUS, THROWS LIKE MARY, ACTS LIKE JUDAS. He managed ten hits, including two homers and eight RBI, in the five-game set.

The reeling Red Sox flew to southern California immediately after the fifth Yankee loss. It was the start of a nine-game West Coast trip. They had lost five in a row and fallen ten and a half games in the standings since July 4. They were playing .500 ball (53-53) against the American League. When they arrived at Angels Stadium, Francona decided it was time for a meeting.

"Typically, that's when a manager will stand up in front of everybody and scream," he said later. "Well, there wasn't any reason to scream. I just told them, 'Hey look, if we're gonna go down, I'll go down with you. We'll do this together.' I felt good about it. Some of the seeds were set for next year for some of those guys. I liked our team. It was part of the whole idea of 'doing it better,' which we thought after we won in 2004. Except for Manny, guys were still playing hard. What we went through the rest of that year built some loyalty for the guys who were coming back."

Dustin Pedroia was called to the big leagues for the start of the trip and joined the team for the first time in Anaheim.

"I had been playing in Ottawa," remembered Pedroia. "They told me to take a 6:00 AM flight to Anaheim the next day, and I got in at two in the afternoon. I was all excited, but it was kind of a bad time to get called up. Everyone was kind of dragging, it was such a tough time. There was a lot of shit going on. I met with Tito, and he said, 'Just have fun. You're not here to save the team.' Then he had the meeting. I just sat back and was trying to figure out what this was all about. I was out of the loop and just happy to be there."

"Poor kid," said Francona. "He said it looked like guys were going to start crying. Nice first day in the big leagues."

Batting ninth, playing shortstop, Pedroia made his major league debut after the meeting. He lined into a double play in his first big league at-bat, then hit a single to center off Angel lefty Joe Saunders. He finished the night, 1-3, and the Sox lost their sixth straight.

Lester snapped the losing streak the next night to improve his rookie record to 7-2, but he still wasn't feeling right. He had night sweats, was losing weight, and couldn't shake a cold. When the ball club flew north to Seattle, Lester's parents sent him to see his uncle Paul, an internist in greater Seattle.

"My parents and I went to the hospital early in the morning, not thinking much of it, but we were there until four in the afternoon," said Lester. "The emergency room doctor came out and told me I either had testicular cancer or lymphoma."

Lester reported to Francona when he got to the clubhouse.

"It's amazing how quickly you can change gears," said Francona. "Before then, I'd been all about how do we get him so he can pitch into the seventh inning? Suddenly, it's all about this being a kid I care about a lot."

John Henry arranged to have Lester, his parents, and uncle fly back to Boston on Henry's private jet.

"We all have our complaints about the workplace, but in the tough times they'd move the earth," said the manager.

Before taking their only child across the country for a battery of tests at Massachusetts General Hospital, John and Katie Lester met with Francona at Safeco Field.

"We will take care of your son," promised the manager.

Lester is only a year and a half older than Nick Francona, the oldest child of Terry and Jacque Francona. The manager's nickname for Lester is "Junior."

While Lester was at Mass General, things got worse for Francona and the Red Sox. They lost their final six games of the trip, three in Seattle, three in Oakland. Manny toyed with his bosses daily, changing his injury from a hamstring to a sore knee. The Sox sent Ramirez for an MRI in southern California and announced there was no structural damage. Manny's malady was reframed as patellar tendinitis. Meanwhile, Lester and his doctor-uncle were calling the manager to explain the fear and ambiguity of his status at Mass General.

Francona was spitting blood into a towel when he met with reporters after the finale in Seattle. He had to restrain himself from snapping when radio reporter Jonny Miller accidently banged his cane into the manager's knee. The Sox offense had been smothered again, and Francona spit out some truth when the inevitable Manny question was asked.

"If a guy says he can't play, I'm not going to make him play," he said. "Go ask him. He says he can't fucking play. We were hoping a couple of days off would be enough. Where it stands now, I don't know."

The manager's bloodletting was due to him biting his tongue while on blood thinners. He was supposed to keep his INR blood level between 2.5 and 3, but it spiked to 5.1 during the road trip.

"If somebody had punched me then, I probably would have died," he said. "I bit my tongue, and it didn't stop bleeding for days."

A perfect metaphor for the trip, and the season.

Multiple injuries to key players contributed to the train wreck. Francona was forced to use Loretta at first, Youkilis in left, a no-longer-capable Javier Lopez behind the plate, and Kyle Snyder and Kason Gabbard as starters. They went 9-21 in August, the franchise's worst calendar month since 1985 under John McNamara.

"We were just going down like soft shit in the rain," said Francona. "That's when Mike Lowell stepped up and turned into a real leader. We had young kids coming up, and I wanted to make sure we did things right. Lowell gutted it out and played every day. He set an example. Those were things I was trying to hang my hat on."

The white flag went up August 31 when Theo traded Wells to San Diego for catcher George Kottaras. Wells had been asking for a trade throughout the season and never seemed happy in Boston, but he pitched well for most of his two seasons at Fenway.

"I always liked David, and he will always be able to throw strikes," said Francona. "But when we traded him, all I could think was, *We made it through this without anything really bad happening.*"

The message was clear. There was still a month to play in the 2006 season, but the Sox were already thinking about the future. The Wells deal indicated it was all about 2007.

The day after the Wells trade the Sox announced that Lester had been diagnosed with large-cell lymphoma, an aggressive non-Hodgkins disease. The Red Sox connection to cancer cure is well documented. In 1953, after the Boston Braves left the Hub for Milwaukee, the Jimmy Fund became the official charity of the Red Sox. Ted Williams befriended Dr. Sidney Farber, the godfather of modern chemotherapy.

"Dr. Farber always told me, 'We're gonna find a way to save these kids!' Williams later recalled. "And goddammit, he did!"

With the help of the Red Sox, the Jimmy Fund has saved thousands of lives. For more than a half-century, Red Sox players have visited cancer patients at Children's Hospital and the Jimmy Fund Clinic. An annual Red Sox flagship telethon raises tens of millions of dollars. Sox pitcher Bob Stanley was a champion of the Jimmy Fund, then benefited from the clinic's research when his own son, Kyle, was successfully treated for a malignant sinus tumor. Mike Andrews, second baseman with the 1967 Red Sox, was chairman of the Jimmy Fund for more than three decades after he retired from baseball.

In the days and weeks after the disclosure of Lester's cancer, Francona protected Lester's privacy furiously, admonishing any reporter who attempted to reach Lester's family. The manager guarded the pitcher and his family as he would have protected his own family. Before starting treatment in September, Lester visited Francona at Fenway. Then, after a single treatment at Mass General, Lester went home and was treated at the Fred Hutchinson Cancer Research Center in Seattle.

"Tito was great with me," said Lester. "I don't think with any other manager I could have had that comfort level."

Lester, Papelbon, and Pedroia were the core of the Red Sox future. They all were scouted, drafted, and developed by Theo Epstein's baseball operations staff. It was important that they do well when they got to the big leagues. Papelbon was an established major league closer (35 saves in 2006) by the time Lester went on the shelf with his life-threatening illness. Pedroia was another matter. The kid struggled in his first days as a big leaguer. In 31 games at the end of the 2006 season, Pedroia batted .191 and committed four errors. He looked a little out of shape.

Still, he had his moments.

One such moment came on a sleepy Saturday afternoon at the Rogers Centre in Toronto. The Red Sox were playing out the string and had call-up Pedroia batting leadoff against Toronto's highly touted A. J. Burnett. Pedroia led off the game with a home run, his second big league home run. He circled the bases quickly, trotted back to the dugout, flung his helmet to the floor, and screamed, "Ninety-six miles per hour coming in, 196 miles per hour going out!"

"Simmer down, Napoleon," said Lowell.

"That got our attention," said Francona. "I think we were half-asleep at that point, but he woke us up. It got us to thinking that we might see some good things out of this kid, even though we were struggling and he was struggling. The kid was unbelievable."

Manny played in seven games in September, batting .211 with one homer. Ortiz hit seven homers in the final month, finishing with a Sox record of 54 homers, but he missed the protection of Ramirez. Ortiz was not happy with the Ramirez disappearing act. Big Papi wanted Manny hitting behind him in his quest for 60 homers, and he was annoyed that he was constantly asked to speak for Ramirez. Fans and media often twinned them as brotherly Dominican sluggers, but Ortiz and Ramirez were not particularly close.

The 2006 Red Sox finished 86-76, in third place, 11 games behind the Yankees. They were not the Idiots, which was okay with the manager. Seeds had been planted for the harvest of 2007.

"We had some guys that hung in there through some tough times, and I was proud of that," said Francona. "We were running some unusual pitching out there, bringing guys up every day, and it was hard. A lot of guys didn't stop playing. I remember thinking, *This is gonna help us,* and it did. Guys were really being professional. I was so proud of that team."

# CHAPTER 10

# · 2007 ·

## "They were all into getting the trophy and they didn't even know I was there"

**T**HE MANAGER AND THE cancer-stricken pitcher cut a deal. Terry Francona didn't need a lot of information or detail, but he wanted to hear from Lester after every treatment.

The text messages would come every three weeks.

"Hey, Tito, just checking in," Lester would type. "Had my treatment today. I'm okay."

That's the way it went until the first week of December 2006. Francona was at the winter meetings with Epstein in Lake Buena Vista, Florida, when his phone rang. Jon Lester calling. No text. The manager was alarmed. Something might be wrong.

He picked up and heard the voice of Lester on the other end.

"I'm cancer-free," Lester told his manager. "You're my first call."

"That's so great, Junior," said Francona. "Thanks so much for calling. Now go get ready for spring training."

Ending the call, Francona put down the phone and started crying. He'd been holding on to his emotions for months. He hadn't talked to anybody about it.

Then, and now, Lester made it a habit to talk about Francona as a paternal presence in his professional life. The lefty was fond of saying his manager had been like a second father to him.

"That made me feel good, but I didn't ever want to say something like that," said Francona. "You only have one father. But you see these kids come up through the organization and you get close. You care about them. You see them grow up. You know more about them than you know about a guy you sign as a free agent who comes in to help you win games."

"I think that phone call caught him off guard," Lester said later. "I don't have any brothers or sisters. I wanted the Red Sox family to know that I was cancer-free. I was ready to come back and pitch."

"It was the most important thing that could have happened at the winter meetings," said Francona. "The meetings were already a success."

Lester was out of the picture for the 2007 rotation. With Wells gone and Schilling slowly breaking down, Theo Epstein looked to the Far East to find pitching help. John Henry was in a spending state of mind, and Theo knew how to get some pitching without giving up any players or draft picks. Thus began the expensive courtship of Daisuke Matsuzaka.

Matsuzaka made his bones in the famed Koshien high school tournament in Japan, pitched eight seasons with the Seibu Lions, and won the inaugural World Baseball Classic, beating Team Cuba in the championship final in San Diego in the spring of 2006. He was targeted by Red Sox international scouting director Craig Shipley and Pacific Rim scout Jon Deeble.

The complex Japanese "posting" system was the first obstacle the Red Sox encountered. Major league teams interested in Japanese players were required to file sealed bids for exclusive rights to negotiate with a Japanese player.

After reading the reports on Matsuzaka, Francona told Epstein, "It sounds like this guy is exactly what we need. It would be a hard check to write, I understand that, but we don't have to give up any players. This is a chance to get a pitcher without losing anybody."

The Red Sox shocked baseball by making a posting bid of $51.1 million for exclusive rights to negotiate with Matsuzaka. The runner-up New York Mets bid slightly less than $40 million. The ever-cool Epstein was sweating. The Sox were serious.

The next step was the courting process. Matsuzaka was represented by Scott Boras, who had a well-deserved reputation for holding out longer than anybody else, then getting a record-breaking deal. Lucchino's secret weapon to counter Boras was Daniel Okimoto, a widely respected political science professor at Stanford who'd gone to Princeton with Lucchino. A Japanese American, Okimoto had many contacts in Japan and understood Japanese culture far better than Lucchino, Francona, or the Red Sox owners. Okimoto had been Bill Bradley's roommate at Princeton when Lucchino was a backup point guard on Princeton's NCAA Final Four team. During Lucchino's San Diego years, Okimoto served the Padres as an unpaid adviser.

It was Okimoto who suggested a lavish get-to-know-one-another dinner at the Pacific Palisades home of Tom Werner. Okimoto convinced Lucchino that a quiet dinner in a private home would be a perfect way for the Sox to introduce themselves and make their pitch to the Japanese legend.

"I was convinced that if Matsuzaka saw the commitment, the excellence, the successes, of the Boston Red Sox senior management team — ownership and senior management team — he would be enormously impressed and would leave the house wanting to sign with the Boston Red Sox and thinking that he would be a Boston Red Sox," Okimoto told author Ian Browne in *Dice-K: The First Season of the $100 Million Man.* "I said, 'If you show this side to yourself of basic honesty and decency and excellence, no one including Scott Boras will be able to demonize the Red Sox.'"

Theo told Francona to pack for a quick trip to Los Angeles. And to bring a gift for Daisuke Matsuzaka.

"We were warned that he was going to bring a gift for us," said Lucchino.

Unable to think of anything special, Francona drove to Dick's Sporting Goods and hastily purchased a New England Patriots number 54, Tedy Bruschi, jersey. Then he went to the airport.

The gathering at Werner's home included Henry, Werner, Lucchino, Epstein, Francona, Boras, Matsuzaka, Okimoto, and a pair of interpreters. They dined on pan-seared Chilean sea bass. After dinner, Okimoto called for the gifts to be brought out. The Sox had presents for Matsuzaka's wife and child, who were not in attendance, gifts for Boras, and gifts for the pitcher. They also brought a replica of the 2004 World Series trophy for Matsuzaka's inspection. Matsuzaka's wife was given a children's book written by Red Sox wives and tour books of Boston. Boras unwrapped a clock, a token of appreciation designed to remind the agent that time was wasting.

Toward the end of the gift session, Matsuzaka unwrapped Francona's Bruschi jersey and politely smiled his approval.

"I think I undergifted," Francona said. "It was a little embarrassing. They were giving him diamond-encrusted pens and things like that. Then I had my Bruschi jersey, and Theo was like, 'What the fuck are you doing? He doesn't know who Tedy Bruschi is.' I guess maybe I could have at least gotten him a Tom Brady jersey."

"Tito is a great guy to have dinner with," said Epstein. "When he's comfortable, he's so much fun to be around. But when you put him in a situation he's not comfortable with, it's painful to be around him. It was hard to be comfortable at that dinner, and Tito was so miserable. You can always tell, looking at his body language, when he can't wait to get the fuck out of somewhere, and that was a prime example. It was just too formal and too awkward. I think he might have dropped a few F-bombs at the dinner, a little out of place. But it was funny. Tito doesn't put on airs for anybody and doesn't respond to people who do."

It turned out it didn't matter. The money did the talking. Lucchino went to Japan two weeks after the dinner, and on December 14 the Sox closed the deal, signing Matsuzaka to a six-year, $52 million deal. Bookended with the posting fee, it was a $103 million commitment for a pitcher. But it did not cost the Sox any players, and that made Francona and Epstein happy.

The Red Sox weren't done spending and they weren't done with Boras. At the end of January in 2007, the Sox signed J. D. Drew to a five-year, $70 million contract. (Confident that the Red Sox were interested, Drew walked away from the Dodgers' three years and $33 million to get his free agency.) The Sox also signed bulked-up free agent shortstop Julio Lugo to a four-year deal worth $36 million.

All this spending came within six months of the night Epstein stood on the Fenway lawn in the middle of the Yankee massacre and said, "We are not the Yankees. . . . We have to do things differently. That's the reality. It's going to occasionally leave us short. It's going to leave us short every time there's a player who's available in a bidding war."

Not anymore. The Sox blew every team out of the water to get Dice-K, and they threw money at Drew when no one else was bidding. At the same time Henry worked out a deal to join Jack Roush's NASCAR racing group for a tidy sum of $50 million. It would not be Henry's final foray into a sporting venture other than baseball.

The manager had mixed emotions about the Drew signing. Drew had a reputation as a talented player who would not play hurt. Francona was aware that Tony La Russa had made harsh comments about Drew in Buzz Bissinger's *Three Nights in August*. Fans and media were also suspicious of Drew as a solution in right field for the Red Sox, but none of it mattered to the stats-crazed young men in Boston's baseball operations department. Drew was the Bill James/Theo Epstein prototype. He was a hitter who worked deep into counts and knew the strike zone better than any umpire or pitcher. He got on base. And he could play defense. He played without passion or emotion, but he was a *Moneyball* warrior, and Epstein wanted him badly.

"If I sign him, will you make this work?" asked Epstein.

"Yes," promised Francona.

Lugo inspired more consensus. John Henry loved Lugo. Bill James loved Lugo. And Terry Francona loved Lugo.

"When we got him, we were all excited, and nobody was more excited than me," said the manager. "He was always a guy who could turn on Schilling's fastball. He used to kill us."

Things were lining up nicely when the Sox gathered in Fort Myers in late February. On the second day of organizational meetings, before any of the players arrived, there was a lot of talk about Pedroia. The September sample had been underwhelming, but Theo and his baseball operations people wanted the manager to stick with the kid at second. No matter what.

Francona valued the opinions of the organization scouts. Returns from his off-season letter to the scouts gave him the information and opinions he needed to kick-start camp. The reports created conversation, and perhaps just as important, they made the scouts feel heard and included. There were plenty of disagreements, but it was a healthy exercise in organizational harmony and team-building.

"They kept talking about Pedroia and saying, 'You're going to see that swing and you're gonna laugh, but stick with him,'" said the manager. "I did, and they were right. Over the course of the eight years, the guys they liked, they were right on. Papelbon, Lester, Ellsbury, Youkilis, Pedroia. I remember early on when they were having an argument about whether Lester was better than Papelbon. I hadn't seen Lester pitch yet, but the fact that they were arguing about it was amazing to me. I was thinking, *If this kid can be anywhere close to Papelbon, holy shit!* But they were right."

Manny didn't get to spring training until February 26, but by 2007 Manny's tardiness was barely story-worthy.

"By this point, I wasn't losing much sleep over what day he came," said Francona. "The veteran players all understood. I'd have those get-togethers with them, and they'd always say, 'We understand it, but we want Manny to play.' I think David got more tired of it than anybody. Gene Mato, one of Manny's agents, was part of the problem. He was always telling Manny how he was getting disrespected.

"I could always tell when Manny had been talking with Gene, because he'd put shit in Manny's ear. Theo and Ben had Gene on the speakerphone one day in Ben's office, and I got wind of it. I was in full uniform — which always makes me more aggressive — and I went in there. Gene started saying some shit, and I let him have it with both barrels.

"You motherfucker!" Francona screamed into the speaker as he leaned over the desk. "You're half the fucking problem here, telling Manny all that shit about being disrespected."

Somewhat nervous, Epstein and Cherington looked at one another . . . then elected to let their manager keep venting.

"None of the shit you're telling Manny is true!" Francona continued. "Cut that shit out! Everybody here respects Manny. Knock this shit off!"

"It went on and on," Francona said later. "I fucking buried him. I think it was all the shit that Theo and Ben wanted to say, and they let me say it. I crushed him. When I walked out of Ben's office, all the secretaries looked away. I don't think they'd ever heard me curse like that. But I think Theo and Ben were secretly amused."

There was considerable debate regarding how to use Papelbon. The GM wanted Papelbon to start. The manager wanted him to finish. Theo had his way at the start of spring training, but dissent from the manager's office was stronger than usual. When Lucchino learned of the Theo-Tito dispute, he gleefully rubbed his hands together, saying, "This is what I like, a good, heated difference of opinion. We need more of this!"

Ten days before the start of the season, with Joel Pineiro, Brendan Donnelly, Mike Timlin, and Hideki Okajima failing to seize the job, Papelbon went to Jason Varitek and told the Sox captain that he wanted to close. When Francona walked past the veteran catcher and kid pitcher, Papelbon said, "If you want to give me the ball in the ninth, I want it." That was all Francona needed to hear. Papelbon would be the closer.

Lester was another story. Francona started the '07 spring with Lester just as he had in 2005 — telling the lefty that he needed to relax. *Don't be in a rush. Work your way back.* For his spring training debut, the manager sent Lester to Hammond Stadium in Fort Myers to pitch an inning of a "B" game against Twins minor leaguers. It was a baby step. Lester threw eight pitches in a one-two-three inning, which is exactly what the Sox wanted. Lester was still ten pounds underweight and far from baseball-ready. He was not part of the plan for the start of the season, but the manager knew the lefty was eager to get back to the majors. Francona called Lester's dad.

"We're going to have a meeting with your son and really piss him off," said the manager. "He's not ready to do this. We have to take care of him. And it's going to be slow."

Francona called Lester into his office.

"Jon, you're not going to start the season with us. We understand how hard you've worked, and this is not an indictment on your work. But you're going to get hurt if you try to come back right now. I don't want you to get discouraged. You just need to go slowly here."

The spring stories of Ramirez, Lester, Papelbon, Drew, and Lugo all paled in comparison with the attention lavished on Matsuzaka. One

Japanese television station had 19 people on the ground for 24/7 Dice-K coverage. Red Sox fans were told they could no longer park near the training site; they were instructed to drive to City of Palms Park and take shuttle buses to the workouts.

"This is different than any other spring training in baseball," observed veteran Schilling. "There are 200 media people here just because of him, but he gives off the impression that he doesn't want to be an inconvenience to people. He's a good kid. He's an ace in the making. He's like Pedro in a sense because he has multiple strikeout pitches."

"I got tired of it pretty quickly," said Francona. "It was too much for me. The Japanese media was relentless. Every time he ran out of the clubhouse they would all pounce and take a picture of his every move. One day when he came out and they started in on him, I heard somebody yell, 'Hey, Dice-K just took a shit.'"

Unless he was on the mound, Matsuzaka had an interpreter by his side whenever he was working at his baseball job. The Sox supplied a translator, and Dice-K brought a second one from Japan. The Sox initially ruled that Matsuzaka could not have his interpreter with him on the field during practice at spring training, but they quickly relented.

The presence of Matsuzaka's interpreter at team meetings was a problem for the manager.

"I was up in front of the room addressing the team, and I heard this fucking mumbling and I started getting the red ass," Francona recalled. "I almost flared on somebody. Everybody knows it's a pet peeve of mine if somebody is talking when I'm talking. Then I realized it was Dice-K's interpreter telling him what I was saying. That took a while to get used to—hearing somebody in the back mumbling every time I talked to the team."

"We knew there were going to be challenges," said John Farrell, Francona's former Indians teammate who was serving as the Sox pitching coach for the first time in the spring of 2007. "I took language classes with Dice-K and Okie three times a week in spring training. The instructor would teach them a phrase in English, and they would teach me in Japanese. I thought it was a way for a coach to better understand their culture and their structure so we could connect in some other ways. To what extent it helped, I don't know. But I wanted them to know I was making every effort to ease the transition for them."

Matsuzaka was on his own throwing program, a rigorous regimen that often frustrated Farrell. When Farrell was late getting to the dugout for the start of a spring training game, he often explained that he'd been occupied watching Dice throw an extra-long bullpen session. The Sox worried about Matsuzaka losing strength in his shoulder.

"He was strong-minded, and that was challenging in a number of ways," admitted Farrell. "And it was the first time I ever had a press conference after long toss or a side session. That seemed a little out of the ordinary, but this guy was the Michael Jordan of Japanese baseball."

Dice-K's final two spring starts were televised on ESPN, and he was on the cover of *Sports Illustrated*'s annual baseball preview issue when the Red Sox got to Kansas City for the 2007 opener. Royals Stadium hadn't seen massive media coverage since the 1985 World Series against the Cardinals. After all the hype, Schilling was routed and the Sox were beaten, 7–1. Lugo struck out in his first three Red Sox plate appearances (a harbinger), and Okajima surrendered a home run on his first big league pitch. On the plus side, Pedroia had two hits in the opener. Two days later, Matsuzaka made his major league debut and struck out ten hitters in seven innings of a 4–1 win over the Royals.

The learning curve was steep for the manager. He was informed that Dice-K viewed a visit to the mound as disrespectful. It wasn't done much in Japan.

Matsuzaka's second big league win came on Sunday night, April 22 (Francona's 48th birthday), in a 7–6 win over the Yankees. But Matsuzaka was not the story. For only the fifth time in major league history, four consecutive batters hit home runs: Ramirez, Drew, Lowell, and Varitek. Back to back to back to back. All off Yankee rookie Chase Wright. It was remarkable to watch, and ESPN was there. The feat had been achieved in 1963 by a Cleveland Indians foursome that included Tito Francona. ("My dad told me about it at least a million times," said the manager.) In 2006 four Dodgers hit consecutive home runs in an inning against the Padres. Drew was part of the Dodger barrage in '06, but if it meant anything to him, he hid it well. While Red Sox players slapped one another on the helmet and a national television audience gushed over the remarkable surge of power, Drew was overheard in the dugout tunnel talking about his latest hunting venture. Francona just shook his head.

Five years later, after Drew's unremarkable and expensive run in Boston, Francona said, "It wasn't as hard for me as people thought, because I knew going in what to expect. I knew J.D. got beat up by Tony, and I knew it bothered him because J.D. and I talked about it when he came over here. I told him, 'J.D., your personality is your personality, but the more you stay on the field, the better team we are. Even if you're making outs, you're a great right fielder and you're a presence at the plate. For me, he was easy. He's the nicest guy ever. He just . . . whether it was pain threshold or whatever, he came out of the lineup for a lot of things, things that maybe other players wouldn't come out of the lineup for. He had a hard time staying in the lineup. There was nothing malicious ever. Sometimes he just couldn't play. The one thing he took great pride in was his being in the right place on defense. Whenever we'd move him in the outfield, we'd notice him kicking the dirt a little. It pissed him off if we didn't think he was in the exact right spot."

The 2007 Red Sox went 33-12 after a 2-3 start. It was a 119-win pace. They swept an April series against the Yankees. By June 2, they had a nine-game lead in the division and an 11½-game lead over the Yankees.

"When people say you can't lose the division in April, they're guys who never won the division before," said Pedroia. "We ran away with the division in the first two months of the season."

Winning routines were established early. Francona, Farrell, Pedroia, and Lowell played cribbage on spring training bus trips while *Night Shift* played on the luxury coach video screen. ("It was always *Night Shift* with Tito," said equipment czar Tommy McLaughlin. "Even though none of the players were old enough to remember the movie.") The cribbage games could be competitive and costly. Francona and catcher George Kottaras picked up $800 from Pedroia and Farrell when the Sox bus got stuck in traffic on Tampa's Skyway Bridge after an exhibition game.

"It was a lot of laughs and cash for him," said Farrell. "One hand after the next. It was an ass-kicking."

Once the season started, Pedroia and Lowell played cribbage every day after batting practice. After hitting, they would come into the clubhouse, grab a sandwich and their game board, and retreat to the upstairs players' lounge for cribbage and early dinner. Francona eventually invited them to play in his office, and the Francona-Pedroia cribbage games remained a staple long after Lowell left the Red Sox. It made sense: Francona and Pedroia were usually the first two uniformed personnel at the ballpark.

With all the attention on Matsuzaka, Francona enjoyed the work of Okajima. The Japanese southpaw had an unusual, almost-violent north-south delivery; he was facing the ground by the time he released each pitch, but he had tremendous success in his first trip through the American League. It took a while for the Sox staff to determine how Okajima might best be used. In the third week of the season, on a night when Papelbon was unavailable after saving back-to-back games, Farrell asked Francona before the game what the Sox would do in the late innings against the Yankees.

"Who's going to save this game?" asked Farrell.

"Okie's going to go right through the middle of the order," said Francona, chuckling.

Farrell gathered up his reports and said, "All right!"

Seven hours later, when the manager summoned Okajima to protect a 7–6 lead against the Bronx Bombers — Jeter, A-Rod, and Abreu due up — Lowell approached Francona and asked, "Where's Pap?"

"Relax, big boy," said Francona. "Don't worry about it. We'll be all right."

After Okajima pitched a hitless ninth, holding the lead for his first big league save, Francona looked at Farrell and said, "Well, that worked out pretty good."

Okajima wound up representing the Red Sox at the 2007 All-Star Game in San Francisco.

Along the way, Francona learned one Japanese phrase: *Ii kanji*. It loosely translates to "way to go."

When Matsuzaka or Okajima came to the dugout after getting out of a tough jam, the manager would say, "Ii kanji, bitch."

After hitting .182 in April, Pedroia was over .270 by the end of May. He peaked at .331 in mid-June. He was on his way to winning the

American League Rookie of the Year Award. Still, he was annoyed that his manager sometimes chose to run for him late in games. This came up more than once during the cribbage matches.

"Why do you run for me?" Pedroia would ask.

"Because you're so fucking slow," the manager would answer.

Most professional athletes regard their speed or lack of speed as God-given. Like a player's height, it is not something that can be changed. Pedroia was different. He couldn't make himself taller (he insists he's five-nine), but he found a way to get faster. In his mind, his speed was a variable — something he could improve.

On baseball scouts' 80-point scale, Pedroia was a 40 to 45 runner when he played in college. Working out at Athletes' Performance Institute in Arizona, Pedroia learned to convert side-to-side motion to forward motion. He improved his fitness, strength, and technique. He made himself into a player who ran faster at the age of 28 than he did at 24. He improved to a 50 on the scouts' 80-point scale. He became a high-percentage base-stealer.

"That's almost impossible," Epstein told *Sports Illustrated* in 2011. "I can't ever remember it happening. It always goes in the other direction."

"Somebody put their ass on the line scouting that kid, and they deserve a raise," said Francona. "It's hard to sit here now and say, 'I knew it.' I ended up sticking with him when he was struggling, but it wasn't because I was this great baseball mind. I wish I could say that. The organization was pretty adamant that he could play, and we were playing well enough where we could handle it. I also loved the way this kid handled himself. He wasn't throwing helmets or letting anything impact his defense."

Epstein approached Francona in midseason and asked if he wanted to talk about a contract extension. The manager was in the second year of a three-year contract that was paying him $1.5 million per season. Francona said he was agreeable to discussing an extension, but Epstein did not follow up. Later in the summer, with the Sox rolling toward the division title, Francona reluctantly reintroduced the topic.

"Theo, I hate to bring this up, but you sort of broached this and then it went away," said the manager. "Are we still going to talk about a contract extension?"

"Ownership wasn't into it," said Epstein. "It's a company rule that we don't really do things early. Let's wait until we get to the end of the year."

"I appreciate the honesty," said Francona. "And just so me and you are on the same page, if we win, all bets are off."

"Okay," said Epstein.

"It wasn't something I was thinking about every day," Francona said later. "It never really bothered me much, but I was glad we had an understanding about how it was going to go later, especially if we won."

In late July, with the trading deadline nearing, Epstein explored the possibility of acquiring Texas closer Eric Gagne.

When Epstein told Francona he was close to completing the deal, Francona said, "Theo, I'm worried about this. We need to talk to Pap."

"Okay, then," said Epstein. "Let's go see him."

Three hours before the trading deadline, accompanied by Farrell, bullpen coach Gary Tuck, and Epstein, Francona drove his Cadillac Escalade to Papelbon's apartment in the Back Bay to talk to the closer about yielding his role to Gagne. They parked in a tow zone in front of Abe & Louie's restaurant on Boylston Street. Dressed in his full Sox uniform, Francona went inside the restaurant, grabbed a menu, and took it outside to put under his windshield wiper.

"Tito, are you fuckin' kidding me?" asked Farrell. "You're walking around downtown Boston, on game day, in your uniform, and you're worried about a parking ticket?"

The four men walked to Papelbon's brownstone, buzzed his apartment, and met with him in the boiler room of his building. Papelbon had memorabilia all around him — stuff he'd agreed to sign for a fee. When Epstein went into his pitch about acquiring Gagne and perhaps moving Papelbon into a setup role, the Sox closer stopped signing and frowned.

"I don't like it," said Papelbon. "I'm the closer."

"Okay," said Epstein. "I understand. Let me go to work on it."

"Part of me was happy Pap said what he said," Francona said later. "But I also knew it made Theo's job harder."

Folks on Boylston Street couldn't believe it when they saw Francona, wearing his game white pants and ubiquitous red fleece, walking back to his car alongside Tuck and Epstein. Boston's meter maids had no regard for the trading deadline and were not faked out by the restaurant menu. There was a $40 parking ticket under the windshield wiper.

Despite Papelbon's objections, Epstein made the trade, acquiring Gagne for outfielder David Murphy, lefty pitcher Kason Gabbard, and outfielder Engel Beltre. The deal proved to be a bust. Gagne went 2-2 with a 6.75 ERA and zero saves, then earned multiple mentions in the exhaustive report on performance-enhancing drugs in Major League Baseball released after the 2007 season by former senator George Mitchell, the chief investigator.

The Mitchell Report was of no concern to Francona in August of 2007. Nor was the parking ticket. He was dealing with MLB's uniform police. Baseball wanted him to wear his standard-issue uniform top when he worked in the Red Sox dugout.

Francona was assigned number 16 when he first got the Boston job in December 2004, but number 16 had been worn by Joe Kerrigan during the train-wreck season of 2001, and clubhouse attendant Tommy McLaughlin suggested that Francona switch to number 47. Numbers meant absolutely nothing to the Red Sox manager. He'd worn number 11 in high school and 32 (now retired) at Arizona. He wore number 16 as a player with the Expos, and also wore 10, 24, and 30 in the big leagues. Terry's dad wore ten different numbers in his big league career, none of them 16 or 47. The best number 47 in recent Sox history had been lefty pitcher Bruce Hurst, who came within one strike of being MVP of the 1986 World Series.

Red Sox fans never knew Francona's number because he always wore a red or blue fleece top, or a Sox jacket over his uniform jersey. "That started in Philadelphia," he said. "I wore my fleece all the time, and it pissed people off. There was a tunnel behind the dugout where it got cold, and I'd wear my fleece. When I found out people didn't like it, I got stubborn, so even when it got hot I'd keep the fleece on. I would sit in the dugout, sweating my ass off, saying, 'Fuck those people.' I wore number 7 with the Phillies, but nobody ever knew it."

During his years at Fenway, the red top was as much a part of Francona's identity as his bald pate. There are very few photographs of Francona wearing his standard uniform jersey top. Unless it was opening day, he was always wearing a red or blue fleece.

"When he walks to the mound in that thing, he looks like he's coming out to change the oil in your car," wrote a *Globe* columnist.

The Valvoline top became an issue in New York in the summer of 2007.

"It was a little bit of a running feud with the league," Francona said. "I started getting fined, and they were sending me letters all the time. Bob Watson was in charge of that for Major League Baseball. I told him to come watch me get dressed. I put my socks on. I put on those two leg sleeves. There's a reason they help your circulation — they're tight and they're hot. Over the top of those I would wear a pair of runner's tights. Then I would put a pair of socks over those, up to my knees, and then I'd put my pants on. I was triple-layered. There were some days in the dugout, I was like, 'Fuck, you've got to be kidding me. They want me to tuck my shirt in?' By the time I put my whole uniform on, I was a little bit claustrophobic. I can't tuck in one more fucking thing. That's why when I'd get on our charter, I'd always take my shoes off and untuck my shirt. It all goes back to my circulation issues. Bob asked me if I could cut off my jersey at the waist so I could wear it and they'd be able to see it under my top. I did that, but I still didn't always wear the uniform top underneath. Anyway, he came to our game in New York one night, and we were laughing about it in the dugout, and I pulled my top up and had the uniform top underneath. Then the game started and it wasn't so funny."

It was a marquee pitching matchup: Beckett versus Roger Clemens. In the second inning, with the Sox on the field and Jeter taking a lead off second, Francona heard a New York City police lieutenant (approved by Major League Baseball), Eddie Maldonado, calling his name from the dugout tunnel behind his back. His first thought was, *This must be bad news.* He was worried about his family. He went back to see what the man had to say.

"I need to check to see if you are wearing your uniform top," said Maldonado.

Francona was wearing his game jersey under the fleece. He was also livid. Jeter was still taking his lead off second, but the manager didn't care anymore. He followed the lieutenant up the tunnel.

"You !@#$%^&*(," said Francona, chasing the lieutenant.

It was loud and ugly. The *New York Post* reported the incident, erroneously stating that Francona had yelled at Watson. The incorrect account enabled Francona to address the issue at his daily presser before the next day's game.

"Terry was within his rights to be upset," admitted Major League Baseball executive Jimmie Lee Solomon.

"It was silly," said Francona. "'Nobody wears that thing. Ozzie Guillen wears a hood. When Joe Torre took over Watson's old job years later, I knew I wasn't going to hear from him about it."

With Francona dutifully wearing his number 47 game top under the fleece, the Red Sox were swept in the series. Francona was ejected in the seventh inning of the finale. Yankee rookie Joba Chamberlain got everyone's attention by throwing a couple of pitches at Youkilis's head, earning an ejection in the eighth inning.

On Saturday night, September 1, rookie Clay Buchholz made his second big league start, against the Orioles. Buchholz has spectacular stuff. Watching closely from his seat behind home plate, Jeremy Kapstein called his friend Bill Wanless, Pawtucket's public relations vice president.

"This kid is going to throw a no-hitter right now," said Kapstein.

Buchholz no-hit the Orioles that night, but there would be no room for the young righty on Boston's postseason roster. After 148 major league and minor league innings that season, Buchholz was losing strength in his throwing shoulder, and Boston's baseball operations people were maniacal about not overextending young arms. Francona agreed with that policy.

The Yankees made a run at the Red Sox in 2007, but the Sox held their ground, finishing ahead of New York for the first time since 1995. Drew salvaged a subpar season by hitting .342 with 18 RBI in September.

When the Red Sox clinched the American League East on September 28, the highlight of the postgame celebration was the sight of Papelbon dancing a jig on the infield while wearing a Miller Lite 12-pack box on his head. It would not be Papelbon's final Yawkey Way Riverdance of 2007.

"You set out to win, but I never wanted to sacrifice our chances of winning the World Series," said Francona. "I always tried to keep the big picture in mind. My job was to put us in position to win a World Series."

The Red Sox steamrolled the Angels in the American League Division Series. Beckett, baseball's first 20-game winner in two years, won the opener, 4–0, retiring 19 consecutive batters after surrendering a single to the first hitter he faced. He threw only 108 pitches. In the 2007 playoffs, he was 4-0 with a 1.20 ERA, 30 strikeouts, and two walks. The Sox won his postseason starts by an aggregate 34–5.

"I had a lot of confidence going into it," said Beckett.

"Every time Josh was on the mound, we knew we were going to win," said Pedroia. "He could have gotten anybody out. Barry Bonds in his prime? It wouldn't matter."

The Game 1 loser was John Lackey, a man who would play a major role in Francona's final two seasons in Boston.

Ortiz and Ramirez hit .533 with four homers and seven RBI in the Sox sweep of the Angels. They were at the height of their powers and had agreed on Ortiz hitting third and Manny at cleanup. The manager let them think it was their decision.

"Early on, both of 'em wanted to hit third," said Francona. "And then both of them wanted to hit fourth. I would bring them in together, and we'd talk about it. I'd say, 'I know you both want to hit third. You can't.' I knew the way I wanted it to work, because I thought Manny protecting David was good. Manny would never go out of the strike zone, so if he hit fourth, it was good. But I wanted them to embrace it. Buddy Bell told me once that I'm sneaky like that — that I can get my way without making it seem like I'm getting my way. And this worked with Manny and David. It was better if they had input and thought it was their idea. They felt like they were part of it, and they were both happy. If you can get the players to think that they're the ones making the decisions, you're way ahead of the game. The idea is to create an atmosphere where the players want to do the right thing. It's like dealing with your kids.

"David batting third and Manny fourth was perfect. I can't imagine what it was like for other teams when they scouted us. How do you get through both of those guys? That's how good they were."

"Manny was more complicated than me," said Ortiz. "To me, it didn't matter. I thought Manny was more of a cleanup hitter than me. He was more patient and chased bad pitches less than me. He would take his walks. I was more aggressive. Sometimes he wanted to hit third, and I told him, 'Go ahead.' It was easy between him and me. Tito let us decide where we wanted to be, and that wasn't a problem."

After dismissing the Angels, the Sox faced Cleveland. Beckett won the opener with another dazzling effort, but then the Red Sox lost three straight to the Tribe.

Ortiz called a players-only meeting after the Game 4 loss in Cleveland.

"I just wanted everybody to be positive," said Papi. "I told them that it wasn't over. Lots of times it's better to have those meetings between players, without coaches. When a player calls you out, it's different than when a coach calls you out. Those meetings weren't about calling people out. It was to remind us to stop chasing pitches and make better pitches and execute the right way in the right situation."

With the Sox down 3–1, Francona was taking a lot of heat. He was criticized for keeping Pedroia (hitting .188 through four games) atop the lineup and for sticking too long with Crisp.

His decision to start Wakefield instead of Beckett in the fourth game was roundly criticized. Francona had asked Beckett about the possibility of pitching the crucial game on three days' rest. Beckett told the manager that he would pitch on short rest, but would be much better on normal rest.

"Tito and I had an agreement," said Beckett. "I just pitched when he told me to pitch. I didn't like that to be in my hands. I don't really tell somebody whenever I want to pitch. If he had told me to pitch earlier, I would have pitched earlier."

"Beckett was ready to be Beckett," reasoned Francona. "That's all I needed to hear. I never could figure out why people made such a big deal out of it. I wasn't giving up Game 4, but we need to win four games. Using Beckett early wasn't going to help us win four games."

Francona went with Wakefield in Game 4, and Wake pitched into the fifth inning, when Cleveland scored seven times. The Tribe's 7–3 victory put the Red Sox on the brink of elimination. Francona was roasted in print and on the air.

Postgame, after facing the cross-examination of the media, Francona went back to his manager's office at Jacobs Field and started getting dressed for the bus to the team hotel. He was surprised to find ESPN's Tim Kurkjian sitting on the couch in his office. He didn't know Kurkjian particularly well.

"How's it going," said Kurkjian.

"Okay," said a suspicious and agitated Francona. "What can I do for you?"

"I just wanted you to know that I agree with what you did, starting Wakefield," said Kurkjian. "There are people who would have liked it if I ripped you, but I didn't want to do that because I didn't believe you were wrong."

Francona felt a little better. He knew he was going to have a rough day and a half before the start of Game 5 in Cleveland.

"I understand fans and media second-guessing," he said. "I'm a fan. I do it. I just wish people would remember that we know things about the team that they maybe don't know. I have more information than anybody. And it's my job to know my team."

Beckett didn't need any extra incentive, but the Indians provided it when they produced one of his old girlfriends, country singer Danielle Peck, to sing the anthem before Game 5.

"I was out in the bullpen with him while he was warming up, and she was on the field, getting ready to sing," said Farrell. "Fans were hanging over the rails, teasing him about his girlfriend. Just when it got quiet and she started to sing, Josh looked over at me and said, 'You know, for the record, I broke up with her.' I was pretty sure then that he was going to be okay."

Beckett beat the Tribe, 7–1, to send the series back to Boston.

"That was Tito at his best," said Farrell. "He believes in a natural cycle to the game, and he really took exception when there was talk about changes in the rotation. He fought that and felt strongly about it. I think it goes back to our relationship when we were teammates in Cleveland. He got to understand the mind-set of a pitcher. The five-day routine, how ingrained they are in their routine and what day of the week it is, and he was a staunch believer that everyone who goes to the mound is going to have to pitch well if we are going to win. Moving Beckett up went against every fabric of him as a manager and how he views pitching. He was dead set against moving a guy up. He felt strongly about never disrupting a rotation."

Games 6 and 7 were Red Sox blowouts, 12–2 and 11–2. The highlight of Game 6 was a first-inning, two-out grand slam by J. D. Drew.

In the eyes of many Sox fans, Drew saved his season with the slam.

"We were getting ready to let them off the hook," said Francona. "We'd had the bases loaded and nobody out, then there were two outs. It was nice having J.D. in a situation like that. You knew it wasn't going to overwhelm him. He wasn't going to lose any sleep over it one way or the other. When he hit it, zero turned into four. It changed everything about that game."

The Fenway lawn party after Game 7 was long and wild. It would be the final Fenway postseason celebration of Francona's eight years in Boston. It was Papelbon's Riverdance II; this time the Big Galoot wore swimming goggles to spare his eyes from the champagne spray.

"That was a real celebration," said Francona. "It was legit. We beat a good team. When I went out to the field the second time and saw Pap out there doing his thing again, I knew it was time to go home. It was cold, and we were all sticky from the champagne. Theo and all of his guys were in my office. On my way out the door I said, 'Pookie [veteran clubhouse worker Edward 'Pookie' Jackson], if I come back here tomorrow and my office is torn up, I'm blaming you.'"

When Francona met with his staff before the World Series, there was a lot of debate about Coco Crisp versus Jacoby Ellsbury in center field. Crisp had made a sensational catch to close out the American League Champion Series, but injured his knee on the play. He hit only .200 in the ALCS.

"That was a hard one for me," said Francona. "Sometimes you are just not sure. It looked like Ells was going to get hot, and there was a lot of push to have him play. I felt like starting Coco was the right thing to do. I had DeMarlo Hale call Coco, and Coco said his knee hurt, so that was that. By the end of the Series, Coco was telling us he was ready to play, but Ells was on fire by then. We couldn't take him out."

Ellsbury had been called up to the bigs in July and hit .353 in 33 games. A 2005 first-round pick out of Oregon State University, he was one of the blossoming stars of the new regime. A rare combination of speed and power, he was successful on nine of nine stolen base attempts in his abbreviated 2007 run in the majors. He was also believed to be the first Native American of Navajo descent to make it to the majors. When they were thinking of drafting Ellsbury, the Cleveland Indians had asked him if he was offended by the "Chief Wahoo" logo that dotted every team cap. Ellsbury said he was not offended.

Epstein and baseball ops were thrilled to have Ellsbury contributing at such a big moment. Francona felt the same way. The Sox were on a roll, just as they had been in 2004, but this time they were doing it with eight homegrown talents, players who presented as thoroughly professional.

The Colorado Rockies had won ten straight and 21 of 22 when they came to Boston for the 103rd World Series. It didn't matter. There was no stopping the Red Sox in October of 2007. Pedroia set the tone with a second-pitch, leadoff homer off Jeff Francis in the bottom of the first inning to start a 13–1 Game 1 victory at Fenway Park. Schilling, a shell of his former self, barely cracking 88 miles per hour on the radar gun, was still able to win on brains, savvy, and his amazing control of the strike zone. He won Game 2 (2–1) with five and a third precise, artistic innings.

Francona knew he was sitting on a winner. But he still worried.

"It's a fine line," said the manager. "I can't remember ever going into a game thinking we weren't going to win. But you don't sit there thinking, *We're going to wipe them out.* You're thinking about everything."

The key play in Game 2 was Papelbon's pickoff of Matt Holliday in the eighth. Assist to Mills. The Red Sox knew that the Rockies would take big leads on Papelbon, who was not known for keeping runners close. With Holliday on first and two out in the eighth, Mills signaled to Varitek to have Papelbon throw to first. Holliday was caught off the base, and the inning was over. It was the first pickoff of Papelbon's career.

"Dana LeVangie [a Sox scout] had told us Holliday would try to get a running lead," said Francona. "Millsie was all over it. He had made Pap aware. As soon as it happened, I was yelling, 'Fucking way to go, Millsie.' It was a great moment. Millsie is such a hard worker, and as a bench coach you don't get a lot of that, so I made it a big point to talk about it with the media after the game."

The Red Sox and Rockies resumed the Series Saturday night, October 27, at Coors Field in downtown Denver. Once again, Pedroia set the tone.

Early Saturday afternoon, getting to the ballpark seven hours before the first pitch, per usual, Pedroia was stopped by a Coors Field security official working outside the visiting players' entrance.

"Excuse me, where are you going?" said the officer.

"I'm going into the clubhouse, man," said Pedroia.

"You can't be in there. You've got to leave."

"I'm a player, man."

"No, you're not. You've got to go."

At this point, Pedroia reluctantly produced his identification. Every big league player has an official MLB ID card.

"Anybody can make one of these," said the officer. "You've got to go." That was it.

"Hey, man," barked Pedroia. "Go ask Jeff Francis who I am! I'm the fucking guy that leads off the World Series hitting a homer!"

With that, Pedroia gained entry. Once inside, he made a beeline for Francona's office.

"You'll never believe what fucking happened to me!" he started.

Everybody loved the story. Then and now.

"He had a full lather going," said Francona. "Probably a few Red Bulls sprinkled in there. But we all knew he was ready to play."

Dice-K threw a three-hitter for seven innings in a 10–5 Red Sox win in Game 3.

"Ii kanji, bitch," said the manager.

Hitting leadoff for the first time in the postseason, Ellsbury went 4-5 in Game 3. Francona was on fire.

The Series was effectively over. Only one team in baseball history ever came back from a 3–0 deficit to win a playoff series, and that was Francona's 2004 team. By 2007, Red Sox fans were so fat and happy, some were hoping to see the Sox lose a couple at Coors so they could win a World Series on Fenway soil.

No manager would entertain such a thought, but Francona felt a twinge of sadness when his team assembled in the visitors' locker room before Game 4 on Sunday, October 28.

"It was weird," he recalled. "Me and Pedey and Mikey Lowell were playing three-way cribbage, and I got to thinking. . . . We were on the cusp of winning the World Series four straight, but I found myself thinking that this might be the last time we played cribbage that year, and it made me a little sad. That shit was so much fun."

Lester got the Game 4 start. Coming back from lymphoma, he'd made 18 starts in the minors, pitching in Greenville, Portland, and Pawtucket. He pitched only three games for manager Gabe Kapler at Greenville, but he met his future wife while pitching for the A-ball affiliate ("Everything happens for a reason," said Lester). He returned to the majors July 23, pitching six innings in a 6–2 win, but experienced elbow trouble and went back to the minors for a start at Portland in late August. By October, he was strong. He was not on the Sox roster for the ALDS and was used only in relief in the Championship Series against the Indians.

In the final game of the World Series, Lester smothered the Rockies for five innings while the Sox scored twice off Aaron Cook. When Lester walked Garrett Atkins on a 3-2 pitch with two outs and nobody aboard in the sixth, Francona walked out of the third-base dugout to pull his 22-year-old starter. Manny Delcarmen was ready in the bullpen. The manager found himself resisting the urge to hug Lester, right there on the mound with the whole world watching.

"Junior, that's plenty," Francona said, taking the ball from the reluctant lefty. "You did great."

Typically, Lester said nothing. He put his head down and shuffled off toward the dugout.

"It was like he had climbed Mount Everest," said Francona. "I know the game wasn't over, but you have to appreciate a moment like that. I knew his mom and dad were there. It was emotional. Having him be a part of that game meant as much to me as anything else."

With the Sox leading, 4–1, Francona pulled Manny Ramirez for defensive purposes before the bottom of the eighth. When Atkins hit a two-run homer off Okajima to cut the lead to 4–3, Francona started to fret. His heart stopped when Rockie flyweight Jamey Carroll hit a rocket to deep left off Papelbon in the ninth.

"As manager, you are paid to be a worrier," Francona recalled. "We were up three games to none and had a lead in Game 4, but I started thinking of all the things that could go wrong. Okie was on fumes.

We'd used him a lot. When Carroll crushed that ball, I was thinking, *Fuck, this game is tied. What are we gonna do?* It's amazing how you worry about everything. It's not that you're not confident in your team, but it's your job to see that nothing goes wrong."

The game was not tied. Ellsbury, who'd moved from center to left when Crisp came in for Manny, ran back to the wall and made a leaping catch.

Sitting on the bench, happy he wasn't screwing up the final game of the World Series, Ramirez yelled over to Francona.

"Nice move, taking me out," said Manny.

Francona just smiled.

There was another issue in the Sox dugout at the finish. Francona had gone to the bathroom when the Rockies made a double-switch in the eighth, and he was absent when home-plate umpire Chuck Meriwether approached Mills with lineup changes made by Rockies manager Clint Hurdle. When Francona got back to his position, Mills handed him the lineup with the changes. As the ninth inning unfolded, it was clear the Sox lineup card was wrong. When Seth Smith came up with two outs in the ninth, batting in the leadoff spot where pitcher Manuel Corpas had been placed during the double-switch, Francona and Mills started arguing.

"The lineup card is fucked up," screamed the manager.

"Not my fault," said Mills. "Those are the changes he gave me."

"We were a hitter off," said Farrell. "We knew it was wrong. Tito was jumping Millsie's ass."

"I'm all fucked up with this," said an exasperated Francona.

On and on they went. Right up until Papelbon fanned Smith on a 2-2 pitch. The argument stopped and the two friends embraced. They were on top of the baseball world. Again. But the lineup card had to be tossed. It would not look good to have an incorrect document in a glass case in Cooperstown.

"By the time that game ended, the card was useless," said Francona. "Scratches everywhere. But I didn't want to run out there in the middle of the inning and interrupt everything with Pap throwing 100 miles an hour."

Just as the 2004 Sox caught fire when all seemed lost, the 2007 team turned into a steamroller when it mattered most. They never lost a game after Game 4 in Cleveland. The Red Sox outscored the Indians 33–5 over the final 31 innings of the ALCS and demolished the Colorado Rockies, 29–10, in the World Series. They outscored the opposition by an aggregate 62–15 in seven-plus games. Lowell, who hit .415 in four games against the Rockies, was named World Series MVP. Francona became the only manager in hardball history with an 8-0 record in the Fall Classic. He finished fourth in voting for American League Manager of the Year.

"It was kind of crazy," said Pedroia. "At the end, it didn't seem that far from when we were down 3–1 to Cleveland. When we won that last game, it still felt like we had to come back and play another game. That's what I felt like. It was crazy. Every move and every break seemed like it went our way. That's what you need. Then again, maybe it was just us being good.

"I think it was just the fight in us. Everyone was on Coco all year about his hitting, but he played great defense. Lugo had a tough year, but in the playoffs he made great plays on defense. Everyone thought I stunk in the beginning, and I ended up playing well. We were kind of the underdogs, and we proved everyone wrong, and it was fun."

"That was the height of me and Theo being on the same page," said Francona. "In '04 we were kind of naive and doing it our way. I think that was a little overblown, but by '07 we were working together really well. Theo and I had our blowups for sure. There were times when he was wrong and times when I was wrong, but when the shit was hitting the fan, I knew right where I could turn, and I always did, and he always had a good answer. I always told him that the Yankees' best player was Jeter, but our best player, Manny, when he leaves the batter's box, doesn't even know what town he's in. That had to change and it did. I knew we needed talent, but we needed guys who gave a shit about winning, and I thought we got more toward that in 2007."

Fox TV set up a temporary stage in the Sox clubhouse for the presentation of the World Series trophy. Epstein grabbed Francona to join Henry, Werner, and Lucchino on the stage with Fox's Jeanne Zelasko and Commissioner Bud Selig. As more people hopped on the stage to share and soak in the glory, Epstein and Francona were nudged to the far left of the plywood platform.

"We'd always joked about how certain folks would only show up when things were going really well," said Epstein. "We'd noticed it in '04, so we were wondering how it would go this time. We were both standing at the end of the stage, and there was a lot of trophy-passing, and it seemed like an afterthought when they passed it our way and we were kind of elbowed off a little bit."

When the temporary staging reached its saturation point, Francona fell off the side — a moment of infinite humor and symbolism.

The manager looked up at Epstein, laughed, shook his head, and said, "Theo, I think my work is done here," then went off to join his players.

"It was kind of funny," he recalled later. "They were all into getting the trophy and they didn't even know I was there."

Rarely credited, ever-blamed. It's part of the job for anyone who manages the Boston Red Sox.

# CHAPTER II

# · 2008 ·

## "This will not help us win"

R ED SOX NATION WAS GLOBAL. The Sox were the most popular sports team on the planet. The Sox were hot. They were trendy. It was almost impossible to travel anywhere without spotting a Sox cap, and many of those caps were pink or camouflage. Garb was sold by the truckload, and tickets were impossible to acquire without going through the dreaded "secondary markets," which somehow managed to get into business with the ball club. New Red Sox books hit the shelves in the spring of 2008, but they were no longer stories about a team that overcame insurmountable odds and shucked off an 86-year-old drought. The 2008 book titles were boastful and conceited. Tony Massarotti wrote *Dynasty,* and Michael Holley penned *Red Sox Rule.*

The Sox were no longer the little engine that could. They were not lovable losers. They were not Cub-like. Their fans were not needy. Hubris and arrogance became the trademarks of the Boston baseball fan. The Sox had become like the Yankees. And ownership seized the day.

This is when Red Sox Nation and the Fenway game-day experience jumped the proverbial shark. There was scalding demand for all things Red Sox, and everything was for sale. Spontaneity was replaced by commercialization of just about everything related to the Boston baseball club. In 2008 the Red Sox became the first team in baseball history to provide fans with two hours of daily live television coverage of *spring training drills*. Sox flagship station NESN aired weeks of bunt drills and pitchers' fielding practice.

More than the Freedom Trail or Harvard Yard, Fenway Park became a destination for tourists who touched down at Logan Airport. Tour groups ponied up their dollars and filed through Fenway at all hours. Drivers barreling toward downtown Boston on the Massachusetts Turnpike noticed that the ballpark lights were always illuminated at nighttime, game or no game. It didn't matter if the Sox were out of season or out of town. There was 24/7 renovation and construction going on in the ballpark, and the club-level function rooms were rented almost every night. If you had enough money, you could have your wedding reception or your child's bar mitzvah at Fenway.

Under the direction of Lucchino, the savvy "new" owners cleverly and regularly reached out to Boston mayor Thomas Menino. Patriots owner Bob Kraft and his son Jonathan famously failed to bring the mayor into the loop when they attempted to build a new stadium in South Boston in the 1990s, and it proved costly. The Sox owners were determined not to repeat the misstep. The Red Sox of the 21st century consulted the mayor on every ball-club decision that had even the remotest impact on city business. Nothing was too small.

The relentless courtship of Menino paid big dividends as the Sox rebuilt Fenway Park. (According to the ball club, it spent $285 million on Fenway renovations in Henry's first ten years of ownership.) With Menino's help, the ball club developed a close relationship with the Boston Redevelopment Authority, an urban renewal agency with sweeping powers. The BRA gave the Sox permission to build Green Monster seats that hovered in the airspace over Lansdowne Street. The BRA also took control of Yawkey Way (the avenue outside of Fenway's main gate) on game days, allowing the Red Sox to use a city street for a private-enterprise block party for 81 dates, plus the playoffs. The sweetheart deal enabled the Sox to sell beer, peanuts, and Crackerjacks on city property while fans were entertained by stilt-walkers, face-painters, balloon-makers, and a band. Fans could also purchase beer and wine and a variety of foods, including a Luis Tiant Cuban sandwich. NESN set up a Yawkey Way stage for 90-minute pregame shows.

According to a February 2011 *Boston Globe* editorial, "The team has earned an estimated $45 million in food vending and Green Monster ticket revenue in exchange for average lease payments to the city of about $186,000 a year."

Every Fenway home opener featured something new inside the ballpark. In 2002 the Sox added dugout seats and the Yawkey Way concourse. The next year it was the Monster Seats — named the best seats in baseball by ESPN SportsTravel and *USA Today*. In 2004 the ball club added right-field roof seats. Then came the third-base concourse and the Game On sports bar. In 2006 the Sox gutted the aquarium-like, glass-encased high-roller section (the 600 Club) in the upper level behind home plate and created plush new executive suites — the EMC Club (EMC is a Massachusetts-based corporation that sells data storage products) and State Street Pavilion — on levels 3 and 4 of the ballyard. In 2007 the Sox added the Jordan's Furniture third-base deck. In 2008 the club added "Coca-Cola Corner" and the Bleacher Bar and expanded the State Street Pavilion. Fenway's right-field roof was renovated for 2009.

Mayor Menino told the *Globe*, "People come to see the ballpark, to see the Green Monster, to be close to the players. Boston must balance development growth with the preservation of what makes our city so livable — our history, character, scale and charm. We are distinct from other American cities because we view our buildings as resources, not liabilities."

Lucchino lieutenant Dr. Charles Steinberg put an indelible mark on the game-day experience at Fenway from 2001 to 2008. (Steinberg left the Sox after the 2007 World Series to work for Dodger owner Frank McCourt, then Bud Selig, before returning to the Red Sox for the 100th anniversary of Fenway Park in the spring of 2012.) In 2003 Steinberg hired Ray Charles to perform "America the Beautiful" on opening day — leading to a memorable moment when Tom Werner was observed waving to Charles. In his Fenway years, Steinberg brought back the Cowsills, the Standells, and James Taylor. He borrowed an old Oriole stunt and had a pretty young woman dressed in an old-time softball uniform sweeping bases between innings. He worked with a local band, the Dropkick Murphys, to come up with a modern rendition of "Tessie," which had been a Red Sox anthem sung by the Royal Rooters in the 1903 World Series. Dr. Steinberg's ultimate production might have been the pregame appearance in Game 7 of the ALCS of Irish step dancers on a stage in center field. The curly-topped lasses high-stepped while the Dropkicks played from a platform adjacent to the center-field bleachers. The Cleveland Indians never recovered and were beaten, 11–2.

"Sweet Caroline" evolved into a Red Sox anthem. The song was introduced to Fenway crowds in 1998 when a Sox employee's wife gave birth to a child named Caroline. By 2008 the Neil Diamond ditty was as much a part of a Fenway game as the national anthem or the seventh-inning stretch. It was played before the Red Sox came to bat in the bottom of the eighth, and it kept a lot of trendy fans ("Pink Hats," as they came to be known) in the ballpark long after they lost interest in the actual baseball game. Starting in 1997, the public-address system played the Standells' "Dirty Water" at the end of each Red Sox victory.

Beginning in 2003, Fenway also became a major concert venue when the Sox left town in midsummer. Bruce Springsteen and the E Street Band were followed by Jimmy Buffett and his Coral Reefer Band, the Rolling Stones, the Dave Matthews Band, Sheryl Crow, The Police, and Neil Diamond himself in 2008. Soon to follow were Willie Nelson, Phish, Paul McCartney, Aerosmith, the J. Geils Band, the New Kids on the Block, and the Backstreet Boys. The shows sometimes disrupted baseball activity. The Rolling Stones played two shows in August 2005, and their stage, which was only slightly smaller than terminal D at Logan Airport, ruined the outfield grass. The entire field had to be resodded before the Sox got back from an 11-game road trip.

It didn't stop there. Fenway was the site of the Baseball Beanpot and soccer games, the NHL's Winter Classic (Bruins-Flyers on national television on New Year's Day), college hockey, public skating . . . even a citizenship swearing-in ceremony. Ballpark tours ($16 per adult by 2012) attracted more than 200,000 people annually. Fenway never slept.

Every game Terry Francona managed at Fenway Park was a sellout. The streak started on May 15, 2003, and was at a major league record 712 when Francona and the Red Sox parted ways at the end of the 2011 season. (The streak became an albatross in September 2012 when the moribund Sox played in front of announced "sellouts" that included thousands of empty seats.) The ever-bigger Fenway allowed the Sox to break the 3 million mark for the first time in franchise history in 2008. Fans who couldn't score tickets to a Fenway game had the option, on some nights, of going to a local movie theater and paying $7 to watch a live telecast of the Red Sox game. There was nothing like it anywhere else in major league baseball.

Fans at Fenway were more a part of the action than at any other time in the ballpark's first century. Under Lucchino and Steinberg, whose motto was "We're in the 'yes' business," select fans were allowed access to the field and pregame ceremonies. Ropes were set up on the track from dugout to dugout, and fans could pester the players while they were trying to get their work done before games. There were multiple first-ball tosses, anthem singers, and seventh-inning-stretch singers. Before the start of each game, a fan (usually a child) yelled, "Play ball." For a price, anyone could do just about anything at Fenway except bat cleanup. Even championship rings could be bought. After winning the World Series, the Sox held a charity ring raffle, awarding the precious metal to lucky fans. This did not go over particularly well with some of the ballplayers, who felt that presentation of rings to fans somehow diminished their achievement.

The commercialism and demand of all things Red Sox took its toll in the dugout.

"It was more and more as the years went on," said Francona. "By 2008 I was dealing with players coming to me and complaining about it. They had opened up the triangle in center for people during batting practice. We'd be trying to get our work done, and we couldn't hit balls off the Wall. It's a fine line. It was just becoming more all the time. Players were getting mobbed by people when they were trying to take BP or get ready for the game. Guys don't mind signing autographs, but their pregame routines are extremely important. More and more people were on the field, and it got difficult to get their work done. We'd have fans getting shown into the Monster, and we'd have to stop hitting balls off the Wall. It got to be a bit much. I'd talk to Larry about it and say, 'This is all great, I understand the organization is trying to make money, but let's not let it get in the way of baseball.'

"This is where, for me, I think the organization was starting to change. It was obvious to everybody. And I'm not talking about the players changing. The organization had more requests. There were a lot of things happening on the periphery that were making it harder to get our work done. That stuff's all great, but you can't forget you're a baseball team."

He was alarmed when Lucchino called to discuss a trip to Japan for the start of the 2008 season.

"I was standing in my kitchen in Brookline when Larry called," said Francona. "He started telling me about this trip, and it didn't sound good to me. I said to him, 'Larry, I understand why you want to do this and why it's good for Major League Baseball. But I want to make one thing very clear. This will not help us win. As a manager, I can't tell you strongly enough that this will not help us win games.'"

"It's not a done deal yet," said Lucchino. "But I will get back to you if it becomes a reality."

"He got back to me almost right away," Francona said later. "I'm sure it already was a reality."

Bud Selig was intent on promoting the globalization of Major League Baseball, and the Red Sox — the defending World Champs with a roster that included Daisuke Matsuzaka and Hideki Okajima — were MLB's dream team, slated for two regular-season games and two exhibitions at the Tokyo Dome. Ever-starving for revenue, the Oakland A's agreed to the same deal.

Planning for Japan was complicated. It required the Red Sox to shorten spring training by a full two weeks. The Sox would leave Fort Myers on March 19, fly 6,000-plus miles to Tokyo, play exhibitions against the Hanshin Tigers and Yomiuri Giants, play two regular-season games against the A's, fly to Los Angeles for three more exhibition games, then on to Oakland and Toronto for another six regular-season road games before coming home to Fenway for the home opener on April 8 against the Tigers.

It was preposterous. It was a 16,000-mile, 18-day, three-country, ten time-zone odyssey that would put money in the coffers of the owners and spread the brand of the Red Sox. But it was not conducive to a quick start for the 2008 title defense.

"The Japan trip had been bad for the Yankees [in 2004]," said Francona. "And I was worried it would fuck up spring training, which it did."

The manager was going into the final year of his contract in 2008. His status had been tabled in the middle of the 2007 championship run as the Sox owners rolled the dice, waiting to see how the '07 Sox performed before committing any more years to their manager. Francona had accepted that judgment and quietly won a World Series. When the owners came to him in the spring of 2008 to talk about an extension, he hardly needed to remind them how things unfolded in '07.

In the spring of '08, while the complicated Japan trip was being organized, Henry, Lucchino, and Francona gathered at the Hyatt Regency Coconut Point to discuss the manager's contract extension.

Once everyone was seated, Francona said, "Before we start, I need to ask Larry if he remembers what I said last year."

Lucchino thought for a moment.

"All bets are off?" offered the CEO.

"Thank you," said the manager. "Now go ahead and make your speech."

It was a pleasant session. Numbers and years were exchanged, but there was no agreement. Francona went back to the business of getting ready for the title defense, but the contract extension weighed on his mind. He didn't want it to become an issue. He didn't want it to go public. He knew he was underpaid relative to many other managers, but he also knew he had new leverage. He'd won two World Series in four years.

Epstein played the role of middle man in the ongoing talks, and every few days the GM would come back to the manager with a slightly better proposal. Francona appreciated the way the process was working and kept nudging Epstein to go a little higher.

"Theo, tell me when enough is enough," the manager told the GM. "When you've hit the edge with them, you tell me, and I'll make my decision."

"I wasn't sure I was going to say yes," Francona recalled years later. "It wasn't a guarantee for me that I was going to sign what they finally came back with."

Ultimately, the deal was struck: three more guaranteed years, which would take Francona through 2011, plus two club options, one for 2012 and one for 2013. Francona's salary went from $1.5 million in '08 to $3 million in '09 and up to $4 million by '11. The '12 and '13 options would pay him a combined $9 million if the Red Sox elected to bring him back after the 2011 season. If the club did not bring him back after '11, it would pay Francona $750,000 severance.

"I wasn't making a lot compared to some managers, but I had never touched money like that," said Francona. "I was very happy with the way it was done."

The Sox were darlings of the national media during spring training. They were consensus favorites to return to the World Series, and veteran Kevin Youkilis noted, "I think people expect us to win every single game."

"I suppose a little bit of excess is inevitable," Lucchino admitted. "As long as we don't fall victim to it ourselves."

Spring training was uneventful . . . until the final day. The Sox bags were packed for the Japan odyssey when they gathered at City of Palms Park on March 19 for their Grapefruit League finale against the Toronto Blue Jays. In midmorning word spread that the Red Sox and A's coaches were not in line to receive the same payments (roughly $40,000) that the ballplayers were getting for the trip to Japan. Francona and all of the Red Sox had been told that the coaches were getting the money, but there was a major misunderstanding.

"We had a day off the day before we were scheduled to leave for Japan," said Francona. "I was running on the treadmill and got a call from [Oakland manager] Bob Geren. We had coached together. He told me that his coaches were not getting any money for the trip. I started calling him a dumb-ass and telling him he'd mishandled it. But then I started to wonder about it, so I started making calls to MLB, and I couldn't get anybody to call me back. They were all in Japan already. I was concerned. Then Jack [Red Sox traveling secretary Jack McCormick] called me back and said, 'Tito, someone's going to call you in the morning, and I don't think you're going to be happy.'"

McCormick was right. The coaches were not getting the money.

When Francona got to City of Palms Park on getaway day, he informed his coaches of the misunderstanding, but promised them that they would be paid. Upon hearing the news, Sox veterans, led by Schilling, Varitek, and Lowell, proposed a boycott of the final spring game and told Sox management they were prepared to boycott the trip in support of the coaches. Even Manny Ramirez joined the chorus of dissent. ("I don't think he cared about the coaches getting the money, but it was a chance to take a jab at management," said Francona.)

When the Red Sox failed to take the field against the Jays for the 12:05 start, the story went international. ESPN had the jump on all outlets because the "Worldwide Leader" was at City of Palms Park to broadcast the Sox-Jays spring training finale.

"The players really stood by us," said third-base coach DeMarlo Hale. "I thought it showed strength among the players and the respect that he had for our coaching staff and what Tito had done as a manager."

Epstein and Lucchino were summoned to the Sox clubhouse. Lucchino was hot. Channeling his lifelong mentor, the late Edward Bennett Williams, who had defended the likes of Jimmy Hoffa, Adam Clayton Powell, and John Connolly, the CEO barked at his manager.

"They can't do this," said Lucchino. "They have to play. This is a wildcat strike. There could be ramifications for you."

"Larry, everybody's really pissed," said the manager. "This isn't going to work. This is a lot of money for those coaches. You know that. We were told they were getting it. You can't steamroll these guys. They are major league ballplayers."

"I was disappointed in the coaches," Lucchino said later. "It came up at the last minute, and I think of the coaches as management in many ways, and we were disappointed that the issue was being drawn at the last minute."

"That's why I was making all those phone calls," said Francona later. "That's why I made the call to Geren. I tried so hard to communicate with everybody because I didn't want this to happen."

There were three meetings with players. Francona went across the field and into the visitors' clubhouse to discuss the situation with Blue Jays manager John Gibbons. Like any good baseball man, Gibbons was naturally sympathetic to the plight of the coaches. Bud Selig called to urge that the game be played and the trip commence without interruption. Sox player rep Youkilis got involved. As the minutes ticked off the clock, Major League Baseball and the Red Sox came up with some money for the support staff, but it was approximately $200,000 short of what would be needed, and the Sox players were still on the bench refusing to play.

"Fuck it, I'll write the check myself," Francona told Lucchino. "We need to play this game. Even if it's an exhibition game. This is professional baseball."

"We can't let you do that," said Lucchino. "We'll figure something out."

"Good," said Francona, bolting from the office. "Let's play."

Francona was never forced to pay the $200,000, but wound up forfeiting his share from the Japan trip.

"Tito was great at making sure the guys who worked for him were taken care of," said Farrell. "He is an incredibly giving person, and money is not his driving force. That was part of the environment he created. It was fun, and guys wanted to be there early."

The Sox came out of their dugout one hour late and lost to the Blue Jays, 4–3, while a Japan Air Lines Boeing 747 waited for them at Fort Myers airport. The 382-seat plane easily accommodated the Sox party of 160, which included Terry and Jacque Francona and their 14-year-old daughter Jamie. Seated near Henry, Werner, and Lucchino, the manager was slightly uncomfortable as the plane lifted off the ground, bound for a refueling stop in Chicago.

"I knew they all wanted to wring my neck," he said. "It was a horrendous way to start a trip. Everybody was pissed before we even left."

The plane made the stop at O'Hare, then took off to Toyko just before 11:00 PM for a 12-hour flight.

Sox management planned carefully to prepare the ballplayers for the lengthy trip. Three team physicians were on board, including Massachusetts General internist Larry Ronan. Days earlier, Dr. Ronan had addressed the team regarding physical issues connected to a 12-hour flight. In the interest of hydration, he encouraged players to drink water and avoid alcohol. He also recommended that they attempt to remain awake during the flight. The Wednesday night flight would arrive in Tokyo just before midnight Thursday, which meant an entire day would be lost to travel and time change. It was important to try to stay awake until arriving in Japan, the doctor told them. He also handed out black, knee-high, anti-embolism socks. Pedroia was first to change into short pants and looked ridiculous walking up and down the airplane aisles in powder-blue shorts and black knee-high socks.

Sox players and coaches organized a cribbage tournament for the flight.

"We had 16 guys playing," said Francona. "You'd go into the winners' bracket and the losers' bracket, and every 15 to 20 minutes guys were moving. It took about six hours."

The manager won the cribbage tournament.

The nocturnal Henry prowled around the plane for much of the flight. When the plane approached airspace over Alaska, Henry went to the upper cabin, where the press was seated, to better view the Northern Lights.

Flying over the North Pole, following the instructions of his medical team, a sleepy, hydrated Francona left his seat to stretch his legs. While he was walking toward the back of the plane, he passed a sleeping, reclined Dr. Ronan, who was wearing a Zorro mask and had two empty mini-bottles of wine on his tray table. So much for no alcohol and staying awake on the flight.

"That will always be my favorite memory of the Japan trip," said Francona.

The Sox were given a day to assimilate to the time change, then played the two exhibition games, winning both. There were a lot of appearances for the manager and players. A lot of people bowing, just like in 2005, but this time it was a cultural exchange.

One local custom caught Francona by surprise when his eighth-grade daughter took a bullet train with Sue Farrell and Luke Farrell, the wife and young son of the Red Sox pitching coach. The Americans went to a natural hot springs *onsen*. Terry and Jacque were mildly shocked when young Jamie came home with tales of old Japanese men walking around naked at the spa.

"It's on the city map as a sightseeing destination," said Farrell. "When my wife got there, she was like, 'Oh my God, where have I taken these kids?' They might have had lunch, but the people eating weren't wearing clothes. It was embarrassing."

The day before the official Major League Baseball opener, Francona,

Lowell, Pedroia, Youkilis, and Okajima agreed to a paid appearance at a reception for EMC at the club's Hotel New Otani Tokyo. The manager, who had been told it was okay to dress informally, arrived wearing a sport coat and a pair of Dockers. Lucchino was not happy when some of the Red Sox players arrived wearing jeans. While the manager and his players noshed on sushi, sipped fine wine, and chatted amiably during the meet-and-greet with EMC clients and assorted dignitaries, Lucchino sidled up to Francona and muttered, "Don't you even think about leaving!"

Pedroia overheard the remark and later asked his manager, "Does he always talk to you that way?"

The EMC event was a hit. The Sox players, even the ones in jeans, did a nice job making the fans and sponsors happy. Their reward came at the end of the function when EMC boss Joe Tucci produced a couple of fielding mitts and insisted on a ceremonial game of catch with Okajima. After catching a toss from Okajima, the exuberant CEO uncorked a wild toss that sailed past the outstretched glove of a sprawling Okajima and punctured a massive LED screen that had been showing a loop of the Sox 2007 championship run. Lowell and Pedroia doubled over in laughter. It was then that Francona decided the event had been worthwhile.

Opening day was Tuesday, March 25, the earliest opener in big league history. At lunchtime, when Francona would usually already be at the ballpark, he was part of a 15-person Red Sox delegation visiting the United States Embassy. The Sox were guests of US ambassador Tom Schieffer, the brother of CBS anchor Bob Schieffer and a former owner of the Texas Rangers along with President George W. Bush. Francona was noticeably uneasy at the embassy. The visit had him out of his game-day routine, and he wanted to be in the clubhouse, especially on opening day.

"I started to aggravate everybody," he remembered. "I was at that luncheon and everybody was toasting, and I was standing at the back of the room waiting to get the go-ahead to go to the ballpark. It was two o'clock and I was still at the embassy, and I wanted to kill somebody. I wanted to be at the ballpark. My routine was all thrown off. I'm as big a creature of habit as anybody, and I was all fucked up. And

I knew Larry was pissed. By the time we finally got to the park, it was late for me. I wanted to have our team meeting before the first game of the year, and I didn't want it to be rushed. That's an important meeting. So I was already in a bad place on it, and right in the middle of me trying to talk to the players, a couple of Japanese women came into the clubhouse holding trays of food. I lost it. I started yelling, 'Get the fuck out of here.' Fortunately, they didn't know what I was saying, but they probably knew I was mad."

It was just after 6:00 AM in Boston when Pedroia singled on an 0-1 pitch to christen the 2008 major league season. Matsuzaka got the start and lasted five innings. The Red Sox trailed, 4–3, in the ninth when rookie Brandon Moss — who was starting in right field because Drew (who had homered in both Japan exhibitions) came up lame after pregame sprints — tied the game with his first major league homer. Manny hit a game-winning two-run double in the tenth and was rewarded with a million yen and a Ricoh color copier. Okajima got the win. Everybody was happy, including the commissioner, who'd made the flight to Tokyo several days after the Sox landed.

"I'm sure the sports bars in Boston were filled for this one, even though it started at six in the morning," gushed Selig. "When I left my hotel today, I had to pinch myself. I thought I was in Boston. Everybody in the lobby waiting to come to the game had Red Sox paraphernalia on. It is remarkable. This is part of the internationalization of the sport, and to have the World Champion Red Sox here is really exciting. This is the ultimate."

Game 2 was not as much fun. The A's beat the Red Sox, 5–1, and when it was over, both managers were required to publicly thank the fans. It was custom. Francona's speech to the Tokyo Dome crowd was brief and uncomfortable. He knew his bosses were watching.

"When I talked with the media after that second game, I could see Larry standing off to the side," said Francona. "He wanted to see if I said something negative about the trip."

"The whole spring was fucked up," Francona said later. "We had guys going to Japan who pitched games that count. A couple starters had to be ready quicker than others. Beckett hurt his back and didn't come with us. It was harder than you can imagine. There were rules coming out the yin-yang. Then you play two games that don't count, then two that count, then you're going back to the States to play more games that don't count. I was all fucked up. It was good for baseball, but I was grumpy as shit."

The nine-hour flight from Tokyo to Los Angeles didn't make anyone less grumpy. And the circus was only beginning.

Four days after their final game in Tokyo — which had been a regular-season major league contest — the Sox were playing an exhibition against the Los Angeles Dodgers at the ancient Los Angeles Coliseum in front of 115,000 fans. It was the largest crowd in the history of baseball. It was also something of a freak show. Baseball's Woodstock venue featured 92 rows of seats and a left-field fence that was 190 feet from home plate.

There were no serviceable locker rooms for the Sox at the Coliseum. The Sox gathered at Dodger Stadium after lunch, got into their uniforms, then took a bus to the Coliseum, where they disembarked in the middle of a massive crowd and stood while speeches were made. The players and staff entered the Stadium under the fabled stone arches situated beyond the makeshift right-field fence. The Dodgers, led by old friend Joe Torre, walked alongside the Red Sox.

As they walked down the crumbling stairs of the Coliseum, Torre asked Francona, "How's your dad?"

"Great," said Francona. "How many more times do we need to be introduced before we can get on with our season?"

They both laughed.

It was the first baseball game at the Coliseum since September 1961, when John F. Kennedy was president and Sandy Koufax struck out 15 in a 13-inning, 3–2 win over the Cubs. The hideous dimensions of the Coliseum created a rare baseball moment when Ellsbury was thrown out attempting to steal second base on a play in which the tag was applied by center fielder Andruw Jones. A rare 2-8 in your scorebook. The exhibition featured nine ceremonial first pitches, one before each inning. Given the number of non-sports reporters assigned to the hardball festival, it was not surprising after the game when one of the local television reporters addressed Francona as "coach."

After the cotillion-like game at the Coliseum, the Sox had to play one last exhibition game at Dodger Stadium. Returning to Chavez Ravine inspired a nice memory for Francona. He made his managerial debut at Dodger Stadium with the Phillies in 1997, and Curt Schilling beat the Dodgers for the rookie skipper.

"It looked easy and from there it was sort of downhill," said Francona.

In the spring finale, Clay Buchholz became the first big league pitcher to pitch spring training games in Florida, Japan, and California.

With the spring training schedule finally over, the Red Sox flew from Los Angeles to San Francisco to resume their regular season in Oakland. Boston beat the A's two straight. Then it was on to Toronto for another trip through customs and yet another opening day. The Blue Jays swept the Sox, three straight. The Sox made six errors and were outscored 17–6 in the final two games in Canada.

"On that last day in Toronto, one of their guys hit a one-hopper to Mikey Lowell's left," said Francona. "Mikey took one step to his left and just fell down. And I was thinking, *I'm with you, Mikey*.

"All the baseball people were uneasy about that trip. We were set up to have heavy legs, set up for us to get our ass kicked. It just seemed that everything we were trying to do was secondary to this trip. It was stop, start, stop, start. It was a total circus. We had cement in our shoes. It was the worst road trip in the history of the game."

"It was pretty bad," said Epstein. "I was always trying to protect the integrity of the baseball operation and our competitiveness. As GM, I'm also part of the broader management team, and I understood where they were coming from. In the end, if we missed the playoffs by a game, all this goodwill would be undone. The only way to successfully market a baseball team is by winning. But we're the Red Sox. We're not going to be able to avoid a trip like that. I just wanted us to mitigate the damage. It was what it was."

"I think it was bad," said Ortiz. "We had just finished winning the World Series. I know this is business and the team made good money at the time, but any team was going to make good money. That trip got everybody out of control."

The Franconamen were 3-4, resting in the cellar of the American League East, when they went through customs one last time to get ready for their 2008 home opener at Fenway and yet another series of pregame introductions.

They returned to Yawkey Way to find a new and improved clubhouse, equipped with as much space (the ceiling was raised) and as many amenities as a 97-year-old structure would allow. Architect Janet Marie Smith was at the top of her game, and Henry was still vested in making improvements to the place the Sox had anointed as "America's Most Beloved Ballpark."

Bill Buckner, the goat of the 1986 World Series, came back to throw out the ceremonial first pitch for the home opener. It was a cathartic moment for the graying first baseman, and his warm reception was another sign that the angst of eight decades had evaporated. In 2008 Boston was Titletown — proud and unapologetic. Bill Russell, Bobby Orr, Tedy Bruschi, Danny Ainge, and Johnny Bucyk carried the Stanley Cup, a Patriots Lombardi Trophy, one of the Celtics' 16 championship buckets, and the 2007 Commissioner's Trophy. The Red Sox were introduced individually and lined up on the baseline for the eighth time in less than three weeks. Tom Werner appeared with Neil Diamond on a video version of "Sweet Caroline," and Steven Tyler sang "God Bless America."

Tyler had a moment with the manager before his performance. Spotting the ubiquitous bag of Lancaster, the rocker asked Francona if he could have a pinch.

"Sure, but watch out for this stuff, it's pretty strong," said Francona, a big Aerosmith fan.

"Thanks," said Tyler, no stranger to stimulants and pharmaceuticals. "I've done a lot worse."

Later, there were reports of a green-faced Tyler running through the clubhouse to spit out his chew.

The Red Sox beat the Tigers, 5–0.

The Sox were leading the American League East by a game on May 19 when Lester no-hit the Kansas City Royals at Fenway Park. The manager was among the last to embrace the tall lefty, and a photo of that moment adorns the den of Francona's home.

Interviewed on live television after his on-field celebration, Lester said, "He has been like a second dad to me. He cares a lot about his players. It's not just what they can do on the field."

"That was a private moment that got played out in public," said Francona. "A kid grows up in your organization, you feel a little more paternal toward those guys. Then you take what happened to him with the cancer. The whole story was too good to believe."

Things started to turn ugly with Manny Ramirez less than a month later. Manny was in the final year of his eight-year, $160 million pact. He had shown unusual commitment during the 2007–2008 off-season, working out at Athletes' Performance Institute in Arizona and arriving at spring training in timely fashion. He skipped the Sox trip to the White House in late February ("I guess his grandmother died again," quipped President Bush), but for the most part Ramirez was engaging and upbeat with teammates and reporters.

On June 5, the night the Celtics were scheduled to play the Lakers in Game 1 of the NBA Finals at the Boston Garden, Manny slapped Youkilis across the face while the two argued on the top step of the Sox dugout during a 7–1 win over the Tampa Bay Rays. There was a bench-clearing incident in the second inning after Crisp was hit by James Shields, but the intramural spat involving Manny and Youk received far more attention.

"I was going to the bathroom when that happened," remembered Francona. "I came up the steps, buckling my pants, and I heard something and I said, 'What the fuck? Can't a guy even take a piss?' They were all talking about who hit who and all that. I said, 'If you guys want to beat each other up, go down the tunnel. Either hit each other down the tunnel where nobody can see you or better yet, just go play.'"

Three weeks later, on a Saturday afternoon at the end of June, Sox traveling secretary Jack McCormick was sitting at a computer in a corridor near the visitors' clubhouse at Houston's Minute Maid Field, making phone calls and filling out ticket orders for the Sox interleague night game against the Astros. McCormick, a former Boston police officer, had been on the job during Francona's entire tenure and was one of the manager's best friends in the Sox entourage.

"Jack only got mad at me once in eight years," said Francona. "We were playing cards on the plane and we had our table out in the middle of the aisle, and he didn't see it and he tripped and went face first onto the carpet, and I laughed. He gave me a look, and that was the end of that. I stopped laughing and never had an uncomfortable moment with him again."

McCormick was unusually fit for a 64-year-old man. He had run the Boston Marathon 17 times.

The road rule of thumb for player tickets is six per player per game — four for family and two for guests. Players who need additional tickets are welcome to dip into the allotment of players who are not using their tickets or make a purchase. Four hours before big league games, it's common to hear major league ballplayers yelling across the room to teammates, asking, "Are you using your tickets?"

Manny was in no mood for asking teammates for tickets on June 28 when he approached McCormick.

"Jack, I need 16 tickets for tonight," said Ramirez.

"Okay, Manny. Just borrow them from your teammates and I'll put them into the computer."

"No, just do it," said Manny.

"Manny, you know how it works," said McCormick. "I can get them, but you have to ask the guys and I'll take care of it."

Louder and more angry, Manny insisted, "Just do it."

Sensing that this was not going well, the six-foot-tall McCormick stood up and faced Ramirez, eyeball to eyeball.

"Manny . . ." he started.

There was no more discussion. Ramirez pushed McCormick violently, and the traveling secretary fell back onto a spring water jug that was on the floor by the entrance to the players' lounge.

"Just do your job and get my tickets!" Manny yelled as he stood over the fallen club executive.

Farrell, Papelbon, Youkilis, and Alex Cora were the first to get to the scene and push Manny away. Francona heard the ruckus from his office and came out.

"It took me a few seconds to realize that this wasn't in fun," Francona said. "There were not a lot of guys around when I got out there, and I saw Jack leaning against a table, kind of dazed. I grabbed Manny and said, 'What the fuck are you doing?' I was hoping he wasn't going to hit me."

Francona brought McCormick into his office to hear his version of events. The two met with Don Kalkstein, the Red Sox mental performance coach, who was traveling full-time with the team. After the meeting, Francona called Epstein.

"Theo, we've got a bad problem," said the manager. "We've got to do something. We've got to send Manny home."

Manny was not sent home. Ramirez was brought into a meeting with Francona and McCormick and apologized to McCormick. He started in left field the night of the incident and hit a game-winning home run a day later.

"I knew what was going on was wrong," said Francona. "It didn't sit well with me. I should have held my ground."

Red Sox management had little response when the story leaked. Lucchino said it was "an internal matter." Henry sent an email stating, "Actions have been taken commensurate with what occurred."

"I don't think there was any kind of effort to coddle Manny at all, by that point," said Lucchino. "We were all upset by it. It was outrageous behavior."

"That was so egregious," said Epstein years later. "We were constantly walking a tightrope with Manny. We needed to maintain some discipline and some integrity because we had to manage the other 24 players, but we understood that if we asked Manny to live up to everything we expected from the other players, we wouldn't even get past the opening series of the year. So there was this balancing act. It was a very difficult dance over the years. Tito was right in the same boat with everyone else. There were things we wanted to do over the years, but Tito and the rest of us at times had to bite our tongues and look the other way. It was so hard to do our job in a way that we could be proud of and also create an environment that would allow Manny to play and be productive. It was hardest on Tito."

Once again, Francona had to go before the media and say things he did not believe. He had to talk about handling things internally. He had to bite his tongue, more than at any time in his tenure as Red Sox manager.

"It was one of the tougher situations for me because it went against something I knew," he said later. "I knew it was wrong. It made it harder because it was Jack. Jack didn't want to be the guy that got Manny suspended. He knew how things worked with the Red Sox. He saw how

Francona's college roommate and best friend Brad Mills was by his side for his first six years in Boston. *Getty Images / J. Meric*

Trusted captain Jason Varitek always had the manager's back. *AP Photo / Jim Mone*

The manager with three of his latter-day favorites (left to right): Jonathan Papelbon, Dustin Pedroia, and Mike Lowell. *AP Photo / Elise Amendola*

In an emotional end to a comeback season, Francona lifts Jon Lester in the sixth inning of the final game of the 2007 World Series, less than a year after Lester recovered from non-Hodgkin's lymphoma.

*Tannen Maury / epa / Corbis*

Red Sox owners John Henry, Tom Werner, and Larry Lucchino, along with Epstein and Francona, pose with the 2007 championship trophy after Game 4 in Denver.

*Getty Images / Brad Mangin*

Francona congratulates Jon Lester again on May 19, 2008, after he pitches a no-hitter in Fenway Park against the Kansas City Royals. *Brita Meng Outzen*

The manager in his spartan office at Fenway. Note the old-fashioned saloon door shielding the corner latrine. *© Photograph by Walter Iooss*

The manager shares a laugh with his players. *Brita Meng Outzen*

Red Sox publicist Pam Ganley (left) delivers the manager to a press conference.

*Brita Meng Outzen*

A light moment with youngest daughter, Jamie, in the dugout.

*Getty Images / Joe Robbins*

Proud father Terry Francona at the wedding of his daughter Leah, who married Marine Michael Rice in downtown Boston on January 11, 2011.

*Brian Phillips Photography*

Pitchers John Lackey, Jon Lester, and Josh Beckett figured prominently in the collapse of 2011.
*Getty Images / Boston Globe*

Trying to figure it out late in the 2011 season (left to right): Dustin Pedroia, Jacoby Ellsbury, Francona, and Adrian Gonzalez. *Getty Images / Boston Globe / Stan Grossfeld*

The general manager and manager plot strategy at Fenway in the final days of the 2011 season.
*Getty Images / Boston Globe / Stan Grossfeld*

Francona's final press conference at Fenway. *AP Photo / Bizayehu Tesfaye*

Sox bosses Werner, Epstein, and Lucchino explain the departure of Francona on September 30, 2011.
*AP Photo / Bizayehu Tesfaye*

At the celebration of Fenway Park's 100-year anniversary, April 20, 2012, Francona walks off the field after tossing his jersey to a child in the stands.

*Getty Images / Boston Globe / Stan Grossfeld*

upset I was and told me to settle down, that he would be okay, but it was backwards. Over the years, Jack looked out for Manny as much as anybody. My patience for Manny lessened after that incident. That one really bothered me. I thought we needed to take a stance. I'd always had meetings with our veteran players about this situation, but now it was obvious they were tired of him too."

Ramirez was fined by the team. McCormick downplayed the incident every time he was asked for a comment.

"The only thing that pissed me off about it was those newspaper reports that I was some frail, 80-year-old man," McCormick said.

McCormick had a theory about Manny's motive. A few days before the incident, Houston GM Ed Wade had been attacked by Astros pitcher Shawn Chacon while Wade was explaining Chacon's demotion to the bullpen. The episode was headline news when the Sox got to Houston. In McCormick's mind, Ramirez was trying to shove his way out of Boston.

"I don't know that Manny thinks that far in advance," countered Francona.

For the first time in Boston, some fans were calling for Manny to be moved.

It got worse for Ramirez in Boston a week after the McCormick episode. Pinch-hitting against Mariano Rivera in the top of the ninth of a tie game at Yankee Stadium, Manny looked at three consecutive strikes, got rung up, then sauntered back to the dugout. He never took the bat off his shoulder. The Sox lost 5–4. For a lot of Sox fans, it was the last straw. It looked like Manny had intentionally quit in the at-bat against Rivera.

"Not true," said Francona. "It was never an issue. I believe it to this day that he was trying there. He had been in the dugout tunnel, getting ready for that at-bat. We didn't have to drag him up there. He knew Rivera enough, he knew what he could hit and what he couldn't hit. He didn't get it and he didn't swing. Mariano painted three. I'm not going to say Manny lost sleep over it, but he was ready to hit. That was a tough one because I always stuck up for Manny, and the media probably thought I was just doing it again, but in my mind he was ready to hit. That one was never an issue."

The Sox were still in first place, owning a 57-40 record, when Francona took a Sunday night train from Boston to New York for his second crack at managing in the All-Star Game. Waiting for him in New York was Phyllis Merhige, a Major League Baseball senior vice president who'd known Francona when he was a young ballplayer and she was director of public relations for the American League. For several decades, Merhige has been the moderator of managers' pre- and post-game press conferences at postseason games and All-Star Games. She was the one who'd been waiting for Francona in the makeshift interview room at Fenway when fans were throwing things at him from the bleachers after the 19–8 loss in Game 3 of the 2004 American League Championship Series. Francona was a favorite of the widely respected Merhige. On the last day of games before the break, Merhige spent the day making arrangements for players and managers from all 30 teams coming to New York. She made sure Francona knew she'd postponed a hairdressing appointment while waiting for him to arrive on the train from Boston Sunday night. Unfortunately, Francona's train was delayed and he didn't get to New York until after midnight.

The ever-loyal Merhige was there to greet him at the midtown All-Star hotel headquarters.

"Phyllis, your hair looks like shit," said the manager as he gave Merhige a hug.

Francona selected New York's Joe Girardi and Detroit's Jim Leyland as his coaches. He was honored that Leyland accepted his offer. Leyland managed in the minors when Francona played at Triple A, and the Sox manager considered Leyland the gold standard of major league managers. Francona's favorite memory of the 2008 All-Star experience was sitting in Girardi's office, swapping stories with Leyland and Farrell while hundreds of media members covered the workout on the Yankee Stadium diamond. Leyland wanted to know more about Pedroia (one of seven Red Sox named to the American League squad), so Francona brought his second baseman into the room to meet with the Tiger manager.

The game, an epic 15-inning joust won by the American League, was an homage to the legacy of Yankee Stadium, which was prepped for the wrecking ball at the end of the 2008 season. Forty-nine Hall of Famers paraded onto the field before the ceremonial first pitch.

"That pregame was one of the most electric moments of my career," said Francona. "That was humbling. One of the most fun things I've ever done."

During the second inning of the game, one of the Cooperstown boys — Milwaukee Brewer Robin Yount — snuck back into the first-base dugout to meet with Francona. They had been teammates at the end of Francona's playing career in 1990. Yount had been the Brewer superstar who made sure benchwarmer Francona had a car to drive while he played for Milwaukee. All these years later, Francona was touched that the Hall of Fame shortstop took the time to find him, even if it was in the middle of an inning at the All-Star Game.

Looking at Francona, who'd been the 25th player on the roster when they were teammates, managing an All-Star team in front of 55,632 fans and 49 Hall of Famers at Yankee Stadium, Yount shook his head.

"Tito, I can't believe it," said the Brewer legend.

"I just can't believe it," Yount said again, still shaking his head.

Then again.

"Yeah, Robin, I know," Francona said, laughing. "Pretty amazing, isn't it?"

The All-Star Game is traditionally difficult to manage, and the task became more challenging after the infamous tie game of 2002 when Joe Torre and Bob Brenly ran out of pitchers. The embarrassment played out at Bud Selig's hometown yard, Miller Park in Milwaukee. The debacle inspired a rule change stipulating that the winner of the All-Star Game would give its league home-field advantage in that year's World Series. "This time it counts" was the new slogan.

With Leyland at his side, Francona found himself in a 3–3 tie and running out of pitching in the 13th inning at Yankee Stadium. Concerned MLB officials dispatched Jimmie Lee Solomon — the vice president of MLB's baseball operations who had been part of Francona's "shirt-gate" episode in 2007 — to ask about available pitching for upcoming innings.

"Tito, how are we looking for pitching?" Solomon asked as he approached the manager from the clubhouse tunnel leading to the dugout.

"Unless you have any pitchers with you, leave me alone and let me figure it out," snapped Francona.

When Kansas City righty Joakim Soria walked off and waved to his family in the stands after shutting down the NL in the 11th, Francona said, "Don't be waving to anybody just yet, big boy. You're going back out there for the 12th."

Pittsburgh's Nate McLouth launched a fly to deep right to start the 14th, but J. D. Drew made the catch. Francona walked down to the end of the bench where Derek Jeter was watching.

"It wouldn't have killed me if that ball went out," said the manager.

Jeter looked puzzled. He knew that home field in the World Series was on the line. More than a lot of players, Jeter cared.

"Aw, I think we can win the Series on the road," Francona said with a laugh.

He loved the fact that Jeter stuck around after he was pulled from the game. Many players left the bench, even the ballpark, when they were pulled from All-Star Games. Not Jeter. Not at Yankee Stadium. Not anywhere.

Baltimore's George Sherrill bailed out Francona by pitching multiple innings. Against Tampa's wishes, Rays starter Scott Kazmir came on for the 15th. Francona kept looking at his lineup card.

"You can look at that thing all you want," said Leyland. "There's no more pitchers on there."

When the American League came in to hit in the bottom of the 15th, Francona talked to Drew about pitching the 16th inning, but it never came to pass. The AL won it on Michael Young's sac fly in the bottom of the 15th at 1:38 AM.

Francona and his staff reunited with Sox players in Anaheim after the break. It would be Manny Ramirez's final trip with the Red Sox.

Manny had switched agents and was working with Scott Boras. Like every team, the Sox were wary of Boras, but he was impossible to avoid. Boras and Manny were looking for a contract extension in the summer of '08, and there was more controversy at the break when Manny told the *Boston Herald*, "I want to know what's my situation. I want no more [times] where they tell you one thing and behind your back they do another thing."

Henry termed the remarks "personally offensive."

Manny was a handful on his final Sox trip. He hit a game-tying homer in the first game in Anaheim, but furnished a *SportsCenter* "Not Top Ten" moment when he flopped badly chasing a fly ball, then started laughing when he realized he was sitting on the baseball. Epstein was in the stands and did not look amused when he saw Manny laughing. Two days later, Manny hit an RBI double, but the Sox lost their third straight to the Angels. After the sweep, while Francona was getting dressed quickly for the trip to Seattle, McCormick popped his head into the visiting clubhouse manager's office.

"Tito, Manny isn't getting on the bus. He isn't going to Seattle with us."

"Jack, let's go," said Francona. "Let's get on the plane. I'll just call Theo and tell him Manny's not with us."

Dressed and packed, Francona snapped his carry-on luggage into place and wheeled it down the corridor toward the Sox team bus, which was waiting to go to the airport. He got on the bus, taking his usual seat in the second row behind the driver. He did not check to see if Manny was on the bus. It was time to go.

Manny was in the back of the bus. He made the trip, played the first two games in Seattle, then got a day off before the flight home.

Storm clouds gathered over Fenway on the final weekend before the trading deadline. The tipping point came when Manny took himself out of the lineup for the Friday night return to Fenway against Joba Chamberlain and the Yankees. Ramirez's name was on the original lineup card when Francona met with the media at 4:00 PM. During the press conference, Manny went to Mills and said he couldn't play due to pain in his knee. Henry and Werner were furious. They ordered Manny sent to the hospital for MRIs on both knees during the game. In midgame the Sox issued a press release stating that the MRIs came back clean. The Red Sox lost, 1–0. Ramirez started the next five games while Theo went to work cutting the best deal he could make.

On Thursday, July 31, an off day, the Sox traded Ramirez in a three-deal swap that sent him to the Dodgers and brought Pittsburgh outfielder Jason Bay to Boston. Francona was at Logan Airport, alone in his car, waiting to pick up one of his kids, when he got the call from Epstein telling him that Manny had finally been traded.

"I got emotional," Francona recalled. "It hit me so hard. I lost my composure sitting there in the car, and a cop came up and asked me if I was okay, and it was embarrassing. I told him everything was cool. I guess until that moment I hadn't realized what a toll the whole thing was taking on me, just going through all that."

The next day Francona called Bay into his office.

"J-Bay, great to see you," said the manager. "Here's what you do. I know it sounds simple, but just do the best you can. You're in a firestorm here, but you're going to be received in this clubhouse better than you can imagine."

Bay hit .293 with nine homers and 37 RBI over the final 49 games.

"He was a manager's dream," said Francona. "He loved to play. We went from a guy that we had been trying to keep in the lineup to a guy you couldn't get out of there."

In 53 games with the 2008 Dodgers, Manny hit a staggering .396 with 17 homers and 53 RBI. In the spring of 2009, Ramirez signed a two-year, $45 million contract with the Dodgers. In May 2009, he was suspended for 50 games for violating Major League Baseball's drug policy. It would not be his last positive test for banned substances.

The 2008 Red Sox finished in second place in the American League East, two games behind the Tampa Bay Rays, with a 95-67 record. The Sox went 34-19 after the Ramirez trade, the second-best record in baseball. They went an alarming 1-8 at the Tropicana Dome in St. Petersburg.

The manager loved his team down the stretch. Manny was gone. Bay was a team player. Pedroia was the American League Most Valuable Player. Lester was a 16-game winner, ace of the staff. Matsuzaka went 18-3, thanks to hefty run support. Papelbon saved 41 games, and 23-year-old Justin Masterson became a dependable long reliever, able to pitch every day. Youkilis was a .312, 29–home run producer and an All-Star starter at first base. Alex Cora, Mark Kotsay, and Sean Casey gave the manager the veteran, go-to bench guys he'd had in Kapler and Roberts back in 2004.

"I think our organization should be proud," said Francona before the start of the playoffs. "It's an exciting time for the Red Sox."

Just like in 2007, they were "doing it better," fulfilling a promise Francona and Epstein made to one another after the storybook but sometime sloppy championship of 2004.

Going into the playoffs, Francona was 22-9 lifetime as a postseason manager. His .710 winning percentage was the best of any manager in baseball history who had managed at least 20 postseason games.

The Red Sox beat the Angels in the Division Series in four games, a series that included a five-hour-and-19-minute Sunday night game at Fenway.

During a group press conference before Game 2 in Anaheim, Francona deflected praise from a well-meaning reporter.

"You've been the manager in the World Series, and if you stopped people on the street and asked them who are the five best managers in baseball, I'm not sure people would say your name," said the scribe.

"My dad would," started Francona, as the group laughed. ". . . If I said I don't care, that sounds flippant and I don't mean it like that, but I guess what I care more about is us winning. . . . You go to Beaver County, Pennsylvania, though, and they think I'm pretty good."

The best part of the exchange was the official MLB transcript, which identified it as an interview with "Coach Francona."

Veteran writers could not wait to show the transcript to the manager.

"About six guys pulled hamstrings running down here to show me the thing," said Francona.

The ALCS was a coming-out party for the Tampa Bay Rays and Tropicana Field, the hideous indoor baseball arcade featuring overhead catwalks that sometimes blocked home run shots.

"I actually didn't mind the place," said Francona. "I loved the clubhouse. If they could take that catwalk away, I wouldn't mind the Trop. There were so many rules, you couldn't keep everything straight, but it played fair. The Metrodome in Minneapolis was much worse."

The Sox won the first game in the Trop, 2–0 (Dice-K), but the series turned in Game 2 when the Sox gambled on Josh Beckett. Beckett had a strained oblique muscle and was taking painkiller shots in order to pitch. He gave up three homers and five hits in the first four innings, but when the Sox rallied to take a 6–5 lead in the top of the fifth, Francona and Farrell tried to squeeze one more inning out of their erstwhile ace.

"We went to him and asked him, and he had it in his eyes," said the manager. "He was ready."

Beckett struck out Akinori Iwamura to start the inning, then walked B. J. Upton, surrendered a single to Carlos Pena, then a double to Evan Longoria. Francona came out to get Beckett, but the damage was done. The Sox lost, 9–8, in 11 innings. Francona went through 72 pieces of bubblegum in the five-hour-and-27-minute marathon. He also hurt Varitek's feelings when he sent Drew up to hit for the catcher in the ninth inning against Dan Wheeler. Sticking too long with Beckett stayed with him long after the game was over.

"It still haunts me a little bit," he said. "But I don't think I'd do anything differently. We just wanted Beckett to get through that fifth inning, then get to the bullpen. It got away from us quick. I could tell Theo was upset with me that night. Those were the nights when it was lonely being the manager."

The series moved to Fenway, and the Rays won two more times, taking a commanding 3–1 series lead. Game 4 was a 13–4 Tampa blowout that sucked the air out of Fenway. Beckett, Lester, and Wakefield were hammered for 17 earned runs in 12 and two-thirds innings and gave up eight homers in the three straight losses.

It looked like the Sox were headed for an embarrassing series defeat in five quick games when they fell behind 7–0 in the first six and a half innings of Game 5. But there was one last rally.

Pedroia started the comeback with a two-out, RBI single in the seventh. After two more runners reached, Ortiz cut the lead to 7–4 with a three-run homer to right. The Sox scored three more in the bottom of the eighth to tie the game and won it on a Drew two-out single in the bottom of the ninth at 12:16 AM. The late-inning drama included high-yield, ten-pitch at-bats by Crisp and Youkilis. The only other team to recover from a seven-or-more-run deficit in a postseason game was the 1929 Philadelphia A's, who trailed the Cubs 8–0 in the fourth game of the World Series, then won 10–8. It was a nice way for the Red Sox to say good-bye to Fenway fans for 2008.

"I've never seen a group so happy to get on a plane at one-thirty in the morning," said Francona.

They went back to Tampa and forced a Game 7 with a 4–2 victory, powered by captain Varitek's game-winning homer off James Shields in the sixth. Varitek was 0-15 at the time, had hit only .220 during the season, and was adjusting to the notion that he might not always be

Francona's best available hitter. He was lifted for a pinch hitter three times in the series.

"I couldn't think of anything more appropriate," Francona gushed after the win. "Our whole dugout went crazy. The way it happened, and as hard as he's worked, it meant a lot to everybody."

One night later, Tampa's young righty Matt Garza dominated the Red Sox, holding the visitors to two hits (including a Pedroia first-inning homer) over seven innings of a 3–1 win in Game 7.

The big moment of the game came with the bases loaded and two out in the top of the eighth. Protecting Garza's 3–1 lead, Tampa manager Joe Maddon used five pitchers in an inning in which the Red Sox never scored a run. The Rays' final hurler was 23-year-old, six-foot-six lefty David Price, who'd been the number-one draft pick in the nation when he was selected out of Vanderbilt in 2007.

"In our dugout, we felt we were going to win," said Francona. "We'd done it so many times, and that's the way it always was with that group."

Price worked the count to 1-2 on Drew, then fanned the Sox right fielder with a 97-mile-per-hour fastball on the black. Drew committed, tried to hold back, but was ruled out on a check swing. Drew never looked at the replay to see if he could have gotten away with taking the pitch for ball two. He never checked with home-plate umpire Brian Gorman. That's the way it was with J. D. Watching the replay or arguing with the umpire wouldn't change the call, so why bother?

"I thought the '08 team might have been our best team," Francona said later. "It was a shame we didn't win that one. But we got beat by a good team."

The flight back to Boston was long, dark, and quiet. Francona abandoned his seat at the front of the aircraft and went back a few rows where he could lie down across a row of three seats. He was more tired than he had been in any other season. He had no way of knowing that his best days in Boston were behind him. After two championships and a seven-game ALCS, it was already starting to unravel. He would never win another playoff game as manager of the Boston Red Sox. Nobody knew it yet, but the fall of the franchise was officially under way.

# CHAPTER 12

## · 2009 ·

## "This is like a reality TV show"

THE MANAGER OF THE Red Sox was comfortable.

This is not a statement made lightly. The man working in the corner office of the home clubhouse at Fenway Park is traditionally paranoid, besieged, anxious, and ever-ready to be fired. Comfort and stability are never part of the job. A sample of the manager's mailbag or a Google search would make even the most secure man cringe. It was Terry Francona who opened his mail in his first year on the job and noted, "Some of the things they ask you to do are impossible. I mean, you can only stick that lineup card so far up your ass."

In spite of the never-ending pressures and expectations, Francona was surprisingly peaceful as he got ready for the 2009 major league season. He was making $3 million per year and had four playoff appearances in five seasons, two World Series rings, and a contract that carried him through the 2011 season. He was easily the most decorated manager in Red Sox history and already had the second-longest tenure of any Sox manager since Prohibition. The years of hunger and uncertainty were behind him. Three of his four children had graduated from high school and moved out of the house. He was at peace with the fans of Boston and found himself almost as popular as Bill Belichick or the late Red Auerbach. Nobody laughed when fans or writers anointed him as the greatest manager in the history of the Red Sox.

As much as New Brighton, Tucson, Yardley, Pennsylvania, or Chestnut Hill, Fenway Park was Francona's home. He loved his job, and he loved the people who were part of his everyday routines. He had his best friend Mills alongside him in the dugout for every pitch of every season. He had former teammate John Farrell as his pitching coach. He had traveling secretary McCormick and team doctor Larry Ronan on speed-dial. He trusted a handful of the longtime Fenway clubhouse workers as much as he trusted members of his own family.

"They were like my little brothers," he said.

Fenway Park's clubhouse workers of 2009 were baseball descendants of the men who'd allowed Francona to hang around clubhouses in Oakland and Milwaukee when he was a little kid and his dad was a big league outfielder and who'd let him steal candy from the ballplayers' stash. Often referred to as "clubhouse kids," the Red Sox clubhouse managers were grown men, some with wives and children. They all had bills to pay, and no one lined their pockets more freely than Terry Francona.

Edward Jackson, aka "Pookie," was a Francona favorite. The Red Sox found Pookie (nicknamed by his grandmother) when he moved to Fort Myers from Jacksonville in 1994. Affable and quick with a smile, Pookie was the go-to guy for clean socks, food runs, practical jokes, and red fleece tops. Jackson was an agent of change for the clubbies in 2002 when he temporarily worked as a driver for new owner John Henry. Afflicted with allergies, Jackson coughed and wheezed while he drove Henry around Boston. The caring owner asked Jackson why he didn't have medication for his condition, and Jackson told the owner that the clubhouse workers did not have health insurance. Henry called Fenway and directed his assistant to include the clubhouse workers in the Red Sox employee benefits program. He told Jackson to see a physician immediately and put the bills on the owner's desk until the insurance kicked in.

Jackson was always in the Sox clubhouse.

"Pookie was almost assigned to me," said Francona. "I'd come in some morning and Pookie would be scrubbing my shower down."

Tom McLaughlin and Joe Cochran were another pair of loyal lifers in the clubhouse. Tall and hilarious, Brighton native McLaughlin started with the Red Sox as a batboy in 1986 and served as visiting clubhouse manager in all the years that Francona managed in Boston. Cochran, who grew up on Cape Cod and started with the Sox in 1984, often worked inside the left-field wall for longtime groundskeeper Joe Mooney. Cochran was the home clubhouse manager during Francona's tenure. McLaughlin traveled frequently, and Cochran was on every road trip. They knew all the locker-room secrets and gave up nothing.

"Clubhouse guys survive by not letting anything get out," said Francona. "And they know everything. If we traded for a guy and I knew the clubbie on that guy's team, I would always call him. You could get information from a clubbie that a scout could never see on a field."

Steve Murphy and John Coyne were additional clubhouse assistants and Francona confidants. "Murph" was the one who prepared the manager's chew and bubblegum concoction. Coyne was a Boston College graduate who looked a little like Pedroia and, according to Francona, could not be trusted mailing packages.

"Pookie walked into my office one day with a wedding present that I'd sent a year earlier, and it had come back to the ballpark unopened," said the manager. "I said, 'What the fuck happened with this?' and he said, 'I told you not to let John Coyne send it. He never writes anything down and then he forgets.'"

The routine on home dates was pretty similar. Francona would leave his Brookline home late in the morning, always on the quarter-hour. Fenway was only a 10- to 12-minute drive from his home, and each day he stopped at the Starbucks on the corner of Route 9 and Hammond Street for an egg salad sandwich. He finished his drive heading eastbound on Boylston Street, turned left onto Yawkey Way, and proceeded to the giant green door near Fenway's gate D. After turning into the underbelly of the ballpark, he'd weave his vehicle around the poles under the stands toward the clubhouse door on the first-base side. The clubhouse door was invariably locked because of the early hour. Francona had keys but never remembered the security code, so he'd open the door and let the alarm go off. Then he'd throw his keys to the clubbies and tell them not to ask if they wanted to use his car.

"Go shopping, go for lunch, I don't care," he'd say. "Just don't fucking ask me if you can drive it."

Francona also had an open wallet policy. During the afternoon, the manager's black billfold sat on the corner of his desk, and it was understood that any of the clubbies or coaches could take out a loan. No need to ask. Twenties came and went without conversation or suspicion.

"I went into that wallet about once a day," said Jackson.

Japanese chef Iso Kosaka, a Dice-K favorite, was working for the ball club by 2009 and prepared Francona a daily lunch of rice, vegetables, and noodles. After eating, Francona would go upstairs for a 20-minute swim in the Swim-Ex machine. Mills would often join him.

"That made for a lot of jokes," said the manager. "Here you got two bald men in bathing suits going upstairs to work out in a little 10-by-12 pool. You can imagine how funny everybody thought that was. It was like we were dating."

After his workout and a shower, Francona would begin poring over data on his computer or perhaps summoning video master Billy Broadbent to fix something on his laptop.

The clubbies were ubiquitous in the early afternoon hours before the clubhouse door was opened to the media at 3:30. Pookie, Murph, or John would make runs to the dry cleaners or to pick up prescriptions at CVS for the manager. McLaughlin would often come over from the visitors' room to shoot the breeze while Francona read printouts and scouting reports.

The door that connected his office to the clubhouse was almost always open. Francona did not believe in sealing himself off from the locker room. He wanted his players and coaches to feel welcome to come in and talk at any hour of the game day. Mills and Hale (Hale had known Francona since their days on the coaching staff of the 2002 Rangers) came into the manager's office every afternoon to go over the night's lineup and matchups. The manager and his coaches wanted to have everything done long before the media entered, long before players started getting ready for the first pitch. He wanted to be free for a game of cribbage with Pedroia before the players went out on the field for early stretching.

"Tito needed that time, by himself," said Hale. "He bounced things off me, asked my opinion. He liked to be there early and not be interrupted. He'd ask what I thought about things. Tito wanted time with his bench coach, then his pitching coach."

The manager-GM relationship was stronger than ever.

When Francona interviewed for the Sox job in 2003, he and Epstein talked about the ideal manager-GM relationship. Epstein had admired the dynamic in San Diego between Bruce Bochy and Kevin Towers, who worked together from 1995 to 2006.

"They could talk about anything, anytime, without threats to one another," said Epstein. "There was an implicit understanding that they were in it together. If there was any probing question or issue to be explored, it was for the good of the organization and it wasn't any kind of personal threat whatsoever, and they had a ton of fun together."

By 2009 Epstein felt that he and Francona had evolved to this level. They protected one another when interacting with the media. Francona always went through Epstein with any questions or frustrations with ownership. Epstein had pulled back from some of the day-to-day suggestions and postgame analysis. It was a marriage that worked.

"Theo taught me a lot about the caretaker part of the organization, and I hope I taught him about the player's side of it," said Francona. "We were meeting in the middle on a lot of that, which was good. I knew when things were getting hot for him, and I knew where I could go. We had the ability to yell at each other and we could move on. One thing I wasn't good at: if I had the rotation set up and he wanted to change it, inevitably there would be an argument and I would tell him a day later, 'Theo, if you give me a couple of days, I'll get around to it. Sometimes it takes me a while to get to the right answer, but I'll get there.' And he understood that. We had a real good understanding of each other, and the one thing that I liked — when things were going shitty, I went right to him and he'd fix it.

"There were things I really appreciated. In eight years, not one time did Theo ever come down during spring training and be pissed that we lost or that somebody didn't run a ball out. That's rare. He never, ever yelled at me about somebody not hustling. He knew I cared, but he was smart enough to pick his battles, and I always appreciated that."

"We were partners," recalled Epstein. "We were really connected in a pretty natural way. There were always issues that would come up where we were at opposite ends of the table, and I thought we did a pretty good job of handling those issues with care so the relationship continued to grow. Our relationship was really important to both of us and it grew, but it wasn't this pat thing. I really respected Tito and tried to put him in a position to succeed and tried to always have his back, and I got to know him really well, to know what he's all about. So it was easy to be there for him in ways that were important to him. He's really good at reading people. He could see through some of my gruffness or some of my moods and connect with me in an important way too, so I think we were there for each other.

"We talked so much baseball. We were trying to build an organization, scouting, player development, major league team. His approach in the clubhouse and on the field was fundamental in that, so we had to talk about it and make sure we were on the same page and make sure he was empowered to do what he wanted. He had so much going on, to balance all those constituencies, it helped him to have a touchstone to bounce things off. It really worked in that respect. Tito wants nothing more than to have people around him who don't annoy him, that he believes in, that he won't have conflict with, and that he can have fun with. I could be that guy, up to a point. Ninety percent of the time, I could be that guy."

Adding to Francona's comfort level was the presence of 30-year-old Pam Ganley in her first year as club public relations director. Francona was a dream manager for any baseball public relations department. He was always on time (invariably early), cooperative, and funny, and he never said anything controversial. Through the decades, the Sox had been well served by public relations directors Tom Dowd, Bill Crowley, Dick Bresciani, Kevin Shea, Glenn Geffner, and John Blake, who worked at Francona's side for three seasons before departing for Texas after the 2008 season. This paved the way for Ganley to become the first female public relations director in Sox history — no small accomplishment in the ultimate old boys' club of Boston. Ganley grew up in Burlington, Massachusetts, went to the University of Massachusetts at Amherst, and worked at Dunkin' Donuts in high school. She joined the Red Sox as an intern in 2000. Only seven years older than Francona's oldest child, she was already one of the manager's favorite people around Fenway when she assumed PR director duties. Like the manager, she did not suffer fools. Sifting through requests, maximizing the manager's time, and sparing him frivolous or unnecessary interviews, Ganley made Francona's life easier. Best of all, she was tough. Ganley did not blush at the colorful language of baseball, and she had a sense of humor.

The 2009 Red Sox were a tad boring for media members looking for traditional Fenway hijinks. Five years after the success of the Idiots, the Sox had gone from an everyday outfield of Manny Ramirez, Johnny Damon, and Trot Nixon to . . . Jason Bay, Jacoby Ellsbury, and J. D. Drew. Awesome talent, long hair, and combustibility had been replaced by slow and steady fires. The '09 Sox were built according to the Epstein blueprint, and it did not allow for an abundance of noise and personality. The Sox were still a 95-win team, but they were less interesting to gossip columnists and tabloid headline writers.

"I loved the '04 team, but you can't imitate something else," said Francona. "It goes back to what I always said about Jeter and Manny. Every time Jeter had something to do with the outcome of a game, it made me nervous. Manny could hit .450, but when he wasn't hitting, all bets were off. So I told Theo, maybe we were a little bit lucky, but we were trying to find guys that cared about winning and cared about each other, but when things got tough, we weren't just waiting for somebody to hit a home run. And I thought Theo did a really good job of that. Every team has individual personality. We were integrating guys like Ellsbury, Youkilis, and Papelbon, and I loved that. Jacoby Ellsbury is quiet. That's okay. He's a good player. If they just play the game right, I don't care about the decibel level in the clubhouse."

"I think the Idiot culture can only exist for so long before it starts to create its own issues," said Epstein in the spring of '09. "We've shifted to players who play hard, care about each other, and focus on winning above everything else in a selfless manner. The more players you have like that, the more cohesive a team you're going to have. We don't want to be on TV for non-baseball reasons. That's what happens when you bring in players who are focused on winning. They tend not to sur-

round themselves with too many outside influences or distractions. . . .
I think people who follow baseball like a soap opera might not appreciate the team quite as much, but in the end I think the reason people
like baseball is just the game itself."

All true. But the Red Sox television ratings were on the decline, and
it would be a seismic issue that eventually affected everyone in Red
Sox Nation more than any other factor. Concern over TV ratings led
to the fall of the franchise.

In the spring of '09, the Sox owners were taken to task for failing to
sign free agent first baseman Mark Teixeira. Like any manager, Francona would have liked to have Teixeira, a power-laden switch-hitter,
terrific fielder, hard worker, and positive influence in the clubhouse
who was coming into his athletic prime.

Francona knew Teixeira from his 2002 year as the Texas bench
coach. He loved the young man's ability and work ethic. Early in the
2008–2009 negotiations, Francona and Epstein flew to Washington,
DC, to court the free agent first baseman. In early December, they met
in a hotel room, supplied by Boras. Francona tossed a cautionary flag
at the start of the five-hour session when he reminded Teixeira to be
mindful of Mike Lowell's feelings. If Teixeira came to Boston, it would
mean switching Kevin Youkilis from first base to third, and that would
bump Lowell off the bag. Lowell was nearing the end of his career and
no longer able to cover much ground, but Francona did not look forward to telling Lowell his time was done in Boston. Teixeira told Francona that he appreciated the manager's honesty.

"It was an awesome meeting," said Teixeira. "Tito was great to me
when I was a young player. He helped me out and showed me how to
deal with players and how to act around superstars. Guys loved him in
Boston, so when he came to visit, he was great, and part of me would
have loved to have played for him. Theo is obviously brilliant. The
meeting with those two was one of the best meetings I've ever had.
They did great."

The Sox thought they had a deal with Teixeira right up until December 18, when talks deteriorated at a Texas meeting that included
Teixeira, Boras, Lucchino, Henry, and Epstein.

"That meeting was a little different," said Teixeira. "When you're being recruited, you would hope that you'd be treated well. That didn't happen. I was treated very poorly by Larry at dinner when they were recruiting me to sign me."

Lucchino scoffs at the notion that his temper killed the deal and blames Epstein for perpetuating that version of events.

"It didn't blow up in Dallas," said the CEO. "It ended after we declined to raise our offer to match another offer. Theo at some point thought it was a mistake to ask him [Boras] where the other offers were coming from, but there was no blowup at the table. He [Boras] said we were the lowest of the five offers that he had, and I said, 'Scott, we don't think that's so. If that's so, tell us. Who are we behind? What do we need to do?' . . . Scott wasn't willing to tell us the other clubs. I asked, 'Who are some of the other teams that have blown us out of the water?' Apparently that ruffled Scott's feathers. . . . The real issue came after, when they did nonetheless make a counteroffer and ask us about another offer, and we said we were not going to bid against ourselves."

Werner said, "I was always under the opinion that no matter what our final offer was, Scott would go back to the Yankees and give them the right to match."

Teixeira signed an eight-year, $180 million deal with the Yankees and won a World Series in his first season in pinstripes. The Red Sox, meanwhile, set about building a high-priced team that would fail in every aspect.

The 2009 Red Sox had hopes of getting back to the Fall Classic. The Sox were only one year removed from a World Championship and had advanced to the seventh game of the American League Championship Series in 2008. Expectations were high when fans were greeted at the turnstiles and in the stands by Francona and his ballplayers on opening day at Fenway.

In full uniform, wearing his red top, the manager stood in an aisle in the lower boxes by the Fenway home dugout and shook hands with citizens of Red Sox Nation. Ever-willing to comply with management, Francona thought greeting fans in uniform was a little over the top.

"One of the things I always thought was cool about Fenway was that we sold out every game just on baseball," he said. "I always loved that. We never had to do the hot dog races and the presidents running around the field and all of that promotional stuff. I know we have to reach out to the fans and we are a business, but I thought it was cool that we never had to do the other stuff that teams did."

Boston mayor John "Honey Fitz" Fitzgerald threw out the ceremonial first pitch when Fenway opened in 1912. Senator Ted Kennedy, grandson of Honey Fitz, had talked about making the toss on Fenway's 100th anniversary in 2012, but Kennedy was in failing health in 2009 and everyone knew this might be his final opportunity. The senator's primary care physician was Sox internist Dr. Larry Ronan, and Ronan arranged for Kennedy to appear at the opener. Ronan peppered Francona with stories about the congressional lion. After Kennedy was chauffeured to the front of the first-base dugout in a golf cart driven by Jim Rice, Francona greeted him with a handshake and a hug and escorted him to a spot in front of the mound.

"That was a cool moment," said Francona. "I was fortunate to do some pretty neat things, and that was one of them. The senator was struggling, but we didn't want it to appear that he was struggling. I had instructions. Larry told me to stay close enough, don't hold him up, but be there for him. We took a picture, and it's one of the ones I kept. I'm not a political person. My dad was a Democrat, but he didn't care much. I grew up in a household of batting averages and slugging percentages. We didn't keep track of political polls."

There was no chance of Francona spilling tobacco juice on Ted Kennedy. In the spring of 2009, the Sox manager temporarily quit the chew because of a $20,000 bet with Larry Lucchino.

The Sox CEO was a two-time cancer survivor and warned everyone of the pitfalls of tobacco and mouth cancer. For years Lucchino would routinely brush past Francona, point to the manager's wadded cheek, and say, "You still using that?"

Six years into the job, Francona threw down the tobacco-stained gauntlet.

"Larry, let's put some money on it," said the manager. "If I'm going to do this, we need to make it worthwhile. I say we go for $20,000. If I quit for the whole season, you give $20,000 to Children's Hospital. If I cave, I'll write the check."

"You're on," said Lucchino.

The Francona-Lucchino relationship was never wholly comfortable. Lucchino's temper was famous, and his perpetual concern with the business and marketing of the ball club sometimes conflicted with the priorities of the manager, who was only interested in winning.

"Ever notice how Larry never says my first name?" Francona would ask people. "He always calls me 'Francona.'"

The bet gave the two men some common ground early in the 2009 season. The abstinent Sox manager made it through spring training and into the first four weeks of the season without the wad in his cheek. But he was not good company.

"Players were saying, 'Dude, it's $20,000, you're not going to make it,'" remembered Hale. "'Just pay the man, cuz you ain't going to make it!'"

When Francona screamed at a NESN cameraman during a rain delay in the Fenway dugout, he knew it was time to fold.

"'Pookie, get me some fucking Lancaster!'" snapped the manager.

Later in the day, he wrote the check. Then he called Lucchino and left a three-word message: "Larry, I lost."

Thanks, Francona.

There was a weird pattern early in the 2009 season. The Red Sox could do no wrong against the Yankees. They swept a three-game series against the Bronx Bombers in April, took three more in New York in early May, then beat the Yankees two more times at Fenway in June. It was the first time the Red Sox opened a season with eight consecutive victories against the Yankees since 1912, which was the first year of Fenway Park and two years before Babe Ruth broke into the majors as a rookie left-hander with the Red Sox.

The House That Ruth Built was vacant in 2009. The Yankees had closed down their majestic stadium and moved next door into the "new" Yankee Stadium. There was amusement in Boston when it was learned that a construction worker/Red Sox fan had buried an Ortiz jersey in the concrete outside the visitors' clubhouse. Outraged Yankee officials ordered a jackhammer crew to dig out the embedded Sox jersey. Francona chuckled when he walked past the site, which was ceremoniously cordoned off by Yankee officials.

But he missed the old stadium.

"I was a big fan of Yankee Stadium," said the manager. "You might swallow some asbestos, but I loved the place. The new one was gorgeous, but for me it just didn't touch the old one. Some of the amenities are out of this world, but they lost a lot when they moved out of the old one. It's like what I heard happened in Boston with the old Boston Garden and the new Boston Garden."

Francona didn't have much of a relationship with Yankee manager Joe Girardi, who was in his second year on the job in 2009. The friendly banter he'd enjoyed with Torre was gone. But he'd developed a fondness for Jeter, especially after managing him in All-Star Games in 2005 and 2008.

Like every ballplayer, Jeter believes in routine. He has used the exact same model bat (Louisville Slugger, P-72) for every professional plate appearance over his multiple-decade career. He is famous for stepping out of the batter's box and arching his back between pitches. By 2009, part of his routine in Red Sox games involved acknowledging Francona before his first at-bat of each game. Approaching home plate, Jeter would look over into the Boston dugout and gesture toward the Sox manager with his hand or bat. If Francona was momentarily distracted, Jeter would step back and wait for a response before proceeding with his work.

"A couple of times somebody in our dugout had to nudge me and say, 'Hey, look over at him! He's waiting for you to wave back,'" said Francona.

"I don't remember when I first started doing that with him," said Jeter. "But I've always respected Terry. I know his players loved playing for him, and I got to know him throughout the years. I always admired him from afar. He always seemed to me like he was pretty even-keel. He never seemed to overreact. He never had a look of panic on his face, and I think that rubs off on teams."

Once Jeter established the ritual, coaches in the Red Sox dugout noticed that A-Rod started to mime the gesture. Francona never noticed. He wouldn't intentionally ignore Rodriguez, but he had too much to do and there were usually runners on base by the time A-Rod came to bat.

Red Sox–Yankee games were traditionally lengthy, often running as long as four hours. Hitters on both teams worked into high pitch counts. Every at-bat was contested, and there were multiple pitching changes and commercial delays.

One of the more prominent time-suckers was the seventh-inning stretch appearance of Irish tenor Ronan Tynan, singing "God Bless America." Some players were agitated by the nightly delay, but Francona enjoyed the appearance of the supersized singer.

"I'd met him over the winter when he sang at the Boston Pops," said the manager. "I teased him about how long he took—that he was fucking my pitchers up—and told him if he took too long next time we were at Yankee Stadium I was going to come out of the dugout and kiss him on the lips. So there we were back at Yankee Stadium, and when he got done singing—it was just as long as usual—he looked over at me in the dugout. I think he thought I was coming out there. Some of the players were bothered by it, but I thought it was a nice ritual. The only bad thing about it was, I had to take off my hat and stand there for five minutes. You know how many insults you can hear from the first three rows of Yankee Stadium in five minutes—while we're 'honoring America'? They should have just introduced it with, 'Ladies and gentlemen, now let's honor America and hurl insults at Tito's head.'"

Francona suffered another Bronx indignity when the team bus left without him after a late game at the new stadium.

"That was the greatest thing ever," said traveling secretary McCormick. "We were pulling out, and all of a sudden I hear this banging on the side of the door, and there's Tito, trying to run alongside the bus. He was pissed, but it was funny."

The highlight of Boston's early dominance against the Yankees came on *Sunday Night Baseball,* April 26, when Ellsbury stole home off Andy Pettitte in the fifth inning of a 4–1 Fenway win.

A steal of home is rare and exciting. It cannot be called or planned. Ellsbury was not without information. He was privy to the scouting report prepared by Sox veteran Dana LeVangie. The report told Ellsbury everything he needed to know: Pettitte could be slow and deliberate toward home plate. A fast runner on third had a chance to steal home against the Yankee lefty.

"Ells always talked about it," said third-base coach Hale. "He'd say, 'I can get it, I can get it.' On this night, I saw where he started to inch down, and I was thinking, *Oh-oh.* I could see he was going to have a chance. It was all on him. It was the perfect storm. The kid is fearless. He thinks he can steal off anybody."

After Ellsbury dashed down the line and easily beat the throw to the plate, he popped up, jogged past a stunned J. D. Drew ("Good to have J. D. hitting there, he probably wasn't going to swing anyway," said Hale), and made his way to the Sox dugout, where Francona greeted him with, "Way to go, Jake! Don't ever be out doing that!"

The first two months of the 2009 season presented Francona with a new and difficult dilemma: David Ortiz stopped hitting altogether. Ortiz was only 33 years old. In his first six years with the Red Sox, Ortiz's average season was .297 with 39 homers and 122 RBI. He ranked second in the majors in RBI, trailing only Alex Rodriguez. He had the third-highest slugging percentage, trailing only Barry Bonds and Albert Pujols. There had been signs of trouble in 2008 when Ortiz hit .264, his lowest average since the Twins released him in 2002; he started only 108 games, suffering a torn tendon sheath in his left wrist that put him on the disabled list in June. But the latter part of '08 was nothing like the beginning of '09, when Ortiz was a .200 hitter without a single home run in the first 39 games. It was no mere slump. It was a stunning decline of a skill set, and it put a strain on the man making out the lineup card. Francona was forced to drop Ortiz in the lineup and considered benching him against tough lefties. The prospect of pinch-hitting for Ortiz had once seemed preposterous, but was suddenly in play. Francona was quizzed daily about how long he could keep Ortiz in the lineup.

"That was one of the tougher things for me," said the manager. "This was a guy that had carried us on his shoulders. I had rings because of him. I knew how much it bothered him because he's so proud. I kept trying to look at the big picture. We had to fight our way through it."

When Youkilis came off the disabled list on the night of May 20, he was surrounded by reporters who wanted to know if his return to the cleanup spot would help Ortiz.

"If everyone stops asking questions about David Ortiz and leaves him alone, maybe that will help him out," snapped Youkilis. "It would bother me if everyone was talking negative about me every day. David Ortiz wants to get out of his slump as much as anybody. That's it. I'm not answering any more questions about David Ortiz."

At the end of batting practice, Francona was pulled away from the cage by Sox community relations employees Sarah Stevenson and

Sheri Rosenberg to engage in some small talk with Sox fans. Stevenson and Rosenberg fielded hundreds of requests from fans anxious to connect anyone in a Sox uniform with a soldier or perhaps a young cancer patient. Francona was at ease with the small talk and the potentially awkward interactions. He was able to make the fans feel welcome, and he appreciated the work ethic of Stevenson and Rosenberg. On this night, when he was asked to visit with a small group of strangers, he brought Big Papi along to sign some baseballs.

After the quick exchange with fans, Ortiz asked his manager, "Tito, do you see anything I'm doing wrong?"

"No," said Francona. "I never hit like you. Just remember, you are one of the best players in the game. Just keep it that way and your time will come. Keep that swagger and you will be fine."

A few hours later, Ortiz launched his first homer of the season (first in 149 at-bats) and said, "I feel like I got my confidence back. I feel like a real hitter, not like the Punch and Judy hitter I've been the first 40 games. Swing like a man, not like a little bitch."

"As a player, you have those blinders on, and little things that Tito would do would pick you up," said Pedroia. "He was the best at that. We all struggle. You got through 20 at-bats like that. I'd be waiting to hit and I'd say to Tito, 'What do you got? Do you see anything I'm doing?' And he would say, 'Fuck, man, are you kidding me? I was the worst hitter ever.' It kind of made you relax."

Ortiz's problems were not over after he finally homered. The Sox sent him to the eye doctor. (Francona got an inadvertent laugh when he told the media Ortiz had visited the "obstetrician.") With Ortiz still struggling badly in mid-June, Francona had a lengthy meeting with him when the Sox played a weekend series in Philadelphia.

"We'd had a rocky couple of weeks," said the manager. "I'd pinch-hit for him a couple of times and sat him against a few lefties, and he was really mad at me. We had to clear the air, so I sat him down and said, 'David, we've been together a long time. I understand how you feel. We've got to fight through this.' And I think we did."

While Ortiz struggled and questioned his professional mortality, Pedroia was becoming more cocky and amusing every day. He stood up to everybody. He got into a scrape with A-Rod in a dustup at second base. He challenged everyone he encountered.

Pedroia spent the early part of the season complaining about lack of respect from his own organization after winning the 2008 American League MVP Award. He complained about the night he received the MVP trophy at a banquet in New York.

"It was unbelievable," said the second baseman, laughing while he spoke. "Tim Lincecum was the Cy Young winner, and he had the whole Giants front office there from San Francisco. Even the clubbies flew out for him. I was there with just Pam Ganley. Brian Cashman — the GM of the Yankees! — had to give me my MVP Award! Our owners gave David [Ortiz] a car or a truck for doing I'm not sure what, and I've got nobody there when I get the MVP. All I got was a handshake."

This was Pedroia's theme throughout the early part of the '09 season. He said the same thing to anyone who'd listen. It never seemed serious; nothing seemed serious when Pedroia was talking.

Ever intent on team-building, Francona decided the best response was a gag gift for his second baseman. He put Pookie Jackson on a mission, and when the Sox returned from an early-season road trip, Pedroia found an electrically charged blue mini-scooter in front of his locker. A pink "AL MVP" helmet hung from the handlebar, and there was a phony note from Lucchino that read: "Sorry we couldn't make it to the MVP presentation, this is our gift to you. Congratulations, Dustin!"

Caught up in the moment, Pedroia — who had an apartment very close to Fenway — fired up his scooter and drove it out of the parking lot, towing his luggage behind him. He did not wear the pink helmet. His teammates thought it was hilarious.

"He looked like *Dumb and Dumber*," said Francona.

"I loved that thing," said Pedroia. "It was like a midget scooter. It went something like 15 miles an hour. It was awesome. I rode it to work every day until John Henry saw me and told me not to do it anymore. It was fucking awesome. I think eventually somebody stole it."

Dustin and Kelli Pedroia were looking forward to the birth of their first child in the summer of 2009. Kelli Pedroia experienced complications during the latter stages of her pregnancy and was bedridden at Massachusetts General Hospital for many weeks prior to the birth. The Sox second baseman lived at the hospital with his wife when the team was at home, and the mothers of both prospective parents came to Boston to be with Kelli through the difficult time.

By all accounts, Pedroia gets his combative personality from his mother, and Francona witnessed an up-front-and-personal demonstration of the family dynamic when he and Dr. Ronan went to visit the Pedroias at the hospital before the All-Star break.

The American League MVP was in the middle of a wrecking ball weekend against the Royals. In four games against Kansas City, Pedroia went 8-19 with four doubles, a triple, and a homer. He also struck out swinging once, flailing at a curveball and leaving the bases loaded. This unfortunate at-bat was the focus of conversation when Francona and Dr. Ronan tiptoed into Kelli Pedroia's hospital room after a game against the Royals.

"It was unbelievable," said Francona. "We walked into the room, and poor Kelli was laying there and Pedey and his mom were going at it over him swinging at that breaking ball in the dirt. I looked at Larry and said, 'This is like a reality TV show.'"

"All true," confirmed Pedroia. "My mom just blew me up for swinging at that ball in the dirt. We're all in the hospital room for Kelli, and my mom was motherfucking me, saying, 'What the fuck's wrong with you?' and Kelli's mom was there, and then Tito and Dr. Ronan walked in. If I play bad and the media gets on me, that's a piece of fucking cake compared to my mom and what I have to go home to. I never hear the end of it. So when Tito and Doc walked in and heard all that while Kelli was laying there, I was like, 'Welcome to my world.'"

As Kelli Pedroia's due date neared, Pedroia had Francona's blessing to leave the ballpark anytime he felt it was necessary to support his wife.

Major League Baseball has lagged behind most of America through virtually every social and cultural change over the last 110 years — the sport did not have any black players until 1947 — and big league clubhouses remain the last bastions of the old ways. In 2009 there weren't many other American workplaces supplying employees with chewing tobacco and beer.

Francona's sensitivity to family matters was largely due to his own experiences. Cleveland outfielder Tito Francona was in Detroit with the Indians when his only son was born in Aberdeen, South Dakota, on April 22, 1959. The elder Francona did not see his son for three weeks. Twenty-eight years later, veteran infielder/outfielder Terry Francona was getting ready for opening day at first base with the Cincinnati Reds when Jacque Francona's water broke back home in Tucson. He told manager Pete Rose that he needed to go home because of Jacque's situation.

"That's fine," said Rose. "Just don't come back."

Terry Francona got the message. He stayed in Cincinnati and hit an opening day home run in an 11–5 win over the Expos. His first daughter, Alyssa, was born the next day, and he was not there. It was the only one of his children's births that he missed, and he never shed the regret. Pete Rose was a friend and was friendly with Francona's wife, but none of that mattered in big league baseball in 1987. Francona pledged that he would do things differently if he ever managed, and that's why he was okay going with an infield of Aaron Bates, Nick Green, Julio Lugo, and Youkilis in a 6–0 loss to the A's in the summer of 2009.

Pedroia was voted starting second baseman of the American League All-Star team, but he skipped the midsummer classic in St. Louis and returned to be with Kelli at Mass General. When Dylan Pedroia was born August 18, 2009, his dad was there. The Red Sox beat the Blue Jays in Toronto, 10–9, improving to 67-51, seven games behind the scalding Yankees.

For most of the summer the Sox were without the services of Daisuke Matsuzaka. On the heels of his 18-3 season in 2008, Matsuzaka developed shoulder soreness and was placed on the disabled list after going 1-5 with an 8.23 ERA early in the season. The Sox were losing patience with the training methods of the stubborn righty, and things came to a boil late in the summer when Dice-K, rehabbing in Florida, told a Japanese publication, "If I'm forced to continue to train in this environment, I may no longer be able to pitch like I did in Japan. The only reason why I managed to win games during the first and second years was because I used the savings of the shoulder I built up in Japan. Since I came to the major leagues, I couldn't train my own way, so now I've lost all those savings."

Farrell was livid when he was told of the remarks. Speaking with the media, the physically imposing pitching coach had difficulty masking his anger. Inside the manager's office — gathered with Francona, Matsuzaka, and a translator — Farrell was allowed to speak freely.

"Normally I would try to settle him down, but not this time," said Francona. "Dice had spoken out of turn, and John was fucking irate. He wore Dice out. There was nothing lost in translation. Dice knew just from the tone. I was nervous about it, but it turned out to be kind of a breakthrough day. Dice was contrite, and we didn't have many problems with him after that."

"I probably didn't handle that too well," said a chagrined Farrell. "There were challenges with someone as accomplished as Dice-K. There were differences, and we had concerns about the volume of throws."

"He had to maintain his shoulder program or he was putting his shoulder at risk," said Francona. "He believed all the throwing just made his shoulder stronger. That's where we were banging heads.

"I didn't pitch, but as a manager, you have to pay so much attention to everything regarding pitchers. I knew the best way to ruin your team was to fuck your bullpen up. You have to watch them. And there were certain things I felt. We kept things monitored very closely regarding usage and the intensity of usage. If I got Alan Embree up quickly some night — you could do that with Embree because he could get ready in a hurry — I'd put a little star next to his name to remind me that he got ready in a hurry and might be a little more stiff the next day. I thought the biggest mistakes a manager could make was not the in-game pitching, it was about warming guys up and not paying attention to that. It was easy to see that you keep the pitchers healthy, they're going to be more productive."

Two days after the Farrell-Matsuzaka storm, the Sox took a hit when the *New York Times* disclosed the names of Ortiz and Manny Ramirez on a list of players who tested positive for performance-enhancing drugs in 2003. The tests had been conducted with the approval of the Major League Baseball Players Association as a means of determining the extent of usage of banned substances in baseball. Players had been assured that results would remain sealed and anonymous. Names would never be released and punishments never issued. The union was in the process of destroying the tests when they were subpoenaed by federal authorities in November 2003. One hundred and four players tested positive in 2003, but as of August '09, the only names that leaked were those of Alex Rodriguez, Sammy Sosa, Ramirez, and Ortiz. By this time Ramirez was a confirmed cheater, having been suspended by Major League Baseball in May '09 for 50 games for taking hCG, a women's fertility drug typically used by PED users trying to restart natural testosterone production at the end of a drug cycle. Seeing Manny's name on the list did not shock many folks in Red Sox Nation. The inclusion of Ortiz was another matter altogether. Sox fans were being told that there was no Santa Claus.

Francona called Ortiz into his office the day the story broke. He closed the door. He sat at his desk while Ortiz took a seat on the Pesky couch.

"David, look at me," said the manager. "Tell me the truth."

"I'm okay," said the slugger.

"Good enough," said Francona. "I'm with you. This meeting is over."

Ortiz had even less to say to the media. The Players Association had told him not to say anything. Ortiz told reporters he intended to get to the bottom of things.

And then there was silence for ten days until a Saturday afternoon when Ortiz, accompanied by Players Association director Michael Weiner, faced more than 100 reporters in the sprawling basement interview room of the new Yankee Stadium. Behind Ortiz and Weiner, who were seated at a table, Francona stood off to the side, in uniform, arms folded across his chest, visible to everyone in the room. Sox players had planned to attend the nationally televised press conference en masse as a show of support, but the team bus was caught in traffic, so it was left to the eternally early manager to stand behind David Ortiz in his moment of need.

"It was important to me," said Francona. "I'd been with David since 2004, and I believed him."

A nervous Ortiz admitted only that he'd been "a little careless" when he bought supplements and vitamins over the counter in the United States and the Dominican Republic. He apologized to Sox fans, owners, teammates, and his manager for the distraction. The Players Association and Major League Baseball both issued statements defending Ortiz. MLB's statement acknowledged that "the names on the list . . . are subject to uncertainties with regard to the test results."

It was a stunning show of support, one that had not been available to Alex Rodriguez months earlier when his name was leaked. When it was over, Ortiz walked back to the clubhouse with Francona and thanked his manager for standing by his side.

"Having Tito there meant a lot to me," said Ortiz. "I had nothing to hide, and I didn't want people to look at me like I was hiding. Tito did what a father would do for his son."

"I was really proud of him," said Francona. "First of all, that's a lot of pressure and he's speaking in a second language. I've been in Venezuela, and I know the feeling of people talking at me when I have no idea what they are saying. I have respect for guys who can do that in a second language. I think that whole thing had an effect on David. People were hammering him. I think it's where he started getting a little bit wary of people. I think he felt like he'd built up enough where he should be off limits."

The Sox manager disputed the notion that Ramirez's drug transgressions and Ortiz's appearance on the list of 2003 tainted the Red Sox championships of the 21st century.

"Baseball as an industry buried its head in the sand in the early, mid '90s, and the consequences are that now we're fighting a fight that's not fair for anybody," said the manager. "I wouldn't want to be a writer voting for the Hall of Fame. There's no right answer, and that's the price we're paying for the industry burying its head in the sand. I see how hard these guys work. I think about what it's like for a guy in Double A who wants to make it to the majors. That's a tough one. If I'm a 22-year-old kid who just had knee surgery and I want to be a good player, would I use? I hope my answer would be no, but you want to be good so bad. . . . I'm not saying I condone it, but I understand how it happens."

The Red Sox owner responded angrily to the Ortiz episode. In an email with a *Globe* columnist, Henry termed the episode "a deliberate attempt by someone to harm David and to harm the Red Sox. . . . David is a gentle soul. He is a prince of a man who only gets upset when he doesn't perform. . . . If anyone in Boston's great history of sports deserves the benefit of the doubt, David does."

The summer of 2009 was a happy time for John Henry. Fifty-nine years old and twice divorced, Henry married 30-year-old Lynnfield native Linda Pizzuti in late June, and the couple was seen at numerous social and charity events around greater Boston. Pizzuti had a master's degree in real estate from MIT and was LEED-accredited. With two championships in his pocket and a ballpark renovation near completion, Henry was content to let his baseball people run the Red Sox while he built a dream home in Brookline and explored new ventures for the ever-expanding Fenway Sports Group. He stayed in touch with his field manager through late-night emails, usually asking why certain decisions were made that ran contrary to the imposing database maintained by the baseball operations department.

Henry's army of stat savants detected significant flaws with the Red Sox defense. The Sox could not stop enemy base runners. It was never much of a concern during the Francona years. Joe Kerrigan hadn't cared about enemy base stealers when he was pitching coach under Jimy Williams, and the philosophy was unchanged during the Grady Little and Francona administrations. The Sox didn't order their pitchers to slide-step. Making good pitches was more important than holding base runners. It led to some ugly stats. In 2009 Red Sox catchers caught only 23 of 174 base runners attempting to steal. Varitek threw out only 10 of 108 thieves.

"It was kind of an organizational joke when I got there," said Francona. "We'd be talking about acquiring a pitcher, and Theo would say, 'He's got a 20-2 record, but we can't get him because he doesn't hold base runners.' They were making fun of me because I cared about it, but we didn't teach it much. It was funny until it got to a point where we couldn't stop people at all. It's hard to win like that. In my philosophy, if you're a pitcher and you don't walk people and don't let people run, they have to beat you getting hits. But if you're just letting them steal second and third, it's a tough way to win. It started to get out of hand with us, and it bothered me. We steadily got a little better at it, and our pitchers paid more attention, but we were never good at it."

The defense on the left side of the infield was equally bad, but less quantifiable. Lowell's hip surgery had limited his mobility, and Lugo (nicknamed "Huggy Bear" by teammates) was not the player the Sox thought they were getting when they signed him to his big contract. Lugo was released in mid-July, and Francona went with a shortstop rotation of Jed Lowrie, Nick Green, and Alex Gonzalez (reacquired in August).

The GM gave his team a boost at the trading deadline when he acquired switch-hitting All-Star catcher Victor Martinez from the Cleveland Indians for righty Justin Masterson and a couple of pitching prospects. The Sox were in Baltimore on the day of the deadline, and for Francona the most difficult part of the busy day was calling young Masterson into his office to tell him he'd been traded. Masterson was only 24 years old and had pitched brilliantly in nine postseason games in 2008. He was talented, wide-eyed, homegrown, and smart. He was exactly the type of player you wanted in your organization. But he could not help the Red Sox the way Victor Martinez could help the Red Sox.

"Theo just made a trade for you," Francona started, as the two sat in the manager's office at Camden Yards. "We are not disappointed in you. But we have a chance to win this year, and this is what it takes. We're sorry."

Fighting back tears, the ever-polite Masterson nodded, said he understood, stood up, and shook his manager's hand. Then he was off to join the Indians, and Francona would root for him to succeed in any game he pitched that did not involve the Red Sox. (After being hired as manager of the Indians late in 2012, Francona looked forward to reuniting with Masterson in 2013.)

Martinez delivered for Boston down the stretch. In 56 games he hit .336 with eight homers and 41 RBI. He also forced Francona to make some new and difficult decisions regarding the deployment of captain-catcher Jason Varitek.

In mid-September, with the Sox still chasing the Yankees but looking like a cinch for the playoffs, Francona picked up a newspaper and saw that the Houston Astros had fired manager Cecil Cooper with 13 games left in the season. Francona called Astros GM Ed Wade. Wade had fired Francona in Philadelphia in 2000, but the two were friends.

Francona told Wade he should interview Mills for his managerial vacancy after the Sox completed their playoff run.

The Sox did not catch the Yankees, but they won 95 games and clinched a wild-card playoff berth for the seventh time in franchise history. It was Francona's fifth postseason bid in six seasons. The 2009 Red Sox played .691 (56-25) baseball at home, their best season at Fenway since 1978, when Don Zimmer's 99-win team went 59-23 at home.

Boston faced the Los Angeles Angels in the Division Series. The Sox had beaten the Angels in the first round in 2004, 2007, and 2008, dominating each time (two sweeps and a four-game set in '08). But Boston was not built to win in 2009. It was quite the opposite. The Sox were swept out of the playoffs in three straight games.

"They did to us what we'd done to them all those times," admitted Francona.

Angel righties John Lackey and Jered Weaver beat the Red Sox in the first two games, with Boston scoring only one run in 18 innings. Youkilis and Ortiz went 1-16 with five strikeouts. Aces Jon Lester and Josh Beckett were the Sox losing pitchers.

It was not a good time for Francona. He suffered a bout of food poisoning in Anaheim and disputed the notion that he was panicking when the Sox said they'd consider pitching Lester on three days' rest in Game 4 if they were able to win Game 3 at home with Clay Buchholz on the mound.

The Fenway finale looked good at the start as the Sox roared to a 5–1 lead through five innings on a splendid, crisp New England October afternoon. The Red Sox led Game 3, 6–4, in the top of the ninth with two out, nobody aboard, and the indomitable Papelbon on the mound throwing nothing but smoke. In that moment, Papelbon was working on a string of 27 consecutive scoreless innings in postseason play. The only pitcher in big league history to start a career with more scoreless frames in October was Christy Mathewson, who did it from 1905 to 1911.

"At that moment, I was thinking, *We're going to win this series*," said Francona.

And then it all went away. Erick Aybar singled on an 0-2 pitch. Throwing only fastballs, Papelbon walked Chone Figgins (0-12, six strikeouts in the series) on a full count. Old friend Bobby Abreu fouled off a 1-2 pitch, then banged an RBI double off the Wall. Francona ordered an intentional walk to Torii Hunter to load the bases. The next batter was Vlad Guerrero, and he ripped a first-pitch, two-run single to center to give the Angels a 7–6 lead.

Francona came out to get his closer.

"He knew I wasn't trying to embarrass him," said the manager. "I'm still thinking we can win the game and get him back out there for Game 4. That's how I always felt. I always thought we could come back. But when you put yourself in a hole like that, you make a mistake and you're going home."

"I don't think anything that happened in this series was completely out of the blue," said Epstein. "We saw things that were reflected early in the season."

There was a subtle, almost unreported front-office change in the latter part of the 2009 season. Janet Marie Smith, the popular architect-visionary who oversaw the renovation of Fenway Park, was relieved of her duties without notice or fanfare. A mother of three children, Smith commuted from her Baltimore home during her eight-year assignment in Boston and would stay at a Back Bay hotel while she performed her job at Fenway. She had an office just a few feet down the corridor from the offices of Henry, Werner, and Lucchino and held the lofty title of Senior Vice President/Planning and Development. Smith and Lucchino were the genius behind the building of Camden Yards, which opened in 1992. Camden Yards changed everything about the way ballparks were built. Smith was considered the best ballpark builder-renovator in the country, possibly a future candidate for the Hall of Fame. Her official job description in the 2009 Red Sox press guide stated, "She directs the renovation and area planning for Fenway Park."

And then she was gone. The Sox no longer needed Smith's services. It was curious, abrupt, and quiet. Henry and Werner never contacted Smith to thank her for her service. Only Lucchino bothered to say good-bye. Smith was hurt, but never went public with her objections. It would not be the last time a valued employee who served the Sox well was dismissed abruptly without getting a call from the owner.

# CHAPTER 13

# · 2010 ·

## "We need to start winning in more exciting fashion"

**T**HEO EPSTEIN WANTED to manage expectations for his 2010 ball club, but he knew it was almost impossible to do that in Boston. He knew that in the 21st century it was never okay for the Sox to speak of "rebuilding." Rebuilding seasons were for the Pittsburgh Pirates and the Florida Marlins. The Yankees and Red Sox were never allowed to rebuild. Fans recoiled at any suggestion that the ball club was less than championship-bound. Money and success had created the perpetual expectation. It was never okay for the Red Sox to slide backward while prospects matured. Lucchino believed in this as much as any Boston baseball fan. He was always about "winning today." It was the way he governed the Red Sox. It was one of the many issues that constantly put him at odds with his protégé GM.

Francona never got involved in the Theo-Larry conflict, but he felt good about his place with the ball club as he looked back on 2009 and got ready for 2010. He'd already managed more Red Sox games than anyone other than Joe Cronin, he'd made the playoffs five times in six seasons, he had two World Series rings, and he had two more guaranteed years on a contract that would make him a rich man. He barely noticed when Theo started talking to the Boston writers about "the bridge."

It was December 2009 and the Red Sox baseball brain trust was camped out at the Marriott in downtown Indianapolis for the winter meetings. The ball club was considering re-signing Jason Bay or investing in free agent Matt Holliday. The Sox were trying to trade Lowell. They wanted to tighten their infield defense and get a little younger. Francona was in the Sox suite late in the afternoon when Epstein routinely briefed Boston's baseball reporters regarding Sox activity at the annual hardball convention.

"We talked about this a lot at the end of the year, that we're kind of in a bridge period," said Epstein. "We still think that if we push some of the right buttons, we can be competitive at the very highest levels for the next two years. But we don't want to compromise too much of the future for that competitiveness during the bridge period. We don't want to sacrifice our competitiveness during the bridge period just for the future. So we're trying to balance both of those issues."

Back in Boston, fans and media reacted as if Epstein was giving up on the 2010 season.

Tom Werner, on the Sox radio flagship WEEI, said, "I think he [Theo] made a very rare mistake saying that. . . . I think Theo would be the first to say it wasn't his finest Winston Churchill moment."

"By definition, it's hard for owners to get it," reasoned Epstein. "They don't have to be on the front lines, giving a true narrative, planning for the future. It didn't really bother me. Tom was generally supportive. That wasn't a great demonstration of his support at the time, and it underscores the conflict between the approach we wanted to take in baseball ops and inherent tension with what the Red Sox became — the public image, the expectations, the dollars, the bigger-better-now mentality. It was a truthful moment, and I think it demonstrated how we'd gotten too big."

"I felt badly for Theo on that one," said Francona. "I knew exactly what he meant, and it got taken so far out of context. I completely agreed with it when he said it. It pissed me off the way that got twisted. He was saying that we're going to find a way to win, but we didn't want to commit money to players that weren't worth the money. Theo was always trying to stay young, and I completely agreed. One of my biggest fears was getting old during a season."

The Red Sox had already been burned by free agency. In exchange for a relatively low yield, Epstein had spent more than a quarter-billion dollars on the likes of J. D. Drew, Daisuke Matsuzaka, Edgar Renteria, Julio Lugo, Matt Clement, and Brad Penny. In the wake of the "bridge remark," it was going to get worse before it got better.

A week after making the bridge remark, a week after taking a beating in the Boston media, in a two-day spree that looked like bridge backlash, the Sox signed Angel righty John Lackey to a five-year, $82 million deal, then inked soon-to-be-37-year-old outfielder Mike Cameron to a two-year, $15.5 million contract.

It was staggering. Lackey was a proven starter who'd won the seventh game of the World Series when he was a rookie in 2002. He was fiery and durable. The widely respected Angel manager Mike Scioscia regularly gave Lackey the ball for the first game of any playoff series. Lackey was no $82 million pitcher, but he was the best available starter on the market, and the Sox wanted to show their fans that they were serious about winning in 2010. The Sox forgot what they had been about. They lost their patience.

"The Lackey signing flew in the face of everything Theo believed in," said Mike Dee, who worked with Epstein in San Diego and Boston and was chief operating officer of the Red Sox from 2004 to 2009. "Theo must have been tied down to make that deal. That was all about television ratings. They were panicked about the ratings."

Cameron was coming off a 149-game, 24-homer season with the Brewers. He was a three-time Gold Glove center fielder and had been an All-Star with the Mariners in 2001. He was a clubhouse leader who spent a lot of time in the community and was considered one of the nicest men in baseball. He'd stolen almost 300 bases in the bigs. Francona remembered managing young Cameron in Birmingham when the Georgia native was a rising star in the White Sox system in 1995.

"He had long, graceful strides and was built for center field," said Francona. "He had all the things that you wanted your center fielder to have. He was going to be a good, solid veteran presence for us."

"Cameron was still showing up as a really good center fielder," Epstein said later. "Little did we know—and we probably should have known and should have seen it—he was literally falling apart from the inside. He had hernias in both groins. It was really bad."

Lackey and Cameron weren't the only new starters who came at a cost. The Sox also signed free agent shortstop Marco Scutaro and agreed to a one-year, $10 million ("make good") deal with free agent third baseman Adrian Beltre.

It was a great deal of money, and it was all pitching, defense, and television ratings. Lackey would provide the manager with another top-of-the-rotation starter who could eat innings. Scutaro and Beltre would shore up the left side of the Boston infield. Cameron would play center field. And Sox fans would get the message that ownership was not giving up on 2010.

Putting Cameron in center required moving Ellsbury to left to replace Jason Bay, who was allowed to walk away to the Mets. Ellsbury was only 26 years old, coming off a season in which he batted .301 with 70 stolen bases and only two errors. Ellsbury was clearly a future star center fielder, but Cameron had played only three big league games in left field. The kid was asked to move. Francona and Hale called Ellsbury to inform him of the switch.

"Jake was initially quiet when we told him," recalled the manager. "We told him this was not any kind of indictment, but that we'd gotten a veteran guy who'd only played three games in left field. We told Jake it was short-term. I don't think he was doing any cartwheels, but he said, 'Okay.' He said he understood why we were doing it and was willing to go forward with it."

"He didn't like it, but he accepted it," said Hale.

Everything about the Cameron signing proved disastrous. Ellsbury never openly ripped the decision to move to left field, but he was clearly not enthused and wound up getting hurt, colliding with Beltre under a pop-up in the sixth game of the season. Cameron played only 48 games in 2010 and went to the disabled list twice with a lower abdominal strain. He hit .259 with four homers and 15 RBI. In June of 2011, with Cameron batting .149 in 33 games, the Sox designated him for assignment, then traded him to the Marlins for a player to be named later or cash considerations.

Epstein took the media hit. He was the one who wanted to sign

Cameron. Henry took the financial hit, paying $15.5 million for virtually nothing.

Francona did not hide from his participation in the Cameron experiment.

"It just made sense for our team," said the manager. "Ells was very fast and could outrun some of his mistakes, but Cam was the best when he was healthy. He had a better arm than Jacoby. It was just going to be for a year, and Cam had never played left. We thought Ellsbury in left for a year could have been like what Carl Crawford was in left for Tampa. That's how we viewed it. We were trying to win now, and that would have been a dynamic outfield. But then Cam showed up and was not healthy right away and Ells got hurt. So much for the best-laid plans.

"I respect how hard those decisions were for baseball ops. They were already thinking of Cam as maybe a fourth outfielder in the second year, and when you are in Boston, if somebody gets hurt, you want to have a good player. As a manager, you can't sit there and say, 'I don't think this is going to work.' The manager and the GM have to try to communicate where you get to the same page. Even when we didn't get to the same page, I still understood why we were doing things."

"We overthought it," said Epstein. "We relied way too much on the impact of defense and looked past the simplest explanation, which was that you don't want old guys playing in the middle of the field. They get hurt. That was my fault."

The push for more defense came from Carmine and from the 2009 season summary questionnaire filled out by all Sox field staff.

"You could see the game trending toward Texas, Tampa, Anaheim, teams that wanted to run," said Francona. "If the ball didn't end up where it's supposed to, you were getting yourself in trouble. Back in '04, we were kind of slow and plodding at times, but we had so much offense it didn't matter as much. By 2009, we were extending innings and not finishing plays. We could see it was getting tougher to win that way."

"If we bring back the same pitching staff and the same defense, we are going to have trouble preventing runs," said Epstein. "Last year was bad, any way you look at it. We were really bad turning balls in play into outs. It's remarkable we allowed the third-fewest runs in the league. That's a testament to our pitching staff and some good situational pitching. You saw Jon Lester early in the year hurt by balls that got by the left side of the infield and by double plays that were not turned. It's the same stuff that the numbers will tell you if you look at it."

Defensive metrics — ultimate zone rating, defensive efficiency, plus/minus — were all the rage at the start of the 2010 season. The Seattle Mariners were a trendy division-winner pick by many experts because they'd played sensational defense in 2009. It was a new frontier of *Moneyball* mania. The Red Sox joined the chorus, but rejected the controversial UZR system popular with stat geeks across America.

"We don't use UZR," said Theo. "We use our own data. We have guys who do it internally. It's been refined. There's been a lot of progress in this area over the last few years. I think if you make any decision exclusively based on what some numbers say, you're taking an unnecessary risk. It's the same way we evaluate guys offensively. If we're going to rely on data, we go through lots of stuff to make sure it's pretty accurate. The same way we know a guy's on-base percentage is accurate. If you're going to trust a defensive number, you've got to make sure you know what it means."

As much as he liked Carmine and all the help from the data, Francona was dubious about defensive stats.

"I didn't need those numbers to tell me that balls were getting through the infield," he said. "Trying to quantify defense is a good idea, but it's not correct yet. They haven't come up with a way to do it. There are too many variables. I'd look at something on an outfielder and say, 'Does this take into consideration where the outfielder starts?' And they'd say, 'No.' I'd say, 'Does it take into consideration the quickness of that particular field?' And they'd say, 'No.' And I'd say, 'Then don't send this down to me.'"

O'Halloran understood the skepticism regarding defensive metrics.

"We knew the analytical world wasn't there yet," said the assistant GM. "We absolutely felt there was and is no defensive statistic that will tell you a guy is a plus-fielder. We know the data isn't perfect, and we can't really rely on it. It really is just a piece to the puzzle. It does help, and sometimes it raises questions. Maybe there is a bias because a guy has great hands and looks dazzling, or maybe he was great ten years ago."

Going into the 2010 season, Francona's daily dilemma was the Lowell-Ortiz platoon. Beltre was going to play third every day, and Youkilis would move back to first full-time. This meant that veterans Lowell and Ortiz would have to share designated hitting duties. Both were accustomed to being full-time players. Both had helped win the Red Sox a World Series (two in Ortiz's case). Both were proud and easily wounded by any suggestion of old age or eroding skills. Both were making more than $10 million per year. Both were enormously popular with Red Sox fans.

Thirty-six-year-old Lowell was in the final year of a three-year, $37.5 million pact that was struck after his MVP performance in the 2007 World Series. He'd recovered from his 2008–2009 winter hip surgery to hit .290 with 17 homers and 75 RBI in 119 games in '09.

Thirty-four-year-old Ortiz batted only .238 in 2009, but hit 28 homers and knocked in 99 runs. He expected to be in the lineup every day. He was making $13 million. In the 2009 playoffs, he was 1-12 with four strikeouts. Epstein had called him out after the 2009 season, saying, "If he's going to be the DH on this team, we need him to be a force. We're a different team when he is that force."

Lowell batted right, Ortiz left. On paper, it looked like a strong combination.

It was not. Ortiz looked badly overmatched at the start of the season and hit only .143 with one homer in April. Lowell hit .250 with one homer in April and wasn't moving well. Both players felt disrespected, and it affected their clubhouse comportment.

For the first time in his career, Lowell was difficult to manage. He knew the Sox had traded him to Texas in December, only to get him back when he failed a physical due to a thumb injury. Throughout the 2010 season, Epstein assured Francona that he would trade Lowell. It never happened.

Ortiz, meanwhile, became more self-centered than at any time in his Sox tenure. He challenged scoring decisions, complained about where he batted in the lineup, and recoiled at the notion that his manager would ever send someone up to hit for him.

Things boiled over in Toronto in the 21st game of the season on April 27. The Sox were 9-11 coming into the game, and Ortiz was enduring one of the worst stretches of his career. There was rampant speculation that Big Papi might be all done, and he was agitated when reporters came to him to gently ask him about declining skills.

The Sox and Jays were locked in a 1–1 game with two out and two aboard when Toronto lefty reliever Scott Downs prepared to face J. D. Drew in the top of the eighth inning. Slotted behind Drew as the sixth batter in the Sox lineup, Ortiz got ready to leave the dugout for the on-deck circle. Before Ortiz left the bat rack, Francona put his arm around Ortiz and told him to check with the dugout before going to home plate after Drew hit. The manager wanted his slugger to know that this was a situation in which he might pinch-hit for him, and he did not want to embarrass Ortiz. Papi nodded and went to the on-deck circle. Meanwhile, Lowell put on a helmet, grabbed a bat, and started to get loose in the dugout runway.

When Drew walked on a 3-1 pitch, Ortiz strode toward the plate with his bat in his hand. He never looked back. The bases were loaded. There were two out. The game was tied. Big Papi was going to hit. He was The Man. This was the kind of situation Ortiz thrived on throughout his Red Sox career. He'd carried the Sox with his bat in the 2004 championship run, and Henry loved him so much that he'd had a plaque made anointing Ortiz as the greatest clutch hitter in Red Sox history (a distinction challenged by some Carl Yastrzemski fans).

But Ortiz was not the batter Francona wanted facing Downs in April 2010, not with Mike Lowell in reserve. Francona was hearing a lot of noise from his bosses, especially Henry, regarding Ortiz's struggles against left-handed pitching. With Ortiz already near the plate, Lowell popped out of the dugout, called to Ortiz, and went toward the batter's box. Proud Papi trudged back to the dugout and disappeared into the tunnel that leads to the clubhouse. Then he took his bat to a water cooler and other nearby inanimate objects. Players in the Sox dugout heard the commotion as Toronto manager Cito Gaston summoned righty Kevin Gregg to face Lowell.

Lowell walked on four pitches to force home the winning run, but it was not a happy night for the manager or his Ruthian slugger.

"Everybody on the bench knew what was going on," said Francona.

"There wasn't a lot of sympathy for David because he didn't look. We had that policy the whole time I was there. If we were going to pinch-hit for somebody, we'd ask 'em to just give us a look before going up to hit. You can't yell from the dugout, especially at Fenway because it's so loud. One time in Tampa I wanted to pinch-hit for Mikey, but I didn't tell him to look over, so he got into the batter's box and I just let him hit. I hadn't told him, and I wasn't going to pull him out of the batter's box. That's my fault. He came to me the next day and said, 'I didn't do that on purpose,' and I said, 'I know. I wouldn't pull you out of the batter's box.' But everybody saw us tell David that night in Toronto, and everybody knew Mikey was ready to hit. It was kind of surreal because David was so mad he was beating up the runway. You could hear David's bat splintering when Lowell drew the walk to win the game.

"Prior to that time, if David was struggling, we could hang with him because the team was doing so well. But that was as tough a balancing act as I've ever had. It fractured my relationship with David for a while. Looking back on it now, I probably wish I didn't do it, even though we might have won a game or two. I think it still bothers David."

Ortiz agreed.

"We play in a town where a lot of things come into people's heads," said Papi. "You need to have the determination to know what's going to be better for you in the end. I talked to Tito, I talked to Larry, I talked to everybody. I told them that my whole life has been a challenge, and you're not going to get the best out of me making me uncomfortable."

Four weeks after the pinch-hitting episode in Toronto, there was another uncomfortable moment with Ortiz. The Sox were still under .500 when they arrived in New York for a two-game series with the Yankees. The Yankees routed Matsuzaka with five runs in the first inning of the first game and beat the Sox, 11–9. After the loss, Francona looked at the next night's matchup against lefty CC Sabathia and wavered on whether to use still struggling Ortiz or Lowell as his designated hitter. Right-handed-hitting Lowell was the obvious choice, but Ortiz had uncommonly good numbers against the Yankee ace and Francona was sensitive to Ortiz's pride. The decision was made more difficult by Henry's regular emails questioning the use of Ortiz against lefties. Francona went to assistant GM Ben Cherington. He went to Hale. Who do you play? he asked them. No one had a good answer.

For the only time in his eight years in Boston, Francona allowed everybody to leave the park without knowing who would play the next night's game.

Ultimately, Ortiz got the nod, batting sixth in the lineup against Sabathia. He rewarded the manager's loyalty with a single in the second and a walk in the fourth. The Sox fell behind, 5–0, again, but rallied for seven runs over the final four innings and won it, 7–6, with Papelbon picking up a heart-attack save.

But there was a problem. Batting against Joba Chamberlain, with Kevin Youkilis aboard and two out in the eighth, Ortiz launched a shot to the wall in deep right-center. Youkilis scored easily, but Ortiz was caught admiring his blast and thrown out as he lumbered into second after realizing his drive did not clear the wall. It was embarrassing and unnecessary, and it happened many times while Francona managed the Red Sox.

When the game ended, Francona, as always, went out to congratulate Papelbon and join the receiving line of handshakes as the Sox starters came off the field. He found Ortiz, who had also come out of the dugout to join the conga line, put his arm around the big DH, and said, "Come see me when you get to the clubhouse."

Ortiz never came to see his manager.

Feeling disrespected, Francona left his office, walked across the room toward Ortiz's stall, and confronted his slugger. It wasn't loud, but it was decidedly non-private.

"There were a lot of things going on," said Ortiz. "I lost a little bit of trust on how he was doing things with me. I thought the situation should be the way it had always been, because I was his player and struggling is part of being a player. He misunderstood things and I misunderstood things. I was a little fired up. Things were a little bit crazy that year. In other years he would have come to me later on and said, 'That wasn't right,' and I'd agree and that would be it. But that day he was calling me out in front of everybody, and I didn't like that. He came to me again after the game, and I went off. We talked the next day and everything went back to normal. He was pressing, I was pressing, there was so much going on, but I regret it."

"My point to David was, the loyalty thing has to go both ways," Francona said. "It was a tough time for both of us, and we were fighting our way through it, and I needed to say something to him. He was being a little stubborn and so was I. Even though we won, I still felt it needed to be addressed. I was sticking my neck out for this guy. Come on, man."

Francona's relationship with Lowell also suffered.

"I had called Mikey during the winter," said the manager. "It was the elephant in the room. Everybody knew we had traded him, but now he was coming back. I just told him, 'Mikey, I know there's shit going on, and I don't have an answer for you. That's my answer. I can't tell you something that I don't know.' He wanted to know. I was trying to balance loyalty, but I didn't want to lie to somebody."

He never told Lowell that two years earlier, when the Sox were courting Mark Teixeira, he'd warned Teixeira that he had a loyal veteran who was probably going to be replaced. (Youkilis would have moved across to third if the Sox had signed Teixeira.) He'd been sensitive to what the signing of Teixeira would have meant to Lowell.

No amount of sensitivity can make things right when a veteran player is sitting on the bench while he still thinks he's good enough to play. Lowell did not make things easy for his manager. Fans and many media members lobbied for Lowell, but there wasn't a daily spot for him in the lineup. There were times when Lowell was physically unable to play, but he didn't want his status publicized. The manager would take bullets for not playing a player who was physically unable to play. Francona was okay with this, but he knew playing time was going to be an issue.

"It hurt our relationship," said Francona. "I was in a bind. There was no right decision. I was so mad about the whole situation with both guys. It wasn't fair. We had two respected veterans, and they were both mad. It wasn't working. All you end up doing is aggravating everybody."

"We tried very hard to trade Lowell," admitted Epstein. "The reality is that Mike Lowell went from being a very good everyday player, a very popular guy, and a very influential guy in the clubhouse to a more complementary-type role. I knew it was going to be a challenge for Tito to manage that situation, but that's part of his job. I knew it would be difficult. We simply couldn't trade Lowell. He had a no-trade, and he didn't want to go to Toronto."

Meanwhile, the Ellsbury situation was a mess, and the manager found himself in the middle of confusion with the Red Sox medical staff.

It all started with the April 11 collision with Beltre. The ball club initially said that Ellsbury suffered a mere contusion and would only miss a few days. By April 23, the diagnosis was downgraded to non-displaced hairline fractures in four ribs, and Ellsbury told the media that the Sox had initially balked at his request for an MRI and CT scan. Red Sox medical director Thomas Gill said, "It really is symptom-based. . . . As long as he can swing, hit, run, catch, do anything that he needs to without feeling it in any way, he'll be cleared to play."

Forty days after the collision, Ellsbury returned for three games, went 1-14, then said he was unable to play. He said he came back too soon. On the advice of agent Scott Boras, Ellsbury flew to Los Angeles to be examined by Dr. Lewis Yocum at the Kerlan-Jobe Orthopaedic Clinic. Yocum found a new fracture in Ellsbury's ribs. Dr. Yocum and an embarrassed Dr. Gill issued a joint statement that read, "This fracture, which is in a different area than the initial fractures and which was not present on previous scans, is likely the result of a new injury when Jacoby dove and impacted the ground during his brief return to play."

Rather than rehab with the team, Ellsbury went to Athletes' Performance Institute in Arizona.

"We don't typically send our players elsewhere," sniffed Dr. Gill.

Dr. Gill was not alone. Youkilis openly criticized Ellsbury, telling the Boston media that the outfielder should be with his teammates. The manager tried to steer clear of the controversy, but sometimes seemed to be speaking in code when he praised disabled Sox players who stayed with the team as a show of solidarity.

On June 10, Ellsbury reported back to the Sox while they were playing in Toronto. He met with Francona before addressing the hungry Boston media.

"Jake was nervous," said the manager. "He told me what he wanted to say, and I told him just to be himself. Then I went out to the dugout and saw him there, reading from his notes. That worried me a little."

It was unusual. Reading from a prepared script, Ellsbury spoke for 11 minutes. He again mentioned the ball club's reluctance to send him for thorough testing. He said it was the team's idea for him to rehab at API.

When it was over, the wounded outfielder went back to see his manager.

Francona knew that Arizona was Boras's idea. The Sox had not sent Ellsbury to API for rehab; rather, they had *allowed* it. The team medical staff sometimes waivered, telling Francona and Epstein that Ellsbury could play, that it was a matter of Ellsbury's willingness to play while less than 100 percent.

Ellsbury made one final attempt to play for the Sox in early August, but reinjured his ribs in a game at Texas and was shut down for the season.

"We put him in a tough predicament," said Francona. "There were times when he didn't help himself, but he was a young kid going through a learning process. We wanted him to be okay so badly. It got mishandled on a lot of fronts. They wanted more testing all along, but our medical people were telling us that there was no reason he couldn't be playing. Then Boras got involved, and it made our doctors look like shit. You can't see inside someone's body. If they're not comfortable playing, you don't know what they're feeling.

"I told Jake, 'I have your back. If it hasn't seemed like that, I'm sorry.'"

Was Ellsbury ever in jeopardy of being traded?

"Never," said Francona. "Theo's the smartest guy in town. You don't trade a stock when it's at $1. You trade at $100. We didn't know we had a 30–home run guy, but we knew we had a good player. If you get mad and trade a player out of frustration, that's a bad move. Theo would never do that."

Two years later, Youkilis expressed regret at calling out Ellsbury.

"I made a mistake by saying something," said Youkilis after he was traded to the White Sox in 2012. "I was frustrated. You've got to be with the team. But I went about it the wrong way."

In the middle of the slump and turmoil, the Sox owners called for a meeting with their manager. The meeting was scheduled for noontime on the day of a weeknight game at Fenway. Epstein came to get Francona to take him upstairs to meet with Henry, Werner, and Lucchino.

Francona had been at the park for about an hour and was already in uniform when Theo came to his office to get him for the meeting.

"Because I was in uniform, I was probably in a little more of an aggressive mood," he remembered.

They took the elevator to the third (EMC) level of Fenway Park and walked toward Henry's in-game suite. Werner and Lucchino were waiting when Epstein and Francona arrived, but Henry was not there yet. The owner was invariably late for meetings, a habit that annoyed the always-prompt manager.

*I've got stuff to do too,* Francona thought to himself while the executives made small talk.

Henry arrived. Sandwiches were served. And one by one, the bosses told Francona what was wrong with the Red Sox. Henry talked about Ortiz batting against left-handed pitching. Lucchino, as always, found fault in multiple places. Werner talked about slumping television ratings and whined, "We need to start winning in more exciting fashion."

That did it. Francona started to get up out of his chair, but Epstein grabbed his knee to keep him in the seat.

"A good move by Theo," Francona said later. "When Tom started talking about ratings, Theo knew I was getting ready to flare on somebody.

"They all gave me their versions of how things should go. And they were all different. When they were done, I said, 'Guys, I just listened to three different opinions. All I can tell you is that the best thing I can do is to be consistent down there. We're going to go through ups and downs. Right now is a down period. We'll figure it out.' I remember later that night David hit a home run off a left-hander and John sent me a funny email telling me he'd changed his opinion. I said, 'John, I know you're laughing, but if I managed like this, do you know what would happen here? This place would explode.' All that stuff was starting to bother me. It wasn't that I need autonomy, but goddamn, let me do my job. We needed to circle the wagons, and we can't do that if I start being inconsistent."

Werner was constantly trying to assert his importance. When the Henry group first purchased the Red Sox, Werner hired a public relations firm to get his name in the local newspapers. When stories were written about Henry or the Red Sox, he was known to call writers and ask, "Why didn't you mention me in your story?" The Sox press guide described him as a member of the Television Academy Hall of Fame "for his extraordinary achievements as a creator and producer of many groundbreaking series." Werner's bio also credited him with shows that "launched" the careers of Robin Williams, Tom Hanks, Billy Crystal, and Danny DeVito. His television company won 24 Emmy Awards, and in 2009 he was appointed to the board of the brand-new Major League Baseball Network. He owned two World Series rings with the Red Sox, but his contributions to the success of the ball club were always difficult to measure. Henry owned controlling interest in the ball club and had the deciding vote on everything. Lucchino, according to Henry, was in charge of the day-to-day operation of the team. Epstein was the boy wonder genius/architect, and Francona was the two-time World Series–winning manager. That left Werner, the "Red Sox chairman."

This much was certain: Werner was in charge of NESN. The Sox network's ratings drove the ball club's revenue, and ratings were freefalling in 2010. According to Street & Smith's *SportsBusiness Journal,* the Sox in 2010 plummeted from first to fifth in Major League Baseball's local television ratings. From 2009 to 2010, NESN's Red Sox ratings dropped 36 percent. It was the first time in six years that the Red Sox did not have the best local ratings among all 30 big league teams. Meanwhile, ratings for Sox games on radio flagship WEEI-AM were down 16.5 percent.

During Henry's ownership, NESN's Red Sox ratings peaked at 13.57 (each point representing 40,000 households) in September of the championship season of 2007. By the end of 2009, the season average number was down to 9.38. In the middle of 2010, it dipped to 6.25. The Sox were also losing their drawing power nationally. According to NESN's own data, Boston's appearances on Fox and ESPN were down 32 and 33 percent, respectively, from 2009 to 2010.

Managing the Red Sox was never independent of television ratings.

"One thing the players were always asking for at those roundtable meetings with Larry was getaway day games," said Francona. "The owners would never go for it. They couldn't have more day games because the ratings were already suffering, and that would have hurt worse."

The Sox recovered from a poor April to go 18-11 in May and were wiping out inferior National League competition at the end of June when they capped a stretch of seven wins in eight games with a 13–11 victory at Coors Field in Denver. Pedroia hit three homers and went 5-5 with a walk in the Colorado finale ("Go ask Jeff Francis who I am!").

"He was on one of those tears, getting ready to carry us for a while," said Francona.

The next night, at AT&T Park in San Francisco, batting against Jonathan Sanchez in the third inning, Pedroia fouled a 3-1 pitch off his left foot. After writhing in the batter's box for a moment, he got up and drew a walk. He limped to first base, then came out of the game. Pedroia was on crutches in the clubhouse after the 5–4 loss. He'd fractured the navicular bone.

It was the beginning of the end of the 2010 Red Sox season. One day after Pedroia's season-ending injury (he would play only two more games in 2010), Clay Buchholz pulled up lame running the bases and had to come out after pitching only one inning. In the series finale on Sunday, catcher Victor Martinez fractured a bone in his left thumb.

It was too much to overcome. By the All-Star break, they were playing without Ellsbury, Pedroia, Cameron, Martinez, and Beckett. Varitek, no longer a starter, broke a bone in his foot and played only 39 games. The Red Sox had 11 players on the disabled list on July 15. Youkilis joined the disabled list in early August and underwent season-ending surgery on his right thumb.

Early birds at the ballpark got accustomed to the sight of Pedroia taking ground balls on his knees. Pedroia and Varitek regularly shucked their crutches and sat in folding chairs while their healthy teammates took batting practice.

"Pedroia and 'Tek found a way to impact the team, even on crutches," said the manager. "It's easier to go through it when your best player acts like that. Pedey and 'Tek were great, just being part of it, just being themselves. You can't have enough team meetings to get that point across. But we were going down so fast, and we didn't have answers. At first it's almost fun trying to figure it out, bucking the odds, but it just got to be too much."

The death of George Steinbrenner during the All-Star break inadvertently shed light on tension in the Sox front office. While the Sox called people back from vacation to compose a thoughtful tribute to their worthy rival ("A formidable opponent," said Werner), a *Globe* columnist used the occasion to reprint remarks Steinbrenner had made about Werner and Lucchino in Bill Madden's brilliant biography of the Boss. In Madden's book, Henry recounted a warning he'd received from Steinbrenner when he bought the Red Sox with his new partners. "I'm concerned about you getting into bed with Werner and Lucchino," Steinbrenner told Henry. "Those are two treacherous, phony backstabbers you've got there, John. You're a pal, but I'm telling you, I've got no use for those two bastards."

Thick-skinned Lucchino was amused and mildly flattered by the resurrection of the passage. Werner was furious and said, "That's gratuitous."

While the Lucchino-Epstein relationship continued to deteriorate, Francona's relationship with Epstein was stronger than ever. Epstein had learned to steer clear of his manager in the early minutes after tough losses, and Francona was always careful to back up his GM even when he was not in agreement with personnel decisions or statistical suggestions. The manager and the GM had one another's backs. Francona enjoyed the youthful exuberance of Theo's baseball ops staff. At 50 years old, he was not above practical jokes and locker-room stunts typically reserved for professional ballplayers.

In late August of 2010, young publicist Pam Ganley witnessed a *Bull Durham* moment when she burst into the manager's office between games of a day-night doubleheader and caught the manager mooning Epstein, Cherington, and O'Halloran. The Sox had won the first game of the twin bill, then learned that Matsuzaka was injured and would be unable to pitch the nightcap. The manager and the baseball ops executives were in a good mood after the afternoon win, and it was not unusual for Francona to engage in immature horseplay. Boys will be boys. It is the way of many baseball lifers. When Epstein made a wisecrack, Francona came out from the other side of the saloon door and dropped trow.

Unaware of the hijinks in the corner office and armed with the press release announcing that Tim Wakefield would start in place of Matsuzaka, Ganley entered the office without warning, saw the manager's bare backside, gasped, then turned and left. Francona never saw her. He had his back to the room. But he heard the door slam while Epstein and his guys were laughing.

Pulling up his pants, the manager turned and said, "Was that Pam?"

"She just left," said the young executives, all doubled over.

Francona bolted from his office, walked quickly across the clubhouse and down the stairs to the tunnel connecting the locker room to the first-base dugout. There he found Ganley, sitting alone on the bench, staring out at the field.

"Pam, I am so sorry," said the manager. "We were just goofing around."

"Don't talk to me," said Ganley. "I can never look at you again. I am never going to be the same."

"There were some human resources concerns going on in the back of my brain, but it was pretty freakin' funny," said Epstein.

Fortunately, the moment passed and Ganley recovered nicely. No need to report to HR. Like many of the young employees at Fenway — the ones who worked long hours for little pay — Ganley was treated royally and respectfully by the manager of the Red Sox. Ever-mindful of his upbringing around big league clubhouses, Little Tito always took care of the people behind the scenes.

While folks in baseball ops and the manager's office struggled to win games with a depleted roster, the Sox owners and folks at NESN were trying to identify why ratings were soft. On July 21, while the ball club was on the West Coast, NESN officials met at Fenway with Werner, Lucchino, and a mortified Epstein for an emergency "Viewership/Team Interest Discussion." The young GM with two World Series championships on his résumé was forced to sit through a meeting that addressed "indicators of declining team popularity" and "possible factors to reduced interest." The television executives agreed to hire a market research and consulting firm to "access factors contributing to lower interest in the Red Sox in the 2010 season." The decision to hire the consulting firm "was evidence to me of the inherent tension between building a baseball operation the way I thought was best and the realities of being in a big market and this Monster which had gotten bigger than any of us could handle," said Epstein. "I thought it was evidence that that conflict was as intense as ever and it was probably inevitable that the business of the game was becoming pretty important."

"Theo was good about shielding me from that shit," said Francona. "There were days when he was a little grumpy, and I could tell somebody probably hit him from up above, but he was really good about that. I don't think he liked that shit about ratings at all."

Back in the dugout, there was no saving the 2010 Red Sox season. From July 4 through the final day in October, the Sox were a sub-.500 team. Francona was forced to use some bizarre lineups and continued to rotate shortstops. Infielder Bill Hall played some center field. Lowell was lurching toward retirement, while kids Yamaico Navarro, Daniel Nava, Ryan Kalish, and Felix Doubront were introduced to the big leagues. The Sox finished in third place, 89-73, seven games behind the Tampa Bay Rays. They sent 19 players to the disabled list and lost 1,013 man-games to broken bones and torn ligaments.

The Lackey signing was not a hit. In his first season in Boston, the big righty went 14-11 with a 4.40 ERA.

"It was a matter of coming into this division with smaller ballparks and deeper lineups," said pitching coach Farrell. "With the Angels, he was making 22 of 34 starts at night on the West Coast in Seattle, Anaheim, and Oakland — the three best pitcher's ballparks in the American League. I don't think he was any different as a pitcher. I just think there were some circumstances that changed."

Run prevention was a bust. The Sox finished 12th out of 14 American League teams in fielding percentage. They made 42 more errors than the wild-card-winning Yankees. The Seattle Mariners, godfathers of run prevention in 2009, won a league-low 61 games in 2010.

John Henry was no longer immersed in his baseball team. The owner and his young bride welcomed a baby girl into their lives in late September, and Henry and his Fenway Sports Group were preparing to buy the Liverpool Football Club for $480 million. The Red Sox were no longer Henry's sole passion. Henry and Werner would spend much of the next two years flying across the Atlantic to tend to matters regarding Liverpool and the Premier League.

The final weekend at home against the Yankees was a public relations disaster for the Sox. Friday night's game was rained out, but fans were not sent home until 10:35 PM. Saturday's day-night doubleheader ended at 1:22 AM Sunday morning. Sunday's season finale was Fan Appreciation Day, and Sox players and the coaching staff greeted fans at the turnstiles. Fans were given nifty round magnetic calendars featuring the 2011 schedule.

It was not Manager Appreciation Day. Werner, privy to the results of the $100,000 study he'd commissioned to explore the Sox television ratings slump, passed Francona on the field during pregame activities and grumbled, "What a shitty season."

"That bothered me," said the manager. "We were up against a lot, and we ground out 89 wins. I was so proud of our guys. They played their asses off. I remember thinking, *Fuck, if this was shitty, I don't want to be around here when it really is shitty.*"

# CHAPTER 14

# · 2011 ·

## "I feel like I let you down"

THE 2011 RED SOX SEASON — which would prove to be one of the most disappointing and tumultuous campaigns in the 111-year history of Boston's American League franchise — was launched immediately after the conclusion of the injury-plagued, 89-win season of 2010. The Sox came up with a slogan that promised better days while acknowledging the disappointment of 2010:

"We Won't Rest Until Order Is Restored."

On Tuesday, November 2, just over a month after the Sox season ended, a group gathered at Fenway to review results of that $100,000 marketing research project the Sox had commissioned back in July. With Werner participating on speakerphone, Lucchino met with the bosses of NESN. Epstein, who'd been reluctant to participate in the study, attended the meeting.

The document distributed to all participants stated that the "research objectives" were "(1) to access factors contributing to lower interest in the Red Sox in the 2010 season" and "(2) to understand factors contributing to less viewing of Red Sox telecasts in the 2010 season."

Listed among the reasons for "lower interest" in the 2010 Red Sox:

- Disappointing news and moves in the off-season; not spending the money to get big players
- The team's positioning of itself as "pitching and defense" after not making "big" trades and acquisitions, and the characterization of 2010 as a "bridge year"
- Not delivering on pitching and defense in April
- Suffering injuries and playing with a "no-name" lineup going into and beyond the All-Star break

On page 28, a section dealing with male-female demographics, the report stated: "The women are definitely more drawn to the 'soap opera' and 'reality-TV' aspects of the game. . . . They are interested in good-looking stars and sex symbols (Pedroia)."

The team-sponsored survey concluded that fans were watching less because "the games are too long with disappointing outcomes." At the top of the list of "key take-aways" was the recommendation: "Big moves, trades, and messaging in the off-season are important."

There was little nuance in the survey. No ambiguity. NESN's in-house memo was telling Epstein and his baseball operations staff what was needed to reverse the costly downward trend in Red Sox television ratings: star power.

Epstein was insulted, amused (Pedroia sexy?), and angry as he sat through the session.

"They told us we didn't have any marketable players, the team's not exciting enough," he recalled. "We need some sizzle. We need some sexy guys. I was laughing to myself. Talk about the tail wagging the dog. This is like an absurdist comedy. We'd become too big. It was the farthest thing removed from what we set out to be.

"That type of shit contributed to the decision in the winter to go for more of a quick fix. Signing Crawford and trading for Adrian [Gonzalez] was in direct response to that in a lot of ways. Shame on me for giving in to it, but at some point the landscape is what it is. I didn't handle it well, but that kind of explains the arc of what we were doing."

"Theo never talked to me about any of that, and I appreciated it," said Francona. "I didn't want to know, and it's good that I didn't know."

Pressed by his bosses and the sagging ratings, Epstein went to work to build the transcendent team, a team that could win 100 games and a World Series, a team that would boost NESN's ratings, a team that would cement the legacies of Henry, Werner, and Lucchino as great owners. It would be a team that could make Epstein and Francona candidates for Cooperstown. Three World Series championships in eight years would make any baseball bosses Hall-worthy.

Everyone in the organization knew the Sox were going to lose two of their best hitters. They were not going to compete for free agents Victor Martinez and Adrian Beltre. The Sox did not want Martinez as their everyday catcher, and he wound up getting $50 million over four years from the Tigers. Beltre signed with the American League Champion Rangers for $96 million over six years.

Epstein had a plan. He was going to make a trade for Adrian Gonzalez, a hitter he'd coveted since 2000 when Gonzalez was a San Diego high school senior and Epstein was a 26-year-old assistant director of baseball operations with the Padres. Caving to the pressure of his bosses, Epstein was willing to trade coveted prospects in order to win immediately. He knew he would be allowed to spend freely. He would be allowed to compete for Crawford and Jayson Werth, considered the best position players on the market. He could even lavish $12 million on a setup reliever (Bobby Jenks). He was going to make the moves that would satisfy the focus groups and the talk shows.

Francona was not at the NESN war-room meeting and did not know that Epstein was charged with the task of making the Red Sox more interesting. He did not involve himself with ancillary issues. He barely noticed when John Henry ponied up $480 million to buy the Liverpool soccer team. He was not listening to the radio when Werner went on WEEI, the Sox flagship station, and said, "I'm concerned about perception. . . . I think we are going to sign a significant free agent. . . . We've got our eyes on a couple of people."

The manager was back at Massachusetts General Hospital for another knee replacement, this time the left knee. Boston's non-playoff status gave Francona the opportunity to plan his surgery for mid-October. Because of the manager's history and meds, it was a complex process. He had to wean himself off his blood-thinning medication and regulate his blood levels, which required a week of presurgical hospitalization. Like all patients dealing with anesthesia and pain medication, Francona came out of the surgical process with a gap in his awareness of everyday events. It had happened in 2006 when Epstein resigned, and it happened in 2010 when Theo was planning to build the super team.

"I missed a lot of stuff again," said the manager. "Maybe that was good."

The surgery and the urgency of the off-season coincided with a difficult time in Francona's personal life. By the time he checked into Mass General for surgery, he'd moved out of his Chestnut Hill home. His marriage of almost 30 years to Jacque Lang — the pretty girl from math class at Arizona and the wonderful mother of his four adult children — was dissolving. Too much devotion to baseball, too many nights on the road, maybe too many nights on that Pesky couch in his office, had taken a toll. Nick had graduated from Penn and was a lieutenant in the US Marines. Alyssa had graduated from North Carolina and was working at Boston College, living in her own apartment. Leah was getting married to Marine Michael Rice, one of Nick's best friends. Jamie was a star volleyballer at Brookline High, on her way to a coveted spot in the US Naval Academy class of 2016. Their dad was making more money than he'd ever dreamed he'd make — and had hopes for another two years if he got his contract extended through 2013 — but Terry and Jacque had reached the difficult crossroads of a long, happy marriage. The manager of the Red Sox left his Chestnut Hill home and moved into the Brookline Courtyard Marriott Hotel near Coolidge Corner, where he would live throughout the 2011 season.

By late November, he'd rebounded from the knee replacement and joined Epstein for a recruiting road trip to Houston and Chicago a week before the annual baseball meetings at the Walt Disney World Dolphin Hotel in Lake Buena Vista, Florida, near Orlando. The Sox GM and manager first went to Houston to meet with Crawford. At Epstein's request, Allard Baird, the Sox vice president in charge of professional scouting, had followed Crawford extensively in the second half of the 2010 season and contributed detailed observations to a 50-page report that included extensive statistical analysis and a comprehensive background check. It was widely believed that the Angels had the inside track with Crawford, but Epstein wanted to measure the outfielder's interest in Boston and thought Francona would make a good recruiter. On November 30, Epstein and Francona met with Crawford and his agents, Greg Genske and Brian Peters, in the agents' offices near Rice University. Francona had managed Crawford on a Team USA unit that toured Taiwan in 2001 and, like everyone else, was impressed with Crawford's speed, athleticism, and sincerity. Crawford had run wild on the Red Sox during Francona's tenure in the Boston dugout. The meeting went well.

"Theo and I did some of our best work on those trips," said the manager. "Theo didn't want us to be the team that they used for leverage. The meeting was very informal, but Carl did a great job of presenting himself. I didn't see Boston as a deterrent for him. When we walked away from that one, Theo asked me what I thought, and I said I thought it went great."

Leaving Houston, Epstein and Francona received a message from Scott Boras indicating that Werth, a Boras client, would be available for a meeting at a hotel near O'Hare Airport in Chicago. The Sox bosses arrived several hours earlier than the free agent outfielder and found an old-fashioned steakhouse bar near the O'Hare Hilton.

"We had a great time at the bar," said the manager. "I think that might have been when Theo made the offer to Mariano Rivera. We got about ten deep, and then Werth showed up and we had an even better time. We knew we couldn't match the money he was going to get. [Less than a week after the meeting, Werth signed with the Washington Nationals for $126 million over seven years.] Later in the winter I got the nicest note from Jayson. He said he would have loved playing for us, but he couldn't turn down that money."

After the free agent meetings, Francona went to Fort Myers for the weekend while Epstein returned to Boston. On Friday, December 3, Adrian Gonzalez and his agent, John Boggs, flew to Boston on Henry's private jet. The Sox put them up at the Mandarin Oriental Hotel on Boylston Street.

Epstein needed to accomplish three things: he needed to work out a deal with San Diego GM Jed Hoyer; he needed to make sure Gonzalez was healthy (Gonzalez had undergone shoulder surgery after the 2010 season); and he needed to work out a contract extension with Gonzalez.

Dealing with Hoyer was probably the easiest part. This was the Jed

Hoyer who "ate his ass off" the night Francona first interviewed for the Red Sox job back in November of 2003. Hoyer needed to move his best player because San Diego could not afford Gonzalez after his "walk" year in 2011. Hoyer knew who the best Sox prospects were because he'd been part of the Boston operation. He knew Theo didn't like parting with top prospects, but the pressure in Boston was great. Epstein was ready to deal, and Hoyer had the experienced hitter Boston coveted.

With Gonzalez in town, the Sox scheduled a series of physicals and an MRI. The slugger passed every test. On Sunday, December 5, the Sox met with Gonzalez at Fenway to discuss the contract extension. Talks broke down early in the afternoon, and Gonzalez returned to the Mandarin. Talks resumed Sunday evening, and late that night the ball club released a bulletin stating that it would have a "major announcement Monday at 11 AM."

On Monday, December 6, with Francona and most of the Boston baseball media gathered at the winter meetings in Florida, Epstein — still in Boston — announced that the Sox had acquired the 28-year-old Gonzalez for prospects Casey Kelly (pitcher), Anthony Rizzo (first baseman), Reymond Fuentes (outfielder), and a player to be named (outfielder Eric Patterson). The key to the deal was the ability of the Sox to work out a seven-year, $154 million contract extension for Gonzalez, an extension that, in order to save millions in luxury taxes, would not be announced until after opening day of 2011.

It was the grandest Red Sox player acquisition since Dan Duquette signed free agent outfielder Manny Ramirez to an eight-year, $160 million deal in December 2000. The press conference to make the announcement was held at Fenway on Tuesday, December 7.

Epstein wasn't done. Still in contact with Crawford's agents, he flew to Florida, where Francona waited.

"It was exciting for us," said the manager. "Youk was on board with moving from first base to third, and Gonzalez gave us a legitimate bat in the middle of the lineup. It was a good way to start the meetings."

Two nights later, Epstein closed in on a deal for Crawford. He had to work quickly because the Angels were bidding for Crawford and the Yankees and Rangers were at a make-or-break point with lefty Cliff Lee. The Sox feared that the loser of the Lee sweeps would set its sights on Crawford.

"There were about ten of us in a suite at the Dolphin, and there was a lot of discussion about Carl," said Francona. "Theo was on the phone in another room, and he came out and said, 'We can do this. This is what it's going to take.'"

What it was going to take was $142 million over seven years. Werth's deal with the Nationals had set the bar.

Late at night in Florida, Epstein called Henry and Werner, who were in England overseeing the new Liverpool soccer acquisition. The GM had to wake his owners after 3:00 AM (Liverpool time) to get permission to make the deal.

"I was in bed," remembered Werner. "I thought there was no deal and that Crawford was going to the Angels. Theo said we had a very limited amount of time to make a decision and he felt very strongly about it. I asked him if Larry and John were supportive, and I said that I was if they were."

When he was reached in Liverpool, Henry granted permission (though he would later say he was never in favor of the deal). Crawford's contract package was one of the ten largest in baseball history.

"That ended the meetings on a high note for all of us," said Francona. "I was ecstatic. Carl had always killed us. It was a lot of money, and it made us heavily left-handed, but I thought this guy would put us over the edge to be a team that was athletic. With him and Ellsbury together, holy shit. All that speed was what we had been fighting, and now we were going to be that team."

The Sox outbid the Angels by $34 million in the Crawford deal. Henry had committed $296 million to two new ballplayers. Tickets for the 2011 Red Sox season went on sale Saturday, December 11, the day Crawford was introduced to the Boston media at Fenway Park.

Ten days later, Epstein was back at it, signing veteran White Sox closer Bobby Jenks to a two-year, $12 million contract. It was a curious acquisition. The Sox already had Jonathan Papelbon and a potential emerging closer in Daniel Bard. Jenks had been a World Series–winning closer in 2005, but his skills had waned, he had health issues, and there were whispers of personal problems.

*We Won't Rest Until Order Is Restored.*

"It was my fault," said Epstein. "I fucked up by giving in to that. There was always a tension between the scouting and development approach and what I call 'The Monster.' 'The Monster,' especially after we won the first time, was that we had to be bigger, better. There had to be more, more, more. We had to push revenues. It became a bit of a distasteful, self-congratulatory tone to some of the things we were doing as a franchise. It's hard to take winning and translate that into a day-to-day modus operandi for the club. There was always an inherent tension between what we were good at, what we wanted to do — the long-term approach — and this Monster. Talk about the arc of the decade. I think our group was really good at fighting that Monster and being true to our approach in the early and middle years, then toward the end — and I blame myself for this — we sort of gave in to it. Seeing the reaction we had when I mentioned the 'horror' of seeing young players develop, seeing the impact it might have had on revenues, and having some discussions with the business people. There came a point where we were almost too big and I lost my willingness to cling to that patience and the approach I thought made us good. I thought we gave in and tried to take the shortcut, and I don't think there are any shortcuts in baseball. We tried to take a shortcut by throwing money at some problems, and the irony is that that led to even more problems."

No one foresaw any problems when the Red Sox looked ahead to 2011. There was only more emphasis on revenue streams that would match the inflated payroll and expectations.

In this spirit, Lucchino called on his manager for an unusual errand in the winter of 2010–2011. Henry and Lucchino wanted to move the right-field bullpens closer to home plate. It was a radical idea, given that Fenway's outfield dimensions had been untouched since the bullpens were installed for the 1940 season. Tom Yawkey had built the bullpens after Ted Williams's rookie season of 1939. The owner wanted to boost the Splendid Splinter's home run total, and the new bullpens were immediately dubbed "Williamsburg." Oddly, Williams's home run total dropped from 31 to 23 in the first year after the bullpens were built.

The notion of moving the bullpens closer to home plate was something Henry had contemplated for several years. During Sox home games in 2010, the owner asked guests what they thought of the idea and how they thought it would be received. The official reason for changing the configuration was alleged complaints from bullpen occupants who held that the bullpens were not wide enough to support double-barrel action without endangering everyone in the pen. The unofficial reason was that the Sox owners wanted to reconstruct the right-field area and cram a few more seats into Fenway Park.

Changing anything inside Fenway was complex. The Sox had successfully petitioned to have their ballpark declared a designated landmark, a distinction that earned the owners tax credits but required them to get permission when they wanted to make renovations. In January 2011, at the request of Lucchino, Francona appeared before the Massachusetts Historical Commission to explain why the bullpens should be wider in the interest of player safety. The manager agreed that the bullpens were too small and was interested in making them bigger to accommodate and protect his players.

"I went there and made the case that someone was going to get hurt, but this lady on the commission was having none of it," Francona recalled. "She was having a bad day to begin with and didn't care who I was. As I was telling her, I could tell she wasn't buying it. She said, 'Do you expect me to believe that this is why you're doing this when you're adding 400 seats?' She laughed us out of the room."

Not willing to take a chance on losing tax credits, the Sox decided the bullpens were not particularly dangerous after all.

"I went to Larry, just to break his chops," chuckled Francona. "I said, 'What about the bullpen safety?' And he said, 'Ah, they'll be all right. They can be careful and work around each other out there. They'll figure it out.'"

Francona was entering the final season of a deal with a two-year club extension that had to be triggered within ten days of the end of the 2011 season. Given Francona's success, his seven years on the job, and the expectations for the 2011 ball club, there was remarkably little talk about the lame-duck status of the team's field manager. Folks close to Henry knew that the owner had had the manager in his crosshairs for a couple of seasons. It would be impossible for any field boss to consistently satisfy and support the data Henry loved.

"A New York writer asked me about my contract the first day of spring training," said Francona. "He tweeted that my contract was going to be picked up the next day, and I actually got a little excited about it, but nothing happened. If anyone asked me about it, I deflected it. I had told them I wouldn't bring it up, so I didn't. I wanted them to appreciate that, but I don't think they did. This was when I was starting to feel that maybe they weren't that big on me. When it was all over later, I told Larry, 'I could have made things a lot tougher on you guys.'"

He stayed in a condo at the Miromar Lakes country club in February of 2011. It was his first year away from the Bell Tower's Homewood Suites. Miromar had a pool Francona could use in the dark hours of morning before driving to the Sox minor league complex.

The manager's recently married middle daughter, Leah, came to visit for a few days with her husband and was disturbed when she came across a bottle filled with as many as 100 Percocets. Like everyone else around her father, Leah Francona knew her dad took pain medication. He had considerable history with pain pills and joked about it regularly.

The manager of the Red Sox had undergone an extraordinary number of surgeries in his 51 years. The most recent knee replacement followed the 2006 knee replacement, knee scopes, knee reconstructions, cervical disk surgery, and numerous wrist, elbow, and shoulder surgeries. He'd cheated death during the Christmas season of 2002, surviving a pulmonary embolism on each side of his lungs, as well as subsequent blood clots, staph infections, massive internal bleeding, and the near-amputation of his right leg. He had a small metal device (a Greenfield filter) implanted into his vena cava vein to prevent clotting. He was unable to jog and would be on blood-thinning medication (Coumadin) for the rest of his life. He still wore sleeves on both legs, and still got cold easily. Anytime he sat too long, his legs swelled and needed to be elevated. He had a hard time remaining comfortable. Blood-level maintenance and pain management would be part of his daily life for as long as he lived.

The vial of Percocets had been stockpiled over a long period of time. "I saved 'em up," said the manager. "I had hoarded them."

Francona's daughter was concerned that his pain was not being carefully managed and asked him to consult with Dr. Larry Ronan, Red Sox head team internist since 2005. Francona knew Ronan well and trusted him.

"I had that bottle, and Leah was worried," said the manager. "It was legal, but it wasn't good. I didn't even open 'em. It wasn't under the Red Sox umbrella. She knew what I'd gone through, and she wanted me to go to Larry [Dr. Ronan], so I did. I told Larry the truth and that it was no big deal. I didn't want to lie to him. I told him, 'I have these, didn't open them, but I like the idea of having them if I need them.' I wanted to be up-front. He said he'd keep an eye on me. The next day he said he wanted to tell Theo. He told me, 'I know you're okay, I see your eyes, but I want you to meet with somebody, a pain management guy.' He said he had to document this."

As Epstein recalled, "I got a call from Dr. Ronan telling me what happened and what he thought we should do about it. We talked about how we could handle this in a way that fulfilled our responsibility to the organization, protected Tito, and, most importantly, protected Tito's confidentiality. We had to limit the amount of people who knew about this and get Tito the help if he needed it. We had to alert MLB that something was going on, but do it without mentioning Tito's name. Terry's name was not specified in our report to MLB. It was just reported that there was a staff member who had an issue. And that was pretty much it. I have tremendous trust and faith in Dr. Ronan. He's one of the special people in the world. So I felt like as long as he was the point guy handling it, that Tito would be in good shape and we'd ultimately be covered too."

Francona agreed to participate in the MLB program and see a pain management specialist several times per month during the upcoming season. Still, he was uncomfortable with the arrangement and worried about his privacy.

"It was just Theo, Dr. Ronan, and me with that agreement," said Francona. "I remember Larry [Dr. Ronan] looking at me as a friend and saying, 'Tito, nobody outside of this room will ever know.' And I said to him, "This will come back to bite me in the ass. I know how shit works here. This will fuck me someday."

According to Major League Baseball's executive vice president Rob Manfred, the only people authorized to know the identity of an individual in the MLB drug program would be the three-man MLB drug policy oversight committee (Manfred, MLB drug abuse consultant Dr. Larry Westreich, and Jon Coyles, director of MLB's drug policy) and the "employee assistance professional" at the participant's own ball club. The employee assistance professional for the Red Sox in 2011 was Dr. Larry Ronan. Under the terms of the program, Francona met with a pain management specialist a couple of times a month, usually at Ronan's office.

Lucchino said that ownership was initially unaware of Francona's participation in the program, adding, "Later in the season I became aware of it, but not earlier in the season."

"They weren't supposed to know," said Francona.

There didn't seem to be much else to worry about. Spring training 2011 was a lovefest.

On Saturday, February 19, Henry, Werner, and Lucchino made their way to the Sox minor league complex in Fort Myers for the annual full-squad team meeting. Minutes before the owners walked in for the start of the annual meeting, Francona had a word with Varitek.

"'Tek, these guys just emptied their wallets for us over the winter," said the manager. "They dug deep. Maybe we should show them some appreciation."

"I'll take care of it," said the 38-year-old catcher/captain who'd been with the Sox since 1997.

"'Tek was such a leader," said Francona. "He was our captain before we ever put the 'C' on his jersey."

When the owners walked into the lunchroom, Varitek popped up out of his chair and started applauding. Everyone followed the captain's lead, and Henry, Werner, and Lucchino smiled and bowed.

"I think this was a sort of spontaneous appreciation for our commitment to the organization," said Werner.

"I think they appreciated it," said Francona. "They'd just spent a lot of money, and I wanted to let them know that we didn't take it for granted."

After the meeting, the owners took questions from the media. Werner noted that the Sox had 16 players on their roster who had been All-Stars at one time or another. Henry said, "This off-season was tremendous." Asked about Crawford, Henry said, "He's the right player for us," words he would come to regret. Lucchino said the Sox paid $85.5 million to Major League Baseball's revenue-sharing fund in 2010 and another $1.3 million in luxury tax. The Sox 2011 payroll came in at $161 million, one of the top three payrolls in baseball.

"We come to camp with a sense of high expectations," said Luc-chino. "There's very definitely a sense of confidence, a sense of opti-mism, a sense of what could possibly be, and you feel it when you first walk into the camp."

The players were no less enthused. Beckett said that he'd never been on a team that won 100 games. (The last Red Sox team to win 100 was the 1946 team returning from World War II.)

"I think it's important to remember that we haven't done anything yet," said Epstein. "We didn't even finish second last year. . . . Baseball is such a humbling game. If you get ahead of yourself even for a minute or two, it kind of knocks you right back down on your backside."

"As good as people thought we were, I thought we were going to be even better," Francona said later. "I thought the pieces fit. I was really excited. Adding Adrian Gonzalez. Adding Carl. I thought, *Wow, this is going to be unbelievable.* We were going to be fast and athletic. Defen-sively, the balls weren't going to hit the outfield grass. I was thrilled."

When the spring games started, the manager was presented with the annual problem of building a team while splitting squads for two- and three-hour bus trips in southwest Florida. The Red Sox Fort Myers site was a logistical problem. Many major league teams had migrated to Arizona or the east coast of Florida, and the Sox faced long bus rides anytime they were not playing the Twins, who trained on the other side of Fort Myers.

"I never felt like I could get my hands on the team enough because of where we were in Fort Myers," said Francona. "Every trip is a bitch. You can go two or three days without seeing people, and it's hard. It's a little bit of a disadvantage as far as getting your work done. I used to tell Theo, 'If I take what I am supposed to on the road, it's gonna fuck us up.' Theo was inevitably going to be getting a complaint from the league because we weren't taking a lot of regular players sometimes. He always said he'd handle it, and I really appreciated that. I'd leave guys back to get their work done. Guys are supposed to be getting ready for a season, and when you put them on a bus for two hours for two at-bats, it doesn't help 'em. We had a lot of new guys in 2011, and that worried me as far as getting ready for the season.

"Expectations are great. That's how you're supposed to feel. That's the way spring training is. But there's a process. You can't go from one mile an hour to 100 miles an hour. You've got to go through spring training and prepare and pay attention to detail. I understand expectations are great, and that's okay, but if you skip the process, you got a chance to have some problems. In 2011 I felt just like I felt after we won in 2007. I had to put up a fight to make sure they knew — if we don't win, all that other stuff isn't going to be there. I wasn't excited about talking about winning 100 games. Those are dangerous things to say. You're forgetting about the process."

*Sports Illustrated* picked the Sox to win 100 games and beat the San Francisco Giants in the World Series. The magazine also projected Crawford as 2011 American League MVP. Six of six *Boston Globe* "experts" picked the Sox to finish first, but more unlikely respect came from the *New York Post,* where seven of seven "experts" picked the Sox over the Yankees.

"We've got a World Series kind of team when you look at our 25 guys," said Clay Buchholz, who had won 17 in the "shitty" season of 2010. "There really aren't any weak spots."

The inimitable *Boston Herald* cut through all logic and presented the 2011 Sox with the headline "Best Team Ever."

And then they lost their first six games and stumbled to a 2-10 record over the first two weeks.

Crawford batted in the number-three spot in the lineup in the opener in Arlington, Texas, and went 0-4 with three strikeouts, leaving five runners on base. The next day he went 0-3 with one strikeout. Francona moved Crawford to the seventh spot in the order for the third game. The Rangers outscored the Sox 26–11 over the three games and hit 11 homers.

It did not change when they moved on to Cleveland. The Sox were swept by the Tribe, losing the series finale, 1–0, on a suicide squeeze in the eighth inning. Matsuzaka and Varitek got into a verbal dustup in the dugout in Cleveland. The 0-6 start was Boston's worst start since 1945, when Ted Williams and friends were still at war.

"It can't get any worse than this," submitted Kevin Youkilis.

The 100th Fenway opener was scheduled for Friday, April 8. Epstein and Francona both addressed the team before the players took the field. It was an interesting juxtaposition for Sox players. The Yale-educated GM dazzled players with Churchillian rhetoric, then yielded the floor to the tobacco-spitting manager, who peppered his rant with F-bombs.

"I tried to keep it simple and direct," said Epstein. "I just tried to appeal to the belief that we can and will get back in this and that adversity is an opportunity for us to define ourselves by how we respond."

"Sometimes I wasn't sure how I felt about the GM speaking to the team, but anytime there's a show of support I'm okay with it," said Francona. "Theo started off, and he was really good. He was more comfortable around the players at this point, and he did a good job. Some of it went over my head. A lot of stuff about 'the measure of a man.' He was up there quoting Martin Luther King. It went above me. When he got done, I stood up and I was just up there dropping 'fucks' on everybody. I guess you could say I wasn't quite as eloquent as Theo."

"It was classic," said Pedroia. "Theo just up there talking about historians and I didn't even know what was happening. I needed a five-minute break to put it all together. For players who are uneducated like me, Tito's speech was better."

"Going 0-6 in Boston was unacceptable, and I knew that," said the manager. "I knew they were going to be hearing it. I told them, 'Hey, you're going to look up at the scoreboard for a while and not like what you see. You have to be patient. People are going to be asking you every day, and it's going to wear on us if we let it. We've got to be strong enough to fight through it. Right now, everyone is taking their shots at us. We'll pull through this, but you've got to be patient.'"

The Sox responded with a 9–6 victory over the Yankees. Carl Yastrzemski, the greatest living Red Sox player, a man who played dozens of games against Tito Francona, threw out the ceremonial first pitch and was gone before John Lackey took the mound to face the Yankees. Yaz was famous for throwing out a first ball, then bailing before anyone could bother him.

Lackey picked up the win. Crawford was the leadoff hitter, went 0-5, and saw his average drop to .143. He'd batted in four different spots in the order in the first seven games. Two weeks later, twenty games into the season, Crawford had batted first, second, third, seventh, and eighth and was hitting .135.

"We talked to Carl on the plane and at his locker," said Hale. "Carl was aware, but he was struggling. Fuck, we were losing. We needed to do things to get him going, and I told him, 'When you get going, we'll move you up.' Tito wanted guys to play with confidence and aggressiveness. Don't be afraid to make mistakes. That's how Tito got the best out of players."

"I never called Carl into my office," said Francona. "Like a few other guys, to him, that would have felt like going to the principal's office. Moving a guy around the lineup goes against everything I believe, but this was hard. I asked him where was his favorite place to hit, and he said he hadn't hit leadoff much, but would do it. He was always fine about things, saying, 'It's okay. I'm cool.' He was just tentative all year. He was jumpy up there. I think he was afraid to make a mistake. It was the same in the outfield and on the bases. I'd tell him to go ahead and run: 'If you're out, I'll tell people I sent you.' Fuck, he wouldn't go."

In his only full season with the Red Sox, Crawford hit .255, with 11 homers, 56 RBI, and only 18 stolen bases in 24 attempts. His on-base percentage was a staggeringly low .289. He was never the player who tortured the Sox for eight years.

"He was a little unique in the way he hit," said Francona. "He was always open, but he started to open more. To close he had to rush, and he couldn't get his foot down in time. He started swinging at balls in the mitt, and it snowballed. He was swinging at everything, trying to go two-for-one and three-for-one. You could see he didn't have his comfort zone all year."

While the Sox slowly climbed toward .500, Francona was dealing with physical and emotional issues.

His replacement knee gave him new problems. He'd learned how to manage the swelling and pain, always wearing multiple layers and wraps. (It was not unusual to see Francona with his legs propped up somewhere in the dugout or in the clubhouse at the end of batting practice.) The Percocets helped, but sometimes the pain and swelling were overwhelming. In the first inning of a game at Toronto on May 10, Francona and Sox trainer Mike Reinold walked over to the Blue Jays' clubhouse training room, where a Toronto team doctor drained Francona's knee.

"I looked over and didn't see him in their dugout," said Blue Jays manager Farrell, one of Francona's best friends. "I didn't know he was right behind me in our room. Physically, he went through a lot."

Five days later, Francona had the same procedure done by a Yankee team doctor at Yankee Stadium.

"People that were around knew what was going on," said the manager. "I had my knee drained four or five times in two weeks. I was so hurtin'. They were worried about infection, and I had trouble sleeping. When you have that much blood in there, it hurts."

Twice a month, he dutifully visited with a pain management specialist at Massachusetts General Hospital.

"It was pretty casual," said Francona. "Larry [Dr. Ronan] would be there and pop his head in. I explained to the pain doctor what I'd been through, that I'd had surgery. But the whole thing has taken on a life of its own. I didn't see the pain management guy until we got to Boston, and Larry and I had first talked about it halfway through spring training. If I was having trouble, you think they'd have waited a month?"

Pain and insomnia were not relieved by Francona's living situation. Few people knew that Francona was no longer living at home. It wasn't something he talked about with his ballplayers or Sox officials, but it was a time of personal turmoil in the manager's life. Homestands became like road trips as the manager padded around his hotel room at the Marriott Courtyard, always keeping the television on.

Living with the white noise of television was something Francona did throughout his adult life. And he was not strictly an ESPN viewer. He liked to watch the History Channel and CNN's *Headline News*, but in the summer of 2011 Francona avoided news programs. US Marine lieutenant Nick Francona was serving a six-month tour as a rifle platoon leader in Afghanistan, and Leah's husband, Michael Rice, was also in Afghanistan, dismantling homemade bombs. News reports from the war zone only made the manager worry.

"It was awful," he admitted. "No getting around it. I thought about it all the time, worried all the time. I tried to stay away from the news, but it's always there."

For the first time in his career, Francona kept his cell phone in the dugout runway during games. He'd sometimes check it between innings.

"I don't know why I did that," he said. "Who was going to call me if something happened to my son? It was just an uneasy feeling for me. I'm sure people saw me checking it."

Months later, when he was out of a job, the cell-phone habit might have been used to hurt him when an anonymous club source told the *Boston Globe* that Francona appeared distracted during the season.

It was actually the opposite. Coping with a dissolving marriage, family members at war, and severe pain and insomnia, Francona sought refuge at the ballpark. He went to work earlier than in any of his previous years in Boston. The clubhouse guys — Pookie, Murphy, Joe, Tommy, John, and the rest — became almost like family members. He developed a tighter relationship with bench coach DeMarlo Hale, but missed Mills and Farrell, who were managing the Astros and Blue Jays, respectively, thanks in part to Francona's recommendations.

"I knew he was living in a hotel, but I respect people's privacy," said Hale. "He knew his responsibility. He knew what he had to do on a daily basis.

"Tito was the same in 2011 as he had been every year before that. I remember walking to the ballpark in Pittsburgh with Tito and asking a question about Nick, and I saw his lip tighten and I knew he was worried. It didn't seem like he wanted to talk about it, so I never brought it up again."

More than most managers, Francona relied on coaches as go-betweens.

"He used us to put out some fires," said Hale. "We all knew that was part of the job."

Coaches could play good cop. Coaches could have a voice. Hitting coach Dave Magadan was a trusted instructor with a voice. Third-base coach Tim Bogar was considered a manager-in-training by Boston baseball ops. Francona enjoyed working with both, but the manager and his players missed the strong presence of Mills and Farrell. New pitching coach Curt Young was smart, affable, and experienced, but didn't have the command of the John Wayne–like Farrell.

"Losing John Farrell was huge, and we knew it was going to be huge," said Francona. "We knew we'd better fill this with someone who knows what they are doing. Theo met with Curt in the Fall League and hired him. Curt's good. He's got a different mentality than John. He's not as take-charge as John. You'll never find a guy more upbeat. During the first week of the season, Curt would be the one saying, 'Tito, it's going to be okay,' and I'd be like, 'Fuck, no, it isn't. You don't know this place.' Curt gave our pitchers more latitude than John did, and I think they took advantage. He was a good hire, but he got caught up in it. If guys didn't listen to him, it's an excuse. Shame on them."

The starters were Beckett, Lester, Lackey, Buchholz, and Wakefield. Buchholz admitted he was afraid of Farrell. No one feared nice guy Curt Young.

"I talk to the pitchers at the beginning," said Francona. "We had a lot of veterans that had earned the right to do things on their own. I told them, 'Don't abuse it. I trust you.' I'd tell guys, 'Follow Beckett. He works his ass off.' I'd always tell 'em, 'If you think you're fucking up, you probably are.' For whatever reason, during the year, I wasn't comfortable."

They also had to break in a new catcher. Long coveted by Epstein, switch-hitting Jarrod Saltalamacchia was acquired at the deadline in 2010 and anointed the everyday catcher for 2011. Varitek still caught Beckett almost exclusively, but Salty was the regular, and he could never be expected to match Varitek's game-calling.

"The guy that followed 'Tek was going to be at a disadvantage no matter what," said Francona. "There was a comfort zone pitching to 'Tek. We always called all the pitchouts and slide-steps from the bench, but we never called a pitch in eight years with 'Tek. There were times we'd help out Salty."

The abysmal Red Sox of April clawed their way back to .500 in mid-May and went over the break-even mark May 16, with Alfredo Aceves getting the win.

"Now we flip the tortilla," said Aceves.

Indeed. For 123 games—four and a half months—the Red Sox went 81-42, posting the best record in baseball. They went 20-6 in July, their best month since 1952.

In June, July, and August of 2011, the Boston Red Sox were the monster team that almost everyone predicted they would be. Gonzalez posted MVP numbers. Ortiz found his old confidence and crushed righties and lefties. Ellsbury rebounded from the broken rib disaster of 2010 and made a legitimate run at the American League MVP Award, hitting .321 with 32 homers, 105 RBI, and 39 steals. Batting in the second spot most of the year, Pedroia hit .307 with 21 homers and 91 RBI.

The Red Sox led the majors in runs, hits, doubles, extra-base hits, and slugging.

Francona observed a nice moment in Philadelphia at the end of June. Nice guy Cameron was at the end and had to be "designated for assignment," which is baseball parlance for "We're letting you go."

"Theo handled that, and he was really great at it," said the manager. "It's always a tough thing to do, but Theo was very respectful. He had a way of making the player feel really good about himself, and with Mike it was all legit. It made me think back to when I was released by the Cardinals. That was awful for me, and seeing Theo talking to Mike made me wish I'd had someone talk to me like that back in 1991."

The Sox were in first place at the All-Star break. Taking advantage of three days off, Francona rounded up five of his favorite clubbies and drove to the Mohegan Sun casino in Connecticut. The manager paid for a suite and still laughs at the memory of Pookie Jackson lying on his back in the middle of the floor, snoring, with a slice of pizza resting on his chest.

"That was the Six Stooges," said Francona. "A true reality show."

Perhaps because the Sox were finally fulfilling expectations, there was virtually no discussion of Francona's contract. It was assumed he would be back in 2012 and beyond. Even though Sox insiders often heard Henry grumbling about his manager during Fenway games, a sixth playoff appearance in eight years and a possible third World Series championship would make Francona a lock to return in 2012 and probably longer. The manager's lame-duck status was not a topic for Boston baseball reporters during the summer of 2011. Things were simply going too well.

The impending disaster of September did not happen in a vacuum. There were signs all along the way. The Sox missed the veteran coaches who had commanded their attention. They had a lot of aging players in the final year of their contracts. They had players placing personal rewards above team success. On the bench, Francona didn't have players who could deliver messages for him — selfless types like Gabe Kapler, Alex Cora, and Eric Hinske. These Sox were looking out for themselves. They also had inattentive ownership — Henry and Werner were consumed with Liverpool, Roush Racing, LeBron James (FSG represented James) — and a general manager who was rumored to be going to the Cubs. They had a cluttered and nervous medical and training staff. And they had a manager who was working more hours than ever, but feared he was losing his voice in the clubhouse.

"I was worried about it all year," said Francona. "Somebody would strike out and go look at video instead of staying on the bench. We had a lot of guys who wanted to play every day. Scutaro would be pissed if he wasn't playing. Jed Lowrie wanted to play every day, but he kept getting hurt. I didn't want to overexpose Josh Reddick. David was in a contract year. Youk got hurt. There was just a lot of frustration with a lot of things. Without the voices of the coaches and veteran players, I was doing a lot more of that work, and the players were like, 'Fuck, man, where is this coming from?' It catches up with you."

The starting pitchers — particularly Beckett, Lackey, and Lester — were a disappointment. They always seemed to be together — in the clubhouse, in the dugout, away from the park. If Beckett came around a corner, Lackey and Lester were usually nearby. Following the lead of Beckett, Lester took on a tough guy persona, grumbled when he didn't get calls from umpires, and was brusque with the media.

All three of them had achieved ultimate big league success at a young age. All three — Lackey with the Angels, Beckett with the Marlins, Lester with the Red Sox — had been the winning pitcher in a World Series clinching game before the age of 26.

Beckett and Lackey were from Texas. Lester grew up near Seattle and settled in Georgia, but was wearing cowboys boots and speaking with a Texas drawl by 2011.

Of the three, Beckett had the best numbers in 2011. Rebounding from an injury-plagued (6-6, 5.78 ERA) 2010 season, he made the All-Star team and had one of the top five ERAs (2.89) in the American League.

"I think every pitcher looked up to Josh," said Farrell, a year after he'd left the Red Sox. "For anyone that was in uniform in the organization in 2007 and watched what he did, he earned that. He was the go-to guy, the guy that didn't back away from challenges. He relished that, and I think guys looked up to him. I don't think Josh abused that in any form or fashion. He wanted to be the guy, and he'd earned it."

Despite Beckett's 2011 numbers, Lester was considered the most dominating pitcher on the staff. He'd drawn the opening day assignment and was on his way to a 15-win season on the heels of 19 wins in 2010. He was also an American League All-Star in 2011.

"I didn't worry about Lester because Lester worked his ass off," said Francona. "He just wanted to follow Beckett and Lackey. I liked the fact that they were loyal to each other. I just wanted the whole team to be like that. It's how you handle frustration, and this team did not handle frustration. They made excuses, and that was troubling."

Lackey was in bad shape in every way. He had difficulty adjusting to the American League East and was on his way to a 12-12 record with a whopping ERA of 6.41. His wife was undergoing treatment for breast cancer, and their marriage was troubled. He had soreness in his pitching elbow and was bound for Tommy John surgery. He was often demonstrative on the field, waving his arms and making faces when teammates failed to make plays behind him. He was short with the media. Fans resented his $82 million contract. After a bad stretch early in the season, Lackey said, "Everything in my life pretty much sucks right now."

In the summer of 2011, wearing their uniforms at Fenway Park, Beckett, Lester, and Lackey (along with Clay Buchholz) participated in a country-western music video shoot for Kevin Fowler's "Hell, Yeah, I Like Beer!"

"We looked into it [video], and we found out that it was just a courtesy to them [the pitchers]," said Lucchino. "It was [approved by] someone who didn't have the decision-making authority on this. It was all very casual and never went up the chain of command, and we weren't aware of it until after this issue developed."

"I never even knew about it," said the manager.

Beer and baseball have walked hand in hand since the game was invented. Through the decades, many teams (St. Louis Cardinals, Toronto Blue Jays, Baltimore Orioles) were owned by beer companies, and no one ever objected to the name of the Milwaukee *Brewers*. Baseball players, coaches, and managers are on the road for 81 games, most of which are played at night. There is beer in the clubhouse, on the plane, on the bus, and in every major league city hotel. Baseball hours encourage the drinking life. Games are late, and participants can sleep late the next morning. Under Hall of Fame manager Earl Weaver, Baltimore Oriole ballplayers were not allowed to use the hotel bar. The hotel bar was home to Weaver and his coaches. Weaver reasoned that banning players from the hotel bar eliminated all possibility of late-night encounters between inebriated, disgruntled ballplayers and their manager. The Sox did not have any such rule.

Rules and traditions changed in baseball and across America in the 1990s and in the new millennium. Drunk driving fatalities, DUI arrests, and liability concerns prompted many major league ball clubs to limit or ban alcohol in the clubhouse. By 2011, 18 of 30 major league baseball teams had officially banned beer from the clubhouse (though it was well known that clubbies had access to beer, even in the "dry" ballparks). The Red Sox of 2011 were not among the clubs with a club-house alcohol ban.

Francona understood the clubhouse culture as well as anyone and believed in treating his ballplayers like adults. Privileges were extended with the understanding that the players would not take advantage of the system. The manager was proud of the fact that the Red Sox did not have a single DUI arrest in his eight seasons in Boston. He knew some of his players drank beer in the clubhouse after the game, and if one of the pitchers had a beer in the late innings of a game in which he was not involved, the manager wasn't going to make a big deal of it.

He remembered the words of the late Chuck Tanner, a teammate of Tito Francona and a World Series–winning manager with the Pirates in 1979. When Tanner made the trip to New Brighton for Birdie Francona's funeral in 1992, young Terry, then a first-year minor league manager, asked the senior skipper how he knew when to step in and when to look away when ballplayers misbehaved on charter airplanes.

"If I didn't turn around, I didn't see it," Tanner told 33-year-old Francona.

"When would you know to turn around?" asked Terry.

"If the plane was going down, I'd turn around," said Tanner.

Point taken. If you wanted to stick around as a big league manager (Tanner managed the White Sox, A's, and Braves in addition to the Pirates), you sometimes looked the other way. Tanner was not going to make a big issue out of a little beer in the clubhouse.

"I'm not saying it was right," said Francona. "But if somebody was drinking, they weren't drinking a lot. It was more disturbing for me to think that they would not protect each other, or one guy wouldn't tell another guy to knock it off. I wanted them to protect each other ferociously. If somebody felt that strong about it, fuck, they could have told me. Most of all, they needed to care enough about each other to stop it, because that's what good teams do. The chicken and beer didn't bother me, but I wanted them there together in the dugout, being all in.

"You noticed it on Sundays when football started. You'd lose the pitchers. I caught Jonny Lester one day. He came bounding into the dugout during our game, and I said, 'What's the score of the Jets game?' He was mortified. It just crushed him. That's all I had to say to him.

"They knew the rules. I always told them, 'The day you pitch, you're fully vested. The other days, there are gonna be days you don't feel like coming down to the dugout. But kick yourself in the ass, because that's the day the position player is going to need a little help.' That's what bothered me with this group. There was too much sitting around and not enough caring on the days they weren't pitching."

"I feel like I was in the dugout just as much as any other year," grumbled Beckett. "I have a hard time grasping what everybody else thinks. Maybe I'm just not smart enough to figure that out."

Francona knew his dugout was crowded late in the season with the September call-ups. He knew J. D. Drew, close to retirement, ever on the disabled list, was addicted to the hunting video game in the clubhouse. He didn't mind players occasionally ordering fried chicken from nearby Popeye's during games (a tradition started in 2010 by Mike Cameron, perhaps the best teammate and most fit player on the club).

"The fact that they ordered from Popeye's every once in a while, that doesn't matter," said Francona. "The fact that maybe guys were getting heavy, that does matter."

Late in the summer, he pulled Beckett aside for a talk — something he'd done just about every year Beckett was with the Sox.

"JB, don't forget the leader you are," said Francona. "Come on, man. I expect a lot from you. I know you are not going well now, and I want to make sure you are okay. You're not going late in games. You seem mad at the world. Let's get going now."

"A week later, I was looking at the same things," said Francona later. "That's where I thought I lost my ability a little bit. That talk normally got him back on track. I used to be able to get to him, and it wasn't working anymore."

"I don't think I took advantage of Tito," said Beckett. "He may think I did, but it's probably not coming from something that I did to him. It's coming from somebody else telling him that's how he should feel. I thought I was prepared to pitch every time I went out there."

Poor conditioning shed light on the ball club's training and medical staffs. Rising star Ellsbury was among the players who did not trust the Sox medical team, and the ball club was constantly overhauling systems of caring for the players. Dr. Bill Morgan, who performed the crucial surgery on Curt Schilling, was removed after the 2004 season. Trainer Paul Lessard was let go after four seasons in 2009. There would be a complete overhaul of the medical and training staffs after the collapse of 2011.

"We had conversations with Theo about what was going wrong before the last day," said Werner. "And we had conversations with Terry. . . . There were certain things that I was surprised about at the end, but we were aware of certain issues, like conditioning issues and medical issues."

"Our medical was all fucked up," said Francona. "There were more egos on the medical staff than there were on the team. Without Larry Ronan, it wouldn't have worked. Without him, we wouldn't have made it. People don't know how many fires we put out there. We tried to set this mold, and we broke it. We kind of outsmarted ourselves. It was terrible. I remember being really pissed when Paul Lessard was let go. Now they know. Some of us paid a price for that."

Late in the 2011 season, Henry talked about the methods of Liverpool soccer strength and conditioning coaches, hoping the Red Sox could adopt similar policies.

"They were explaining how their athletes punch in every day," said the manager. "The players recorded what they ate and how many hours they slept. John wanted to do that here. I said, 'John, are you shitting me? You can't do that with these guys. Just be happy they show up at the ballpark.'"

On August 3, Ortiz stepped up to the plate with runners on second and third and drove a single to left. Cleveland left fielder Austin Kearns did not field the ball cleanly and Youkilis scored from second.

Kearns was charged with an error, and Ortiz received only one RBI. The Red Sox won, 4–3, improving to 68-41 and holding their first-place lead over the Yankees.

The next day Francona was conducting his pregame press conference with the Boston media at Fenway when Ortiz burst through the door to the left of the stage, pointed at Francona as he was met by Ganley, and said, "I'm fucking pissed. We need to talk this out, you and me."

"I'll be with you in a minute," said Francona.

Ortiz turned and left, yelling, "Okay. Fucking scorekeeper always fucking shit up."

"I was fired up," Ortiz admitted later. "It wasn't like I was mad at Tito, but I needed to talk to him. I wanted him to talk to the scorekeeper because he's my manager. They knew they screwed up. I wanted my RBI."

Ortiz was correct in his assessment. Youkilis had never broken stride on his route from second toward home plate and would have scored on the hit regardless of Kearns's error. Ortiz was particularly sensitive about his numbers after he had finished with 99 RBI in 2009. Ganley petitioned the league and had the decision changed.

Video of the press conference incident went viral.

"I'm glad I brushed my hair that day," said Ganley.

Back in the clubhouse, Francona found Ortiz, and they met in his office, Ortiz taking a seat on the Pesky couch.

"David, if you want an RBI, I'll try to help you, but you did wrong," said the manager.

Pleading his case, Ortiz started, "But . . ."

"No 'buts,' David," said Francona. "You did wrong."

The moment came to symbolize the 2011 Red Sox. They would be remembered as individual stars who cared about themselves more than their team or their manager.

"It wasn't the end of the world, but if that bothered him that much, that was a warning sign for me," said Francona. "Guys were starting to think about themselves. It was a concern. I took a lot of pride in having teams where guys got mad if we lost, not if they went 0-4."

By this time the roundtable meetings involving Lucchino and the players — a staple of the Sox structure for most of the eight Francona seasons — started to ring hollow. It became difficult for Jack McCormick to find willing participants.

"I think the players were starting to think it was a little transparent," said Francona. "By the end, I was asking Jack to get a couple of coaches because you couldn't get players."

There was a long-distance player-management issue when the Sox played in Texas late in August. The Texas series concluded a stretch in which the Sox had played 14 of 17 games on the road with stops in Minneapolis, Seattle, and Kansas City. The schedule called for the Sox to return from Texas for a three-game weekend series against Oakland at Fenway, but Hurricane Irene was scheduled to slash through Boston by the end of the weekend. There was almost certain to be a rainout Sunday. Baseball ops wanted management to absorb the rain and reschedule any lost games for September 29 after the end of the regular season. The Sox did not want to lose the gate. (Lucchino was obsessed with getting all 81 home games played every season.) Lucchino called his manager and proposed that the Sox move up the Sunday game and play a day-night doubleheader on Friday.

"That was a tough one," said Francona. "We were not going to get home from Texas until six in the morning on that Friday. I was against it all the way around. I told them, 'This doesn't help us. I know you have other concerns and there are revenue things to think about.' I wasn't trying to be a fly in the ointment, but they asked me a baseball question, and I told them. Daniel Bard got into it because he was the player representative and they wanted to have a vote on it. The players were feeling like they wanted some help on that one. We were beat up. We were tired. We were on fumes. For me, a rainout wouldn't have been the worst thing in the world. I know it costs them some money, and I respect that, but we had a decent lead and couldn't give the guys a blow."

"Larry Lucchino called me when we were in Texas," said Pedroia. "I said, 'Larry, I don't give a fuck. I'll play whenever you tell me to play, but I've got to talk to the guys.' It taxes the team. I would have been dead the rest of the year if we'd done that doubleheader Friday. We didn't want to do it. We wanted to play it at the end of the year if we got rained out."

Lucchino said, "We'd made a major investment around 2004 to

change the drainage and the turf around the ballpark so that there would not be so many rainouts. We've had pretty good luck doing it. We'll tell our players we're going to play 81 home games and try to get them all in."

The compromise was a Saturday day-night doubleheader. Sox players were angry, but won both games. It would be the last time in 2011 that they would win two games in a row.

When Henry got word that players were upset about the doubleheader, he offered them a night on his boat (no children allowed) and had a set of $300 headphones delivered to the stall of each of the players. Sox coaches were not invited on the boat and did not receive headphones.

"There was this growing sense of disconnection," said Epstein. "The players were pissed at the world, including management. There was a sense of discord in the clubhouse. Ownership felt disconnected, and there were these nice headphones lying around. We'd done it in the past. In 2004 we gave them iPods."

"I never got the headphones, but about 20 of us went on the boat, and that was pretty sweet," said Pedroia. "But we started playing bad right after that doubleheader."

"There was a conflict between pitchers and position players," said Youkilis. "It shouldn't be like that. Guys weren't getting along, and we didn't address it as players. Players wanted the manager or the general manager to do something. It just became a rift. You can point fingers all you want, but players have got to play. When you come in the door, you have to be accountable for your actions and what you are doing. It just sucked for Tito. We didn't take care of what we needed to take care of. I wasn't accountable when I could have been accountable."

By the end of August there was considerable speculation that Epstein was leaving. He was under contract through the 2012 season, but Chicago Cubs owner Tom Ricketts had fired general manager Jim Hendry and recklessly identified Epstein as the man he wanted to run his team. Epstein went underground after the rumor surfaced and was careful with his words when he finally addressed the Cub question before a Sox-Yankees game at Fenway.

"I'm completely focused on the Red Sox of 2011," said Epstein.

Werner called the Epstein rumor a "non-story." But Epstein made no verbal commitment beyond 2011 and said nothing to squash the speculation. It was one more distraction leading into one of the worst months in baseball history.

Hours after Epstein's vague remarks, the Sox beat the Yankees, 9–5. Boston held a one-and-a-half-game first-place lead over the Yankees and had a nine-and-a-half-game lead over the closest wild-card contender. The Sox were 81-42 since their 2-10 start.

Everything unraveled in September, and it had little to do with chicken and beer. The Red Sox folded because their starting pitching collapsed. In the month of September, Boston starters did not survive the fifth inning in 12 of 27 games. Beckett, Lester, and Lackey were an aggregate 2-7 in September with a 6.45 ERA. The Red Sox lost 11 of their 15 starts. Beckett gave up six earned runs in each of his last two starts, both against the last-place Orioles, losing leads in both games. Lester, in his last four games, went 0-3 with an 8.34 ERA, giving up 25 hits and 12 walks in 19 and two-thirds innings. Buchholz, who'd suffered a stress fracture in his back in June, never made it back to the mound. Erik Bedard, a lefty acquired at the trading deadline, won only one of eight starts in his two-month tenure in Boston and couldn't get through four innings on the next-to-last night of the season in Baltimore. Setup man Bard, after not giving up a run over 25 consecutive outings from late May through the end of July, grew tired from too much work and lost three games in a single week in September.

Tim Wakefield's exhaustive quest for his 200th victory was difficult for everyone, including the manager. The 45-year-old knuckleballer picked up victory number 199 on July 24, then made seven starts without reaching the milestone. It was a strain on the team and the manager.

"I wanted Tim to get the 200th, but I also wanted us to win," said Francona. "Part of the good thing about getting records is doing it during the normal course of events. Unfortunately, we kept losing his starts. After the fourth try, it was like, *Damn.* You try not to think about it, and I was careful to manage the same way I would always manage, but it seemed like energy was going toward personal things that weren't team achievements. I don't blame anyone for wanting that win, but for a manager, you want your guys to be concerned about team goals."

Wakefield finally beat the Blue Jays at Fenway on September 13, but that was his only win in September. Wakefield's September ERA was 6.45, and the Sox lost four of his five starts.

On Monday, September 5, after an 11-inning, 1–0 day game loss in Toronto, Francona had McCormick get a suite at the Park Hyatt Toronto so that Sox players could hold their annual NFL fantasy draft. Pizza, beer, and sandwiches were served. It was usually a good team-building event.

The night did not go as well as Francona hoped. Francona noticed teammates rolling their eyes when other players were attempting to be funny. This was nothing like 2004 when the Idiots drank beer, played cards, and teased one another on charter flights. It was nothing like 2007 when there was always a friendly cribbage game, or the fantasy draft night in Baltimore in 2010 when players interacted, busted chops, and left the room feeling good about themselves and one another.

"DeMarlo and I were looking about the room and I thought, *These guys just don't like each other like they used to,*" remembered Francona. "It was a different atmosphere. You could tell the guys weren't as close as the teams we'd had in the past."

"It's usually a night with a lot of laughter and fun, and when you came out of it you felt good," said Hale. "But Tito didn't think this one connected like it did in the past. This team was just a little different that way."

Crawford kept to himself and was usually alone in the dugout. Ellsbury was friendly, but seemed to interact only with Jed Lowrie. Youkilis was on the disabled list (sports hernia) during the collapse of 2011. Youk was much easier to have around when he was contributing on the field. When he was on the disabled list, he was not as positive a force in the clubhouse or the dugout. Ditto for Drew, who carried himself like a man who'd already retired. Gonzalez complained about tough travel after too many Sunday night games.

The night after the fantasy draft, the Sox demolished the Blue Jays, 14–0, with Lester pitching seven innings, striking out 11. It would be Lester's only win of September.

Francona met with Varitek early the next day.

"There were things that I was seeing, and I wanted to bounce that stuff off 'Tek," said the manager later. "We spoke for about a half-hour.

I told him I was worried. I valued his judgment. He'd tell me if I was overreacting, but he told me he was seeing the same things I was seeing. That convinced me that what I was seeing was there. It wasn't something I was just imagining. It was weird, because we'd just won that 14–0 game. But I wouldn't just come out of the blue with a meeting."

After the lengthy talk with his captain, Francona alerted Hale that he was going to call a team meeting.

"That kind of surprised me," said Hale. "He told me to get everybody into the room, and I was asking, 'A meeting about what?' and I'd go up to guys and they'd say, 'A meeting about what?' And guys were saying, 'What the fuck for?'"

The meeting was not a success.

"You guys might think it's weird having a meeting after a 14–0 win," Francona said at the start of the meeting. "But I'm seeing things that are bothering me. If I didn't tell you, I'd be wrong. You don't always have a chance to win, and we have a team that can win this year. But if we don't put our best foot forward, that's going to bother me. I see us worrying about too much shit that doesn't mean anything. People are spending too much time and energy worrying about things they can't control. For a team that's supposed to be good, we sure bitch about a lot of stuff. We need to stop bitching about scoring decisions, contracts, personal goals, bus times, getaway days, the media, everything. Just stop. Remember what we are here for."

He didn't single anyone out.

Back in his office with Hale, Francona looked at his trusted bench coach and said, "That fell on deaf ears. All they are doing is wondering why I'm having a meeting when we just won 14–0. Everybody is going their own way."

"We just ain't playing good," Hale told Francona. "We're not scoring runs. We're not getting the two-out hits. We're not defending. We're not picking the ball up. We've tried a lot of things here."

"It wasn't about Tito," said Pedroia. "Some guys were mad for other reasons. They weren't mad at him. He didn't need to change."

"That meeting just seemed weird," added Youkilis. "It was too open. Guys were like, 'What the fuck is going on here?' Some guys didn't know what was going on. You're so wrapped up in doing your thing, it was vague. I knew what he was saying wasn't directed at me, but I think some guys took offense. Here in Boston it's just story after fucking story after story. And drama, drama, drama. Guys that can't handle it go nuts."

The Red Sox lost five straight games after the meeting, including a damaging three-game sweep at the hands of the surging Rays in St. Petersburg. They came home for an off day, September 12, and Henry entertained the players on his boat.

Before the next game, at home against Toronto, Francona and Epstein had a lengthy talk in the manager's office.

"Theo, we're going to win, but it's going to be a little more interesting than we want," said Francona. "Beckett and Lester will dial it up for a couple of games. One of them will throw a masterpiece, and we'll be okay."

Wakefield won his 200th that night, pitching six innings of an 18–6 blowout. The next night they lost, 5–4. This was a pattern. A week later they would beat the Orioles 18–9, just a few hours after losing the first game of a day-night doubleheader 6–5. This was how the Sox managed to rank third in the majors in runs scored during a month in which they went 7-20.

The final homestand was disastrous. The Sox split a pair with Toronto, lost three of four to the Rays, and three of four to the Orioles. In the Fenway finale Wednesday, September 21, Beckett blew a 4–1 lead in the sixth and gave up all six runs before he was pulled in the eighth of a 6–4 loss. The Sox went down one-two-three in the ninth. Fenway fans booed with gusto when Lowrie grounded out to end the game.

After the game, Pedroia went to Francona and asked him to put Crawford into the second spot in the batting order.

"He wants it bad," Pedroia explained.

There was an off day before the Sox went to New York and Baltimore to close the season. They were a full eight games behind the first-place Yankees, but still held a two-and-a-half-game lead over Tampa for the wild-card spot. When they arrived in New York, Francona was greeted with a report that his relationship with Epstein was fractured. Speaking on *The Dan Patrick Show,* Peter Gammons — the veteran scribe Francona visited in the hospital when Gammons suffered a brain aneurysm in 2006 — said, "I'm sensing an increasing disconnect between Theo Epstein and Terry Francona."

It was a telling remark, given Gammons's close relationship with Epstein. By the end of the 2011 season, Gammons was established as a virtual mouthpiece for Boston's baseball operations department. Epstein and Gammons spoke almost daily, and the scribe's public utterance of a GM-manager "disconnect" was an indication that Theo wanted to float the story.

"I was pissed about that," countered Epstein.

According to Francona, "Theo called me and said, 'Tito, that's crazy,' but I knew those guys talked so much. I know Peter reports stuff he probably shouldn't, but I was tired by then and everybody's nerves were shot. When you're losing, we all were worn thin. It was another thing I kind of had to defend, and I didn't appreciate it."

Friday's game was washed out. On Saturday, Lester was pummeled for eight runs over two and two-thirds innings of a 9–1 loss to the Yankees.

Sunday morning, standing in front of the Sox Manhattan hotel headquarters (the Palace Hotel at 465 Madison Avenue), Francona was waiting for Hale when he was approached by a stranger with a foreign accent.

"You must win today," said the well-dressed stranger.

"Hey, asshole, what do you think we're trying to do?" said the stressed-out manager.

Before the astonished intruder could react, a security official got between the two men and explained to Francona that the man he was barking at was a well-meaning diplomat from South America. There was an international convention nearby, and the streets around the Sox hotel were peppered with government officials and security officers from around the world. Francona reintroduced himself to the visiting dignitary, posed for a photo, then got in a cab bound for the Bronx. It was going to be a long day at the ballpark.

Wakefield made the last start of his career in the afternoon game, losing to the Yankees, 6–2, despite a couple of home runs from Ellsbury, who was making a late-season push for Most Valuable Player.

Against the wishes of Epstein, Lackey was Boston's starting pitcher for the nightcap. Lackey's previous start had been rough, like most of his starts in 2011. At Fenway against the Orioles, he was battered for eight runs on 11 hits, and he glared at Francona when the manager came to take him out of the game in the fifth inning.

"I never paid any attention to that," said Francona. "I'm just out there saying, 'Gimme the fucking ball.' Ten minutes after the game, I don't think John even knows he did it. I could have made a bigger deal out of it, but we were trying to get him not to bury himself."

In 2011, Lackey was officially the worst starting pitcher in Red Sox history for a single season (28 starts, 6.41 ERA). He was a pariah to most fans. Sports talk show hosts in Boston made an issue of Lackey showing up his teammates and his manager. Francona and Sox players continued to support the big Texan, but Lackey made things hard on everyone. His pitching elbow was throbbing (he underwent Tommy John surgery after the season), his marriage was over, and he just wanted to get away from everything.

"He was going through a lot," said Francona. "His stuff had backed up, and every mistake he made on the mound, Goddamn did he pay for it. He could give up two-out runs with the best of 'em. He'd get two guys out and then make a mistake. He probably didn't help his cause with the media. He was kind of surly. But he wasn't like that around us. He was hurting at the end, and we gave him every chance to go on the disabled list, but he talked us out of it. I think he just wanted to win so bad, and he couldn't believe what was happening."

Francona had to sell Lackey to his general manager before the final game in New York.

"I know you don't like this, and I'm trying not to be stubborn, but I've got to do what I think is right," Francona told Epstein. "I know what's on the line, but I've got to do what I think is right. If we pitch somebody else and we lose, it's going to hurt our team. If we pitch Lackey, he might not pitch well, but it won't fuck our team up."

"I don't think he should be the guy," said the GM. "I don't think he gives us the best chance to win. But this is your call. You're the manager."

Things got more complicated between games of the doubleheader when Lackey got a text message from a TMZ reporter asking about his troubled marriage. TMZ was getting ready to report that Lackey had filed for divorce from his wife Krista, who was battling cancer.

Lackey was appalled by this invasion of his privacy, didn't believe the ball club was doing enough to protect him, went to Francona's office, and shoved his cell-phone screen into the face of his manager. Twenty minutes before the game, Francona called Pam Ganley into the office to see if she could help with the delicate situation and the media storm that was sure to follow.

In a nationally televised game that lasted 14 innings, covering five hours and 11 minutes, the Red Sox beat the Yankees, 7–4. Lackey gave up three runs in the first inning, then settled down and blanked the Bombers through the sixth. Five more Sox pitchers went to the mound, and Francona used 14 position players to keep the desperate Sox alive. The doubleheader split gave Boston a one-game lead over the Rays with three to play.

Standing in front of his locker after the game, Lackey started things by saying, "Thirty minutes before the game I got a text message on my cell phone from one of you, someone in the media, talking about personal stuff. . . . It's unbelievable I've got to deal with this."

A short flight and a few hours later, Francona walked into the visiting manager's office at Camden Yards and discovered Ganley sitting alone on the couch. She was distressed about the way things had gone down with Lackey the night before in New York. She was also powerless to control TMZ.

It pained Francona to see Ganley upset. Dealing with an epic collapse, no contract for 2012, bus schedules, reports of a "disconnect" with his GM, sulking veterans, and pitchers out of shape and out of control, the manager donned his big brother cap and consoled the loyal public relations employee. It was not the first time he'd come to the rescue of Pam Ganley. In her first year on the job, she'd taken some shots from pitching coach Farrell regarding official scorers' decisions, and Francona had urged that Farrell apologize. Now he had to gently extinguish a brushfire between his struggling $82 million pitcher and a dedicated publicist who logged 100-hour workweeks for less than John Lackey paid in clubhouse tips.

"It's okay, Pam," said the manager. "We all know you are doing the right thing."

A few hours later, the Orioles beat the Red Sox, 6–3, and the Sox fell into a tie with the Rays for the wild-card spot. Boston's nine-game lead over Tampa was completely gone. The Yankees weren't trying very hard to beat Tampa, but nobody begrudged New York's opportunity to rest. Derek Jeter and Alex Rodriguez sat out every inning of Boston's 14-inning victory over New York Sunday night.

In his final start of 2011, Beckett coughed up a 2–1 lead, giving up all six runs in six innings. Batting in the second spot for the fourth consecutive game, Crawford went 0-5 again, which made him 3-21 since Pedroia petitioned the manager to elevate the sensitive outfielder.

"Fuck, Pedey, I can't do this anymore," Francona told Pedroia. The manager moved Crawford to the eight spot in the batting order for the final two games of the season.

The Red Sox and Rays both won their next-to-last game Tuesday night. With Lowrie batting in the cleanup spot for the first time in his career, Sox rookie catcher Ryan Lavarnway hit a pair of homers and Papelbon survived a 28-pitch ninth to give the Sox an 8–7 win.

The last night of the 2011 Major League Baseball season was one of the most exciting nights in the history of the game. The Red Sox, Rays, St. Louis Cardinals, and Atlanta Braves were all playing with the playoffs on the line. Four other teams, the Tigers, Angels, Brewers, and Phillies, played with home-field advantage at stake.

At 10:30 AM on Wednesday, September 28, Francona came down the elevator from his room in the Renaissance Baltimore Harborplace Hotel. He placed his bag behind the velvet ropes in a space set aside for the Red Sox traveling party. He did not have to check out of the hotel or even stop by the desk to pay incidental charges. Trusty Jack McCormick took care of paying the manager's incidentals. It was a managerial perk.

McCormick had a lot of other duties on the final day of the regular season. The Sox were going somewhere after the game. There were four possibilities. They could be going to Detroit or Dallas–Fort Worth for the first game of the American League Division Series. They could be going to Tampa–St. Petersburg for a one-game playoff against the Rays. Or, worst case, they could be going back to Boston for a long, awful winter. McCormick had a Delta 757 jet (Air Red Sox flight number 8884) set to transport the team from Baltimore-Washington Airport to one of those four destinations after midnight.

"No problem," said McCormick. "They have maps to every city."

Francona walked to the ballpark for the final game. The walk would have been intolerable in midsummer, when hundreds of Red Sox fans patrolled the half-mile of streets between the Renaissance and Camden Yards. It was not a problem in September. The manager made his lonely walk and enjoyed every step.

"We were fighting for our baseball lives, and I was excited that morning," said Francona. "For me it felt like we were getting ready to play a Game 7, and that was exciting. We had all those destinations, which was somewhat hilarious, and I still thought we were gonna win. We had Lester on the mound, and he was 14-0 lifetime against Baltimore."

The manager arrived at Camden Yards before 11:00 AM.

There was much to do. Lester was getting the ball on three days' rest for the regular-season finale, but Francona had to think about the possibility of a one-game playoff game the next day in Tampa. There was no logical starting pitcher for the Red Sox. Beleaguered Lackey (weeks away from surgery) was a candidate on three days' rest. Beckett, whose wife was scheduled to deliver a baby any minute, said he'd start on two days' rest. Wakefield, as ever, could pitch anytime. Rookie lefty Felix Doubront, 0-0 in 11 relief appearances, was available. Buchholz, who hadn't pitched since June, was another possibility. Henry was petitioning his manager to pitch Buchholz. Meanwhile, in a bizarre twist, urged on by Henry's nervousness, Epstein explored acquiring Kansas City lefty Bruce Chen for a one-game stint with the Red Sox.

Francona hated the idea of bringing in Chen for such an important game.

"I'm really uncomfortable with this," Francona told his GM. "I want us to do this with our team. These guys are busting their asses. He's not even part of what we are doing."

"I've got a responsibility," countered Epstein.

Epstein was unable to acquire Chen.

The GM had a few things to say about the final lineup. He wanted rookie catcher Ryan Lavarnway, a September call-up, back in the lineup, batting fifth, behind Gonzalez. Francona agreed.

"We were always pushing Tito to play Lavarnway," admitted Epstein. "But where he was in the lineup was Tito's call. That was something Tito wanted to do."

Lester said he was comfortable throwing to a rookie who had less than a month of experience behind the plate.

The finale unfolded in ghastly fashion. Pedroia's solo homer in the fifth gave the Sox a 3–2 lead, which was where things stood when Baltimore rains interrupted the game at 9:33 PM in the middle of the seventh. When the Sox got to their clubhouse, they tuned in to the game in Tampa. Folks watching the Sox NESN broadcast back in Boston were bombarded with virtual ads urging them to watch Liverpool versus Wolverhampton the next day at 4:00 PM, which would have been the same time the Red Sox might play Tampa on TBS. The Yankee broadcast showed the Bronx Bombers leading the Rays, 7–0. It looked like the Sox were going to either Detroit or Texas for the ALDS.

The Sox had been in their clubhouse for more than a half-hour when the Rays came to bat, still trailing 7–0, in the bottom of the eighth. No Yankee team had blown a lead of 7–0 or greater in the eighth inning or later since 1953, but Yankee manager Joe Girardi was not playing to win. Jeter, Teixeira, Curtis Granderson, and Nick Swisher had all been pulled from the game. Tampa scored six times in the eighth, pulling to within a run on Evan Longoria's three-run homer.

Epstein and Cherington watched with Francona in the manager's office at Camden Yards.

"The Yankees had that lead, then started treating it like a spring training game, as we would have," said Epstein. "There was a sense of dread as we saw them chipping away. It was like watching a car wreck in slow motion."

"We started that night worrying about ourselves," said Hale. "But when all that stuff started happening in Tampa, I thought we stopped thinking about ourselves and thinking about them."

The Sox were getting ready to go back on the field when Longoria rounded the bases to make it 7–6.

"Fuck!" barked Epstein.

"Fuck!" barked Cherington.

"Don't worry," said Francona. "It ain't happening."

The Sox were scheduled to resume playing at 10:58. Epstein and Cherington stayed behind in Francona's office when the Sox went back down the runway toward their third-base dugout. At 10:47, Tampa pinch hitter Dan Johnson tied the Rays-Yankees game with a two-out home run off the right-field foul pole off Cory Wade. Johnson was batting .108 and hadn't gotten a big league hit since April.

*At least we're still winning our game,* Francona thought to himself. *There are worse things than having to play tomorrow.*

After the one-hour-and-26-minute delay, the Sox and Orioles went back to work in the bottom of the seventh.

The Sox blew a chance to pad their lead when Scutaro hesitated on the base path and was thrown out at home plate in the eighth. Boston failed to score in the ninth despite having runners on first and third with no outs and the heart of the order due up. Oriole manager Buck Showalter, a known Red Sox antagonist, managed as if it were the seventh game of the World Series and got his team out of a jam by intentionally walking Gonzalez, setting up an inning-ending double-play grounder by Lavarnway.

Papelbon came on for the ninth. The 2011 Sox at that moment were 77-0 in games in which they led after eight innings, and Papelbon had blown only two saves all season. There was cellophane covering the lockers in the Red Sox clubhouse, but Francona was anxious because his ninth-inning man had thrown 28 pitches the night before.

Beyond the pitching mound, directly in his line of vision, Francona could see play-by-play updates of the critical game unfolding in Tampa. The manager knew what everybody knew: a Yankee win or a Red Sox win would vault the Sox back to the playoffs for the sixth time in eight years. The worst case still seemed to be a one-game playoff in Tampa the next day.

Then everything came apart. With two outs and nobody aboard, Papelbon surrendered back-to-back doubles to Chris Davis and Nolan Reimold to make it a 3–3 game. Sox-killer Robert Andino was next and hit a sinking liner to left. It was a classic 'tweener. Crawford hesitated, lumbered forward, but could not make the catch. When Reimold slid across home plate with the Orioles' winning run, the Red Sox ran off the field knowing they could not make the playoffs without playing one more game the next day in Tampa. Disgusted by the Oriole dogpile at home plate, Scutaro flung his glove into the third-base dugout.

While Hale dutifully peeled the lineup card off the wall and gathered scouting reports, Francona left the dugout and headed for the clubhouse. The manager never wanted to linger for the cameras.

He made his way down the tunnel, through the clubhouse, and into his office, where Epstein and Cherington were sitting on the couch. The office television was turned off. It's baseball etiquette. After a loss, the TV goes dark. The manager stood at his sink and started brushing the Lancaster out of his gums. Sitting on the couch, tracking the Rays game on his iPad, Cherington saw Longoria's home run down the left-field line.

"Longoria just won it with a walk-off," said Cherington.

"Fuck," said . . . everyone.

There it was. The Sox were not going to Detroit or Texas for the American League Division Series. They were not going to Tampa for a one-game playoff, which would have been their first one-game playoff since the infamous Bucky Dent game of 1978. They were going home. Their season was over. They were the first team in baseball history to fail to make the playoffs after holding a nine-game playoff lead in September.

"I feel like I let you down," Francona told Epstein.

"This was all of us, and you did the best you could," said Epstein.

Ganley made it into the room and told Francona he would need to do his postgame press conference in the hallway outside the Sox clubhouse. Showalter was using the Camden interview room ("For those two Baltimore reporters," Ganley sniffed), so Francona had no option but to go outside the clubhouse and into the corridor for his session with the Boston media. With his back supported by the cold cinderblock wall, Francona answered every question, referencing "the mess we got ourselves into." It felt like a firing squad. And it was.

Less than an hour later, there were three buses waiting for the Red Sox entourage outside Camden Yards. The traveling party was larger than usual because everybody thought they were going to the playoffs and there were a lot of front-office folk and family members in the group.

Francona came out of the ballpark, saw the three buses, strolled toward the door of the first coach, and walked up the steps, only to find somebody sitting in his seat — driver side, second row by the window. He said nothing and went to find another seat.

*It doesn't matter,* he thought to himself. *It's not going to be my seat anymore.*

# CHAPTER 15

# "Somebody went out of their way to hurt me"

THERE WAS NOTHING FUN or collegial about the bus ride and the last flight home from Baltimore. No Texas Hold 'Em, no "Mississippi River Rule," no back-slapping or chop-busting.

Red Sox manager Terry Francona didn't get to his room 421 of the Brookline Courtyard Marriott until 4:00 AM on the morning of Thursday, September 29, 2011. He was scheduled to meet with Theo Epstein later at Fenway for the perfunctory "postmortem" press conference in the second-floor media room, and it was not going to be pleasant. He knew his future was in doubt. The Sox were holding a two-year option for his services — at $4.5 million per season. The option had to be triggered within ten days, and the manager was not in a favorable position with Henry, Werner, and Lucchino.

He was back at Fenway late in the morning. While Sox ballplayers came and went, packing boxes for a winter that was coming earlier than they expected, Francona closed the door to his office, sat down, and looked up at Epstein.

"Do these guys want me back?" he asked his general manager. "Cuz I don't think they do."

"No, probably not," Epstein said softly. "But you need to think long and hard about whether you want to be back. If you really want to be back, I will go to battle for you. We'll make this happen, but you need to go home and think about whether you want to be back. They said they want to talk with you. We have a meeting with them at nine tomorrow morning."

Armed with this information, Francona marched upstairs with Epstein to answer media questions about the collapse. It was part of the job, even when the job was coming to an end. There was always a need to explain what happened, especially after an awful finish. The 33-minute question-and-answer session was broadcast live by NESN.

Asked if he wanted to return as manager of the Red Sox, Francona said, "Theo and I talked a little bit. I think we'll continue to talk tomorrow. Maybe it's best today to stay with where we're at. It's still pretty fresh and pretty raw."

"I was so disinterested in that press conference because I had a feeling we were going to be having another one," Francona said months later.

In the wake of the collapse, there were already multiple media reports citing clubhouse drinking and other player misconduct.

"The way the clubhouse culture has evolved, and this falls on me ultimately as the general manager, we need to be more accountable," Epstein admitted to the press. "In some small ways, we've gotten away a little bit from our ideal of what we want to be on the field and off the field. It's our responsibility to fix it. . . . We're less than 24 hours removed from the end of the season. We need some time to calm down, get objective, and look at ourselves, look at 2011, look ahead and make the best decisions for everybody."

Shortly after the awkward session, NESN, the Red Sox–owned flagship station, cut away from postconference analysis to air a replay of a Premier League soccer match involving Wolverhampton. Rival station Comcast Sportsnet New England stayed with Sox analysis.

In less than 24 hours, Francona would no longer be the Sox manager and "clubhouse culture" would be cited as a major contributor to the Sox collapse. Francona and Epstein forever would point to the woeful starting pitching in September, but Red Sox Nation and the media beast wanted more. Reports of clubhouse drinking, Popeye's takeout, and unprofessional behavior leaked from every crevice of ancient Fenway. "Chicken" and "beer" were destined to stand forever as the bookends of Boston's epic fold.

After the press conference, Francona went back to his office, sorted through some bills and personal mail, and visited with Pookie, Murph, Tommy, Joe, John, and the rest of his favorite clubbies. Then he went to the Courtyard Marriott and tried to keep his television tuned to a Sox-free zone. It was not easy.

He knew John Henry had been critical of him for several years. The owner could not tolerate decisions that flew in the face of the data. No doubt Werner was thinking of this as another "shitty season." Francona knew Lucchino would need someone to blame. They weren't going to sacrifice Theo Epstein, Adrian Gonzalez, Carl Crawford, or Josh Beckett. Not yet. Not after one bad month.

Early Friday morning, when Francona wheeled his Cadillac Escalade into Fenway for his final day as manager of the Boston Red Sox, there were multiple television crews camped out at the intersection of Van Ness and Jersey Streets.

He met with Epstein before going upstairs to meet with the owners.

"Theo, given everything you've told me about them not wanting me back, why are we even having this meeting?" Francona started. "I told you from the get-go, it was not only their privilege but their responsibility to get the right person. And if they don't think I'm the right person, there's not much to talk about."

"I think you need to recharge," said Epstein. "You need to get away. Go reconnect with who you are. If you can do that, you can come back with a new voice. You might be able to come back with a new voice. I don't want you to come back either unless you can commit to doing that. You have earned the right to do that."

At 9:00 AM ("Never be late, never be early," was a favorite Lucchino expression), Francona and Epstein went to the third floor and sat at the large oak table in the meeting room connected to Lucchino's office. Also at the table were Cherington, Henry, Werner, and Lucchino.

It was an awkward, passive-aggressive session lasting almost an hour, accomplishing little. Francona knew the owners didn't want him back, but no one was willing to express this uncomfortable truth. All the men were exhausted, still hurting from the shocking defeat in Baltimore less than 36 hours earlier. The manager said that the players didn't care about one another or protect one another. In Francona's mind, that was the worst part about the locker-room drinking: players were not looking out for one another, and they were telling stories behind one another's backs. He admitted he was bothered by things that hadn't bothered him in the past. He did not tell them that ownership interference was one of the most difficult parts of his daily life, but he was pretty tired of emails about Ortiz batting against lefties.

Henry, Werner, and Lucchino all took turns speaking. None would voice the plain truth that Francona was not being offered the extension. Henry and Werner routinely recoiled from confrontation, but it was unusual for Lucchino to hold back. The CEO traditionally played the heavy in awkward situations and had a Rolodex of enemies to prove it. Not this time. Nobody wanted to be the man who fired the two-time World Series winning manager, not even after the worst collapse in baseball history. They knew they were not going to bring him back for two years for $9 million. He would never be presented with that option. But appearances were top priority for this group, and it would be much easier to sell the story if the popular manager simply asked to leave.

Exasperated with the conversation, Francona finally said, "If you don't know what you are doing about me, why am I here? This is a silly meeting. You guys know me. I've been here eight years. If you don't want me, just tell me."

"We want you to wait and think about it," said Lucchino. "Take the weekend. Sleep on it. See how you feel."

"It was a sentiment that we all felt," Lucchino said later. "That there be an orderly process to this and that we have time to think about it."

"We had not come to any conclusion about whether to move forward with Terry or not," insisted Werner. "I was very clear about it. I thought we needed to have a conversation with Tito about what went on in September and how it happened and how we were going to move forward in the future. It was at that meeting that Tito said that he had lost control of the clubhouse . . . he was very forthright about it, that he was not the right person to continue as manager going forward. I had kept a very open mind about what to do going forward and was hopeful that he would be not only specific about the problems, but how to correct them. He said, 'I'm not the guy to move forward with you.' Given the historic collapse that we had had, you certainly would want a manager who would articulate not only what the problems were, but how he would go about addressing them. You want a guy who is going to go through a wall, and he was clearly of the opinion that he wanted to leave. His body language in that whole meeting was 'I'm not the right guy for this.'"

"I never said I lost control of the clubhouse," countered Francona. "I said I hadn't been able to reach some of the guys. I was just trying to take accountability. But I kind of viewed that meeting as a charade."

"Here's what happened at that meeting," said Lucchino. "We began by saying, 'Wow, what happened?' It was informal. He went through his analysis of how things deteriorated and things that contributed to the decline. Right after he finished, we asked him, 'What should we do about these things, how do you propose to deal with it?' and that's when he said, 'I'm not the man to deal with these things. They need to hear a different voice down there. I'm not the guy.' We were all a little surprised. . . . It seemed to us that a little time and space would be appropriate, but that didn't seem to be the timetable that Terry or others had in mind."

"Down deep, I didn't know if what was said in the meeting was everything," Francona said later. "They basically told me they wanted me to wait and think about it before stepping down."

"I don't think anyone at the meeting felt it went well," said Epstein. "There were more questions — 'Do you want me back?' and 'Do you want to be back?' — than there were answers. It was awkward, to say the least, and we weren't really getting anywhere. Afterwards, Tito and I went down to his office to process the meeting. We laughed at ourselves, at how circular it all was, and how that wasn't exactly the type of meeting you get with people who want to keep working together. As we got serious, we went back to what was the key issue in my mind, the same one we had identified the day before. Could Tito take some time to reinvigorate himself and come back as 'the new voice' we all agreed was necessary to reclaim the clubhouse? I'm sure the lack of an endorsement at the meeting was bothering him, because this time he was definitive: 'I don't think so. It doesn't feel right. You were at the meeting. It's time to move on.' I asked him if he was sure, and he said he was. We reminisced, hugged, and soon Tito went home and I went upstairs."

TV crews were waiting on the sidewalk when Francona pulled out of Fenway, bound for his family's home in Chestnut Hill. He wanted to prepare Jacque and his children for what was coming. Youngest daughter Jamie was still at Brookline High School, and Leah was living locally while her husband was at war in Afghanistan.

Back at Fenway, Pam Ganley worked on a "Statement from Theo Epstein":

> John Henry, Tom Werner, Larry Lucchino, Ben Cherington and I met with Terry Francona this morning at Fenway Park to exchange thoughts and information on the 2011 season and discuss areas for improvement going forward. We all plan on taking some time to process the thoughts expressed in the meeting. There are no immediate plans for an announcement.

The statement was released to the media at 1:25 PM.

Driving toward his home in Chestnut Hill, Francona became aware of an eye in the sky. He was being followed by a local news helicopter, like O. J. Simpson in his white Bronco.

*Holy shit, with all the stuff going on in the world, they have a helicopter following me,* Francona thought to himself. *What a waste of money.*

The manager's cell phone rang. It was Theo.

"I talked to them after you left," said Epstein. "It's pretty clear that the decision has been made."

"Okay, fine," said Francona. "I understand. But let's not put up a charade. Let's just have a press conference and get it over with."

Lucchino and Werner dispute this timetable, as well as the notion that they had made up their minds.

"That's just not what happened," said Werner. ". . . I would take exception to that. Theo can say what he wants. . . ."

"If Theo did that, he did it on his own motion," added Lucchino. "He reached his own conclusions about what he thought, but he didn't clear any of that with us or talk to us about that."

A reasonable person would conclude that the Sox owners wanted their popular manager to quit before he was fired. That was certainly the opinion of Epstein and Francona.

"Before the meeting, I was told they didn't want me," said Francona. "After the meeting, I was told that there was nothing to think about. And I knew how I felt during the meeting."

Months later, when Lucchino was asked if he thought Epstein "played" both parties against one another, the CEO paused for several seconds and said, "I feel in my bones a certainty of certain things that I don't want to say publicly, and this is one of them."

When Francona got to his house—the house he'd moved from almost a year earlier—there were television crews parked on the street next to his driveway. He went inside, spoke to Jacque and the girls, then went back to Fenway. It was surreal. While he drove back to Fenway to pick up some laundry and have a personal tax document notarized in the Red Sox legal department, he heard radio reports about "Francona on his way back to Fenway for another meeting."

He called Epstein again.

"Let's do this," he told the GM. "Everybody's mind is made up. Why the fuck are we waiting? I don't want to wait through the weekend, Theo. This is getting silly. Let's do this now."

"I'll get ahold of them," said Epstein.

"It was just terrible," recalled Epstein later. "I just tried to be as honest as I could be with Tito. He deserved the right to come back if he wanted to come back."

"There were issues with his option not being picked up before we started playing bad," said Josh Beckett. "It kind of looks to me like they didn't want him to come back one way or the other."

"It was Tito pressing for a resolution," said Lucchino. "We didn't. We wanted a little time and space. So much was going on. We wanted to slow the process down and do things in a more orderly way, but Terry was eager to move on."

Was there any possibility that Francona could have been offered the two-year extension to come back?

"We discussed various options," said Lucchino, "all of which we thought would be part of the process that sort of got truncated and never developed."

"It's pretty simple," said Francona. "If they wanted me back, they'd have picked up my option."

Back at Fenway, Francona busied himself, packing a few things in his office, while Ganley worked upstairs on a press release that would serve as the manager's Red Sox obituary. When Francona saw Ganley's first version, he asked her to make some changes. All parties were specific and careful with the wording of the statement.

"It was a scramble, " Francona said. "I didn't want it to be a charade. Pam kept coming down with statements. They were adamant about how it was worded, and I'd say, 'Pam, this isn't how I feel.' She told me they were adamant about certain things [for instance, that the word "fired" would not appear anywhere in the document]. I was starting to get mad. I told her, 'Pam, this isn't how I feel. I know they want it a certain way, but I lost my job.' We went through about four versions before we agreed on the last one. Whatever happened, I don't know. I just know what I was told."

The high-nuanced press release, delivered at 5:19 PM, included expansive statements from the ownership trio of Henry, Werner, and Lucchino, plus Epstein and Francona.

From the owners: "Tito said that after eight years here he was frustrated by his difficulty making an impact with the players, that a different voice was needed, and that it was time for him to move on. After taking time to reflect on Tito's sentiments, we agreed that it was best for the Red Sox not to exercise the option years on his contract."

From Epstein: "Ultimately, he decided that there were certain things that needed to be done that he couldn't do after eight years here, and that this team would benefit from hearing a new voice."

From Francona: "After many conversations and much consideration, I ultimately felt that, out of respect to this team, it was time for me to move on. I've always maintained that it is not only the right, but the obligation, of ownership to have the right person doing this job. I told them that out of my enormous respect for this organization and the people in it, they may need to find a different voice to lead the team."

Polygraphs exploded across New England while the Red Sox announced that Francona would meet with the media at Fenway at 7:00 PM, followed by a contingent from Sox management at 8:15. Adding to the chaos was news that John Henry had been injured in a fall on his yacht and transported via ambulance to Massachusetts General Hospital. The WBZ-TV (CBS) evening news featured helicopter coverage of an ambulance parked near Henry's docked boat at Rowes Wharf behind the Boston Harbor Hotel. Henry walked off the *Iroquois* wearing a neck brace.

On his way to the interview room, Francona remembered the winter day eight years earlier when WBZ's Jonny Miller asked about the notion of every Red Sox skipper ultimately waiting to be fired. He had told himself he wouldn't be one of those guys, he would leave on his own terms.

*Damn, I almost made it,* he thought to himself.

Every local station covered the press conference live. Francona stayed on script for most of his session. He spoke of "a sense of entitlement" regarding veteran players, but took much of the blame:

"I felt frustrated with my inability to reach guys that I've been able to in the past, or affect the outcome a little bit more differently, and that bothers me," he said. "I wanted desperately for our guys to care about each other on the field. I wasn't seeing that as much as I wanted to do. When things go bad, your true colors show, and I was bothered by what was showing. It's my responsibility."

While he spoke, the clubbies stood shoulder to shoulder in the back of the room.

They heard Francona reject an opportunity to refute a *Boston Herald* report that players had been drinking in the clubhouse during games. He stated that leaving the Sox job was "my decision" and that "it was time for a new voice here."

Gordon Edes of ESPNBoston asked, "Tito, what was it about this situation that kept you from going in this morning and instead of saying, 'I think we need to part ways,' instead of saying that, saying, 'Look, I didn't have a good year, but I'm going to redouble my efforts, and I'm going to reach those players next year.' What is it about this situation that . . ."

Interrupting the drawn-out question, Francona said, "Well, some of it may be personal, but I just thought it was time. To be honest with you, I didn't know, or I'm not sure how much support there was from ownership. And I don't know that I felt real comfortable. You've got to be all-in on this job. I voiced that today. Going through things here to make it work, it's got to be everybody together, and I was questioning some of that a little bit."

When Francona was done, he left the room through the back door on the left side of the stage and walked through the empty weight room back toward the stairs that led to his (former) office. Safe in the ancient corner space with the brown saloon door fronting the toilet, he found Pookie Jackson and told him to round up the rest of the clubbies for one final night on the town. Upstairs in the media room, Epstein sat uncomfortably between Werner and Lucchino and took questions about Francona. When Lucchino was asked about Francona's lack of "support" from ownership, the CEO said, "I was puzzled by that remark."

"I didn't understand why Larry said that," said Francona. "I didn't crush them, but they knew. Don't do that to me. They all knew why I made the comment about support. We'd been talking about that all morning."

"Looking back, I can understand his rationale," Lucchino said later. "He wanted a quicker resolution, and if we were really behind him, we would have said, 'Hey you, it was a terrible September, but you're our guy.' We didn't. We sat down and said, 'Let's talk about it. Let's have a process and see how this comes out.'"

After the press conferences, Francona and the clubbies went to the Red Lantern in the South End of Boston, an Asian restaurant owned by Ed Kane, one of John Henry's best friends. When the ex-manager checked his phone messages, he had one from Manny Ramirez.

"Papi, this is Manny," said the faint-but-familiar voice. "I just wanted to give you a call. You were an okay manager. Call me back."

"I could hear someone in the background coaching him through the phone call," chuckled Francona. "I could hear a voice telling him it was okay to hang up now. It was kind of funny. He was getting coached even then. But I did appreciate the gesture."

Francona and the clubbies stayed at the Red Lantern long into the night.

The next morning, while Francona and the clubbies slept it off, John Henry's wife, Linda Pizzuti, tweeted, "Happy John is home. He slipped down stairs, injuring his neck. Kept at hospital as a precaution, but made it home in time for the derby."

Swell. The owner, who'd been unable to attend the press conference announcing the firing of the manager who brought two World Series titles to Boston, made it home in time to watch Liverpool beat Everton, 2–0, in the Merseyside derby. It would not be the final insult Sox management hurled at Terry Francona in October of 2011.

The Sox wanted Francona's club-owned laptop back, almost immediately.

"I asked Brian O'Halloran, who I love, if I could keep it because I had so much on there," said Francona. "It wasn't like I was stealing secrets because the stuff was mine. They said no. But I knew how to run it, so I bought the same model and asked them to help me put my stuff on there, but in doing so they erased a bunch of my stuff. And they disconnected that little air card that I think costs about $6 a month."

Five days after the 2011 Sox season ended in Baltimore, *Boston Globe* sports editor Joe Sullivan assembled a team of veteran baseball reporters in his office and put Bob Hohler in charge of a piece that would go behind the scenes to explain the Red Sox collapse. Hohler was a veteran reporter who served in the *Globe*'s Washington bureau for seven years before taking over the Red Sox beat in 2000. He was the *Globe*'s daily Sox reporter in 2004 and had a good professional relationship with Francona. Hohler left the Sox beat after the 2004 World Series and was an award-winning investigative/project reporter in his post-baseball career at the *Globe*.

Hohler knew all the principals in the Red Sox front office and had additional sources in and around Fenway Park. His sources told him about Sox players drinking beer, eating fast-food chicken, and playing video games in the clubhouse during games. Anonymous team sources also painted a picture of a manager rendered less effective because of his troubled marriage and pain medication. Hohler left Francona a message, but the ex-manager was busy working in the FOX Sports broadcast booth, covering the American League Championship Series between the Tigers and Rangers.

On October 10, Francona was sitting in the broadcast booth with partner Joe Buck before the second game of the ALCS when he was approached by *Globe* baseball reporter Peter Abraham.

"Bob Hohler has been trying to get ahold of you," said Abraham. "He's working on a story, and I think you need to talk to him."

Francona reached Hohler later that night. The ex-manager was surprised and furious when Hohler asked him to confirm details of his personal life and charges that his failing marriage and health issues in some way contributed to the team's demise. Hohler was citing multiple team sources who had requested anonymity.

"I couldn't believe it," said Francona. "Somebody went out of their way to hurt me. You go eight years in Boston and it's coming to an end, which is hard enough. I couldn't digest it because I was fending off all these allegations. I was like, 'C'mon, Bob. What the fuck?'"

Headlined "Inside the Collapse," Hohler's story ran on Wednesday, October 12, on page 1, above the fold, accompanied by a Reuters photo of Francona, looking dejected, standing in a dugout somewhere on the road in 2011. There were multiple examples of ballplayers behaving badly, but the ex-manager took the biggest hit as details of his personal life were uncovered.

"By numerous accounts, manager Terry Francona lost his ability to prevent some of the lax behavior that characterized the collapse," Hohler wrote. "Team sources said Francona . . . appeared distracted during the season by issues related to his troubled marriage and to his health."

Further down in the story, Hohler wrote, "Team sources also expressed concern that Francona's performance may have been affected by his use of pain medication."

There it was. Francona's spring prediction to Dr. Larry Ronan ("This will fuck me someday") had come true.

The ex-manager was quoted through the story, saying, among many other things, "It makes me angry that people say these things because I've busted my [butt] to be the best manager I can be. . . . It [pain medication] never became an issue, and anybody who knew what was going on knows that."

The damage was done. Reaction was swift.

"He got kicked in the balls just like everybody else that ever left Boston," said Beckett. "Why won't they just let you leave? They did it to Nomar. Now they do it to him. He was a great manager, and all they did was kick him in the nuts when he leaves. I hope he buries all those fuckers."

"Does anybody ever leave here happy?" added Pedroia.

"I wonder who the fuck was saying that shit," said Youkilis. "It got put in there that I was a problem. There was a lot of stuff that was written that was unfair. We all have our personal lives. Not one person is exempt. The bottom line on that season is that we were not accountable."

"Wherever that came from, somebody screwed up big time," said Ortiz. "It's a huge impact when you are playing for a guy that has your back. It doesn't matter how tough things are for you, he still believes in you. That is huge. That will turn you into a monster. Pedroia — the reason why he is the player he is, is because Tito gives him confidence. All of us got used to that."

Francona was especially hurt by "team sources" talking to Hohler about his use of pain medication.

"Larry [Dr. Ronan] called me right away, and I think he was as upset as I was," said Francona. "He felt so badly. I took so much less pain medicine this year, and everybody that knows me knew it, and that's what bothered me more than anything. Anybody that knew me knew why I was taking pain pills, but somebody was trying to hurt me. That's what bothered me. I know I'm not perfect, but somebody felt compelled to tell that. I'd like to think that whoever did it, maybe it was a slip of the tongue, that the intention wasn't what it was. It was said to put me in a hole. And Bob Hohler tells me it came from multiple sources."

Larry Lucchino also called Francona.

"I thought it was sad that this story was coming out now," said Lucchino. "There should have been a better, more positive, more orderly process for his departure I didn't want him to think that we — John, Tom, and myself — were responsible for providing any of the adverse information. As far as I knew, I know I didn't talk to Hohler. I don't think anybody talked to Hohler. I think there was a limited email exchange on certain issues with Hohler by John [Henry], who is the email guy. But no one discussed these kinds of issues with Hohler at the ownership level, and I thought he had the right to know that."

"He [Lucchino] was the only one of my bosses who called me," said Francona. "He called to assure me that he hadn't said anything about the pills. Larry wasn't even supposed to know about the pain management system we had put in place with MLB. It was a good conversation, but I said, 'Larry, do you care what they are saying, or are you just worried that people think it's you?' He said he'd look into it. He said, 'Consider it done.' But he never called me back."

"He [Francona] urged us to make public who the person was that did this," said Lucchino. "It was such a transgression. And I did tell him that we would look into it. I did not tell him we would do some Nixonian investigation, because I knew how impossible that kind of examination was. But I did tell him we would look into it, and we did. I talked to people in the baseball ops department in particular. I knew it didn't come from ownership, and I was trying to find out where it could come from. The possibilities were players, who were largely gone at the time, or baseball ops. There may have been others that I hadn't thought of. I inquired among those people and couldn't find anything. I called Hohler to try to get some further information from Hohler, if we could get anything out of him. John and Tom talked to him, trying to get guidance. You know, 'Was it a he or a she?' All that kind of stuff. Hohler initially was a little bit cooperative. He acknowledged privately that it didn't come from ownership, then he sort of backed away from that acknowledgment publicly. We were hoping he would acknowledge it publicly."

"Larry did not contact me after the story was published," said Hohler. "He knows as well as anyone that the *Globe* will fully protect its sources. Larry, Tom, and John each know whether or not they communicated with me for the story. Beyond that, I never privately revealed to anyone in the Red Sox organization who was or wasn't a source for the story."

Francona called Henry multiple times and received no response.

He never spoke with Werner.

When he spoke with Epstein, Epstein said the same thing Lucchino said: *It wasn't me.*

"When all this shit came out, I felt like they were more worried about how they looked than about what happened to me," said Francona.

"I don't have a drug problem, that's pretty obvious. I don't drink that much, but I joke about it a lot. Anybody that knew me knew I had taken more painkillers in '04, because my knees were shot, than I did in that last year."

"The article was unfortunate for a lot of reasons, not the least of which is that it threatened to taint Tito's legacy," said Epstein. "The nature of the article and its timing brought disproportionate focus on issues that were isolated to a difficult period when none of us were at our best and that were secondary, if not irrelevant, considering his greater body of work. Tito's overall legacy, I think, is proving that even in a crazy place, a good man can be himself and still win. He never tried to be something he wasn't. He always put the game — and those who played it — first. He treated people well when no one was looking. He connected with players, fans, and an organization in a simple, genuine way that left us all better than he found us. And he was the on-field leader of an organization that had a decade for the ages. Tito will have a third act — and I have no doubt he will win again — but what he accomplished in Boston was unique, unassailable, and unforgettable."

There were two sets of snitches to chase. While Francona wondered about the "team sources" who spoke about his pain management, Sox players wondered who was talking about chicken and beer. Beckett texted more than a dozen Sox clubbies and team personnel.

Youkilis emerged as a primary suspect regarding the chicken-and-beer leakage. It was a rumor that would follow Youkilis to Fort Myers in the spring of 2012.

"I guess guys thought it was me for some reason," Youkilis said in July 2012, after he'd been traded to the White Sox. "But I never talked to reporters after the season ended. I never talked to anyone in the off-season. I was in California. My agent declined every request after the season."

Well aware that he would become the poster boy for chicken and beer, Beckett said, "I don't want to be remembered as a guy who talks about what goes on in the clubhouse. There's things that go on in the clubhouse that are nobody else's business. That's a sacred place to me. But I'd rather have all this stuff fall on me than some of these other guys. Some of these guys have a lot longer left to play. If that's where it ends up being pointed, I'd rather it be that way. I don't like seeing guys behind me — guys like Jon and Clay — go through any of this. We did everything we could to win. People can believe that or think I'm blowing smoke up their ass, whatever the fuck they want. The fact is, we all wanted to win."

The Red Sox continued to make off-season news, most of it negative. Epstein was still on the job, but everyone knew he was going to the Cubs. Two days after the *Globe* story, Henry burst into the Brighton studios of 98.5 The Sports Hub to defend himself against comments being made by sports radio hosts Tony Massarotti and Mike Felger. He spoke of a "media riot" and made it clear that he'd opposed the Crawford signing. The owner said, "Larry Lucchino runs the Red Sox." Henry said he'd sent an email to upper management during the season in which he suggested picking up the option years on Francona's contract. Henry said there was never a discussion about a replacement manager, not even through the last day of the season. Regarding the *Globe*'s details on Francona's private life, the owner said, "It's ridiculous that people would talk about things like that. . . . If it's someone with the team, and that's what it says in the newspaper, then I'm very upset about it. It's reprehensible that it was written about in the first place."

On October 17, Lester became the first Sox pitcher to admit to having chicken and beer in the clubhouse during games, telling the *Globe*'s Abraham, "People knew how Tito was, and we pushed the envelope with it. We never had rules, we never had that iron fist mentality. If you screwed up, he called you on it. . . . Tito was the perfect guy for this team for a long time, but I think he got burnt out. . . . I should have been on the bench more than I was."

A day later, WHDH-TV's Joe Amorosino, citing two unnamed Red Sox employees as sources, reported that Lester, Beckett, and Lackey had been spotted drinking Bud Light from red cups in the dugout during games. This report prompted an 11:04 PM statement in which the three pitchers, Lucchino, and Francona issued denials.

"I wasn't even manager anymore and volunteered for that one," said Francona. "Pam called me and told me, and I told her that in 34 years of professional baseball, I have never seen someone drinking in the dugout. That was one of the worst things I'd ever heard of. Guys don't come down to the dugout and drink. It just doesn't happen. Lester would come down in the first inning every night with a big cup of Coke, and I would grab it and taste it — not to check on him, but just because it really pissed him off. It was Coke, but it was in a big beer cup. I knew what he was drinking. I don't believe there was beer in the dugout, and I never will believe it."

On October 21, Theo Epstein officially resigned from the Red Sox.

"In a way, I'll never recover from September," Epstein said in 2012. "I don't fully understand it. It was so disheartening. We were playing, I thought, world-class baseball for four and a half months. Clearly we had our vulnerabilities — the pitching attrition was real, and we were walking a tightrope — but for the wheels to fall off the wagon the way we did. . . . For us to lose not only our competitiveness and our place in the standings, but our identity as a team was painful to watch. I'll never really ever get over that.

"When I think of my relationship with Tito and what I'll hold on to, I think a lot of those early years, going through it together when we both had the sense that we were like kids who broke into someplace we shouldn't have been. We were kind of together running this team and organization. We weren't jaded yet. We were just having a great time seeing what we could pull off, having fun together, taking chances, having each other's back, and being a small part of watching this great thing unfold that touched so many people. It was really wonderful, and that's what I really think about. As the dynamic shifted and the Red Sox became too big, it became less fun for everybody. Tito, me. It was time to move on for both of us."

On November 8, Francona went to Cincinnati (home of St. Louis Cardinals owner Bill DeWitt) to interview for the managerial vacancy created when Tony La Russa retired after winning the 2011 World Series. It was a formality. The Cardinals wanted to stay in-house and hired Mike Matheny, a former Cardinal catcher who'd been serving as a special assistant in the organization. Francona hadn't expected to get the job, but it bothered him when they started the interview by asking if he could ease their apprehension about the events of September.

"They started by asking me, 'Given what happened in September, explain why we should hire you or why we shouldn't be afraid of you?'" Francona said. "They didn't ask me about the chicken and beer, but the first thing they brought up was September."

When Pedroia and Ellsbury attended the Gold Glove Awards dinner in New York, Pedroia blushed as Bob Costas told the large audience, "The bar is going to be closed now . . . except for the Red Sox."

"I was embarrassed, man," said the second baseman. "Being a Boston Red Sox, my name is part of that, and it's frustrating."

Francona spent a good portion of November swimming and getting his health back at the Canyon Ranch Hotel and Spa in Miami Beach. He was committed to not managing any baseball team in 2012.

On December 1, the Red Sox named Bobby Valentine the 45th manager in franchise history. Francona's cell phone exploded with texts from Pedroia, Beckett, and many other Sox players. Five days later, ESPN announced that Terry Francona would join the network's *Baseball Tonight* crew and replace Valentine as in-game color analyst on *Sunday Night Baseball*. Francona and Valentine were interviewed together by ESPN's Karl Ravech in the lobby of the Hilton Anatole at the annual winter meetings in Dallas.

"The only advice I can really give [you] is that in the bathroom there is a mouse," Francona told Valentine. "It's in there every day."

The interview was funny and cordial. When it was over, Francona stood up, took off his microphone, looked around, and said, "Is the camera still on?"

Motioning toward Valentine, clearly joking, Francona smiled and said, "I fucking hate him."

Everybody laughed.

"When people ask me if I left the Red Sox on my own or if I was fired, I don't even know how to answer that," Francona said later. "I really don't. But I know I spent a lot of my time in Boston putting out fires. I didn't get to do as much baseball stuff as I would have liked. I love the baseball stuff. It's a lot harder to just let these guys play the game in Boston because of all the other stuff that's going on. You have to deal with it. Even if stuff wasn't there, people put it there. We fought for eight years to keep stuff in that clubhouse. I tried my ass off to help put the team in position to win, and I worked my ass off that last year more than ever.

"Our owners in Boston, they've been owners for ten years. They come in with all these ideas about baseball, but I don't think they love baseball. I think they like baseball. It's revenue, and I know that's their right and their interest because they're owners — and they're good owners. But they don't love the game. It's still more of a toy or a hobby for them. It's not their blood. They're going to come in and out of baseball. It's different for me. Baseball is my life."

# CHAPTER 16

# · 2012 ·

## "I guess we should have won a third World Series"

**T**HE 2012 RED SOX were a disaster. The way things ended in 2011 carried into the following season, impacting the ball club and Red Sox Nation with disturbing regularity. In September of 2012, John Henry admitted to *Sports Illustrated* (via email): "What appeared to be an outlier month in September 2011 turned out to be a harbinger instead."

The once-beloved team morphed into one of the most disliked units in the history of New England sports. The '12 Sox were a rare blend of entitlement, underachievement, and sloth. The Sox finished in last place for the first time in 20 years, compiling Boston baseball's worst won-loss record (69–93) in 47 seasons. It was systemic failure from top to bottom.

The season-long collapse was unexpected. The Red Sox made multiple changes after the disappointment of 2011, but none of them improved the product.

Before the season began, the Sox issued a press release on January 19, announcing "restructuring of major league medical staff." Dr. Thomas Gill was relieved of his duties as team medical director. The

Sox also parted ways with assistant trainer Greg Barajas and strength and conditioning coach Dave Page. Trainer Mike Reinold was demoted from head trainer to head physical therapist.

Tim Wakefield retired February 17. Jason Varitek retired two weeks later.

When the Sox arrived in Florida, every media outlet wanted to talk about chicken and beer.

"People gotta eat, whether it's chicken or steak," said Adrian Gonzalez.

An angry Beckett complained about a clubhouse snitch and said, "Somebody was trying to save their own ass, and it probably cost a lot of people their asses." Multiple reports fingered Youkilis as a chicken-and-beer source, and there was a Beckett-Youkilis showdown in the early days of spring training.

"I don't know why I was the one put out there," said Youkilis. "We dealt with it in Florida at the start. It was confronted and it was over with. If anyone asked anyone, I think I would have been the last one to say anything."

While Beckett demonstrated little contrition and looked for leaks, Lester admitted, "We've got to earn the trust back from the fans."

Lackey, who'd undergone Tommy John surgery, said, "Guys having a beer after their start has been going on for the last 100 years. It's not like we were sitting up there doing it every night. It's not even close to what people think."

In a February 18 column in the *Boston Herald*, Francona told Michael Silverman, "I called John Henry seven or eight times. Never heard from him. I have not talked to John since the day I left. It makes you kind of understand where you stood."

This got Henry's attention. The owner called his ex-manager and said he was concerned that Francona was angry with him. He repeatedly asked the manager why he had talked about ownership not having his back.

"Like I always said, it's not only your right, it's your obligation to get the right manager," Francona told Henry. "I understand that. And it's not me anymore. But if you're hearing what I heard before our meeting, during our meeting, after our meeting — and then reading that article — how would you feel if you were me? Instead of caring about me and my reputation, you start running into radio stations to make sure it's not about you. I wanted you to care about me."

Ten minutes after they hung up, Henry called Francona again.

"Tito, how about if you come back and throw out the first pitch for us on opening day?" asked the owner.

"No thanks, John," said Francona.

"It was the same stuff as before," the ex-manager said later. "It was like when they made us play the doubleheader and John thought he could make up for it by giving the players his boat and giving them headphones. But the one thing that did come out of that — John promised me he would get back to me regarding that Hohler story. He said the same thing Larry said, but to this day I never heard back."

When Henry met with the Boston sports media in Fort Myers on February 25, he explained, "I wasn't avoiding him [Francona]. . . . We had a long conversation after I found out he was trying to get in touch with me. It was a great conversation, one that we should have had prior to this. We were able to clear the air. I think there were points of view that we both had about last year when he left, so it was good."

Later the same day, Bobby Valentine announced that the Sox would no longer have beer in their clubhouse at Fenway Park.

"It doesn't matter," said Ortiz. "We're not here to drink; we're here to play baseball. This ain't no bar."

Appearing on ESPN Radio, Francona said, "I think it's a PR move. I think if a guy wants a beer, he can probably get one."

Valentine shot back with, "Remember, you're getting paid there for saying stuff. You get paid over here for doing stuff. I've done both."

Francona felt badly and promised "to be more careful with my words."

Red Sox spring training spawned dozens of accounts of Valentine's innovative and busy preseason camp. The new manager's training regimen was described as "creative" and "sophisticated." Bobby V loved the bunt and hated the windup. He did not communicate much with his coaches. He had a video director put together theme-driven highlight reels — a one-hour loop of pitchers making good fielding plays, or perhaps 60 minutes of nonstop relays from the outfield — and ordered the tapes played throughout the day in the clubhouse. Valentine did not delegate to his coaches, several of whom were team-appointed.

Francona tired of hearing about the new "clubhouse culture" and the new way of doing things in the Sox spring camp.

"When somebody takes over, it's always going to be bigger, better, newer," said the ex-manager. "I grew weary of that. I wanted to put rings on both fingers and say, 'We were a little above average.'"

The Francona-Valentine contrast could not have been more clear. Francona was a communicator, ever-mindful of the feelings of his players. Valentine made little effort to extend himself to his players. Francona hated it when reporters' cell phones went off during press conferences. Valentine smiled when his own phone rang while he was addressing the media. Francona had kept everything in-house. Valentine was always likely to blurt out something that would ignite a media storm.

Francona returned to Fort Myers for the first time when the ESPN crew visited JetBlue Park for a much-hyped exhibition game between the Yankees and the Red Sox in March. The former manager skipped the customary pregame information session in Valentine's office, leaving Dan Shulman and Orel Hershiser to carry out the chore.

"It was strange to be there that night," said Francona. "I was glad that the first Red Sox game I was doing was spring training and not regular season. The one thing I didn't want was pictures of me standing in the middle of the clubhouse looking awkward. So I went on the field and saw some guys and retreated upstairs and got ready for the game. I didn't want to be the focal point of somebody's story."

Despite his efforts, Francona was corralled by a number of Boston baseball reporters. He said he hadn't heard anything about Fenway's 100th birthday bash in April, adding, "I'm not quite ready for the hugs yet. I'm still trying to stop the bleeding."

Later that night, when Valentine ordered a suicide squeeze in the bottom of the ninth, Francona correctly predicted that Yankee manager Joe Girardi would take his team back to Tampa even though the game was tied after nine innings.

Francona was in the broadcast booth for the Red Sox opening day loss at Comerica Park in Detroit, but once again he avoided the Sox clubhouse. Given the traditional pomp of every opener, it was easy for him to stay under the radar. The game unfolded in a fashion remarkably similar to the final night in Baltimore in September 2011. The Tigers beat the Red Sox with a cheesy walk-off single to left in the bottom of the ninth.

A week later, Francona was in a phone store in Tucson with three young Verizon employees, learning to use his new iPhone. While the phone was on speaker, he took an unexpected call from Larry Lucchino.

"Tito, this is Larry," Lucchino started.

"Hey, Larry," said Francona. "Just so you know, I'm in a Verizon store learning to use my new phone and we're on speakerphone here."

"Fine," said Lucchino. "I was just following up to make sure you know we'd love to have you on hand with all the other ex-Sox players and managers when we celebrate Fenway's 100th on April 20."

"Larry, you know what my answer is, don't you?" said Francona.

"Yeah, you're not ready to hug everybody," said Lucchino. "I read all about it."

"That's right, Larry," said the ex-manager, "I said the same thing to John. I told him I don't want to be included in anything to do with the Red Sox until he gives me a decent answer on who fucked me in the newspaper."

"It wasn't fucking me!" insisted Lucchino. "And it wasn't fucking John!"

"That's fine, Larry," snapped Francona, aware that folks in the phone store were starting to look at him. "I believe you. I'm just telling you how I feel. I don't want anything to do with the Red Sox until you care enough to find out who said it. Call me when you got a better answer!"

*Click.*

Later in the day, Francona tried to call Lucchino and left a message with the CEO.

"I thought it was a respectful message," said Francona. "I wasn't emotional. I just wanted him to understand why. I didn't want it to be a fight."

When Lucchino returned the call, Francona started to ask if the CEO had received his message, and Lucchino snapped, "I didn't listen to it."

"That led us to round two," said Francona.

"That was a bit of a blowup," admitted Lucchino. "He was mad we hadn't publicly identified the person who had leaked this story, and I told him how hard it was and how frustrating it had been my whole career, and you just can't keep turning your organization upside down and expect that you're likely to find who had done it. I never had any great success."

"I never asked them to publicly identify the person," said Francona. "I just wanted to know who it was. And Larry told me if he found out, it was a fireable offense. And that he'd call me back. But he never did. Both Larry and John said they'd get back to me after their attempts at discovery, and neither one of them ever did. I think that's what bothered me the most."

Nine days before Fenway's 100th celebration, news of Francona's rejection of the Sox invite hit the front page of the *Globe*.

"It's a shame," Francona said in the article. "I'm sure they'll have a great event and I was part of a lot of that stuff there, but I just can't go back there and start hugging people and stuff without feeling a little bit hypocritical. . . . I just feel like someone in the organization went out of their way to hurt me and the more we talked I realized we're just not on the same wavelength. They're probably better off going forth and leaving me out of it. . . . Until I'm more comfortable with some answers on what happened at the end of the year, I don't want to have much to do with the organization and that's a shame. With all the good things that were accomplished, I just feel pretty strongly about that. . . . When I spoke to John he made me think they were going to make an effort. John and Larry made it clear to me they weren't responsible for what was said. I thought they owed it to me to get to the bottom of it a little bit."

"I understand how strongly he feels on this matter and I accept that," Lucchino said in the *Globe* story.

After disclosure of his refusal to attend Fenway's 100th, Francona received phone calls from friends urging him to return for the celebration. Nomar Garciaparra called the ex-manager, no small deed given the acrimony that had accompanied Nomar's departure from the Red Sox. Commissioner Bud Selig called to deliver the same message: Francona owed it to the fans.

The ex-manager started to have second thoughts. He called his 78-year-old father in New Brighton, Pennsylvania.

"Dad, you've always been the voice of reason," said Terry Francona. "They want me to come back. What should I do?"

"Tell them to shove it up their ass!" said Tito Francona.

"Thanks, Dad, I just needed to hear you say that."

Then he got a letter from Red Sox vice chairman Phillip H. Morse, a limited partner in the ownership group who had been close to the manager during his eight years in Boston. Francona trusted Morse and valued his opinion. Morse informed Francona that if the ex-manager did not see fit to return to Fenway, Morse would also skip the celebration.

"That really floored me," said Francona. "Phil was one of my favorites, and his letter made me think hard about it. I wanted to acknowledge the fans. My strong feelings about what happened with the organization didn't change, but I was making myself the story and I wasn't comfortable with that. I didn't feel good about myself. So I decided to go."

He called Pam Ganley two days before the event to tell her that he'd changed his mind. He didn't want to interact with his old bosses, but he wanted to be there for the fans.

Francona was in Connecticut, working for ESPN, on the morning of Friday, April 20, the 100th anniversary of the inaugural Boston–New York baseball game at Fenway. After the morning broadcast, ESPN senior vice president Jed Drake drove the ex-manager from Bristol to Fenway High School on Ipswich Street outside the ballpark, where more than 200 ex-Sox players, coaches, and managers were gathered for the pregame ceremony. Francona was issued Red Sox uniform top number 47, a jersey fans rarely saw during the eight years he worked in the Boston dugout.

Departing from his modus operandi, Francona arrived at the holding area across the street from Fenway just a few minutes before the group boarded buses for the ceremony. He was happy that he was too late to hear the welcoming remarks delivered by Henry, Lucchino, and Werner. Ganley arranged for him to wait in a private room in the high school where he could escape unwanted attention.

Lucchino still found him.

"I opened the door to his room and said hi," recalled Lucchino. "We shook hands. We didn't talk. It was just a greeting. He was kind of cold and distant. . . . I would hope that the passage of time would allow him to appreciate the gigantic contributions he made here. There has to be a way that there can be a rapprochement that will allow us to accord him the kind of respect and gratitude — we owe an enormous debt of gratitude — and there has to be a time when we could do that."

On the short bus ride from Fenway High to gate C behind center field, Francona sat with Dave McCarty, one of his 2004 warriors. He didn't know what to expect when he got off the bus and walked into the ballpark underneath the center-field bleachers.

The scene under the stands had the feel of a livestock auction. Fenway's 100th was a complex ceremony that required planning and organization normally reserved for a presidential inauguration. Every Sox alum was issued a number, and the men were lined up in four rows of fifty. While he waited, Francona visited with old-timers Gary Bell and Tommy Harper, who asked him about his dad. When Garciaparra and Lou Merloni came over for a visit, the anxious ex-manager thanked Nomar for his phone call, but reminded both ex-players that he was not ready to extend an olive branch to the Red Sox owners.

"I'm not in the mood for any public hugs," said Francona.

The beloved World Series–winning manager was not one of the first to appear in the 20-minute parade of alums. Jim Rice, Dwight Evans, Bill Buckner, Frank Malzone, Jerry Remy, Luis Tiant, and other gods of the 1960s, '70s, and '80s were in the initial wave of ex-greats, while 92-year-old Johnny Pesky and 94-year-old Bobby Doerr sat in folding chairs positioned outside the first-base dugout. None of the individuals' names were announced to the crowd, but each man's image appeared on the center-field scoreboard as he walked in from the imaginary cornfield beyond the warning track.

When Francona finally emerged from the shadow of the doorway, a wave of noise and love washed over the Fenway lawn. The ex-manager held his hand over his heart as he walked toward the infield amid chants of "We want Tito!"

His was the loudest reception.

"Sounded like a Learjet," said Millar.

"It felt good, but I didn't know if they were clapping for me or the next guy," said Francona. "Mostly, I just wanted to find my place on the field and get it going. I was trying to be a little bit inconspicuous."

Standing in the Fenway infield with 200 other Sox veterans and present-day players, Francona visited with Pedroia and the trio of pitchers who broke the rules back in September: Lackey, Lester, and Beckett. He got emotional when he saw Nate Spears, a 26-year-old utility player, who'd been in professional baseball for nine years.

"He was a great kid, and we'd taken him on every spring training road trip," recalled Francona. "When we sent him down in 2011, I had told him, 'Kid, you're going to play in the big leagues.' For some reason, seeing him there that day really got to me."

Sox bench coach Tim Bogar saw Francona getting emotional.

Pointing to Spears, Bogar asked Francona, "Are you all misty-eyed because of him?"

"Yeah," said the ex-manager, shaking his head, wiping his eyes. "All this fucking shit and I'm crying because of him."

Spears was optioned to Pawtucket six days later.

After the ceremony, Francona marched shoulder to shoulder with the likes of Randy Kutcher, Jody Reed, and Buddy Hunter toward the center-field triangle. After he reached the warning track, the ex-manager peeled off his Red Sox jersey, rolled it into a ball, spotted a young girl in the front row of the bleachers, and hurled the shirt over the wall toward the girl. Typically, the gift was intercepted by an adult, but Francona got the man's attention and directed him to give the jersey to the young girl.

Then the ex-manager raised his left fist in the air and disappeared under the stands. He walked toward the ESPN compound, where he was picked up by a driver and escorted to the Langham Hotel in Boston's Post Office Square.

Terry Francona accomplished the impossible; he got out of Fenway faster than Carl Yastrzemski.

"They took me right to the car and got me the heck out of there," said Francona. "I was probably halfway to the hotel by the time everybody got off the field. I was just uncomfortable and didn't want to linger."

Had he stayed at Fenway a little longer he would have noticed personnel changes at security stations around the ballpark. There was a new guard outside the Sox locker room. Inside the clubhouse, veteran clubbie Joe Cochran had been banished to the visitors' side, replaced by Tommy McLaughlin, who'd been working in the visitors' room for more than a decade. There was also paranoia upstairs, where Werner had a veteran Boston sports columnist evicted from the EMC level.

The Yankees beat the Red Sox, 6–2. A day later, on national television, the Sox blew a 9–0 lead, losing 15–9. Valentine was booed every time he popped his head out of the dugout as fans continued to chant, "We want Tito!" After the stunning loss, Valentine said the Sox had hit "rock bottom." Little did Valentine know, but there would be many more rock bottoms in 2012.

The ESPN *Sunday Night Baseball* game was rained out, saving Francona an awkward evening of dodging his old bosses on his 53rd birthday.

Lucchino understood that Francona felt betrayed and persisted in his efforts to find the leak. But there were limits. As a Yale law student, Lucchino had worked alongside classmate Hillary Rodham on the Senate Watergate impeachment committee. Lucchino rejected the notion of conducting what he termed a "Nixonian investigation."

"That would entail a special prosecutor and literally calling people in," said Lucchino. "I've never done this, tempted as I've been in the past to have people come in and take a lie detector test. I've been frustrated enough about leaks that have been damaging to me and the organizations I've been with and the other people in the organization, and I know how hard it is to try to identify that person."

Lucchino believed the primary source was someone who had already left the organization.

"The people who actually do know aren't saying it," said Lucchino. "So I'm not sure the responsibility falls on those of us who don't know."

"That's interesting coming from someone who promised to find out," said Francona. "Maybe this will help people understand my frustration."

The Red Sox piled up dozens of injuries and underachieved throughout the season. Valentine got himself into a jam in mid-April when he criticized the commitment of Youkilis, claiming the veteran was not "as physically or emotionally into the game" as he had been in the past. Pedroia reacted, saying, "Maybe that works in Japan or some-thing . . . but that's not the way we do things here." Cherington came to the rescue of an agitated Youkilis, and Valentine apologized to the veteran.

Valentine was not happy with the constant presence of Cherington, O'Halloran, and other members of baseball operations in his office be-fore and after every game.

"Other teams don't do that," the new manager grumbled as he sat in the Fenway dugout before a Sunday afternoon game in May. "It's just one meeting after another around here."

Watching the Boston circus from Chicago, Epstein responded to a reporter's query with an email that read (in part), "Too bad for you nothing is going on in Red Sox land to capture a cynic's attention. Wow. It's even stranger to watch from afar than it was to be in the middle of it."

When the last-place Sox visited Wrigley Field in June, Epstein looked back at the path not chosen.

"We joked about it all the time in the front office," he said. "We'd say, 'Wouldn't it be great if we could just say, "Screw free agency altogether. We're going with a purely homegrown lineup. We're going with old-school, Branch Rickey–style, pre–free agency, pre–draft, whatever." [Will] Middlebrooks at third, Lowrie or [Jose] Iglesias at short, Pe-droia at second, [Anthony] Rizzo at first, Lavarnway catching, Ells-bury in center, Reddick in right, Kalish in left.' Wouldn't that have been fun? We kind of clung to that in the back of our minds, knowing it was impossible, recognizing that there was an inherent tension between that approach and bigger business. I kind of kick myself for letting my guard down and giving in to it, because that might be a better team in some ways and resonate more with the fans than what we ended up with."

Why, then, did he waste all that money?

"As far back as '04, I kept hammering everyone internally," said the ex-GM. "I'd say repeatedly, 'We can't forget what we are. We're a base-ball team. We can't get too big. We can't promise things that aren't going to happen. We have to be patient.' . . . We did fight that battle. We protected ourselves in baseball operations. We were insulated. We were in our own little environment. We did well for a long period of time, but we became too big, and then I fucked up and kind of gave in to that and didn't execute it well, and for a period of time we lost part of our identity, and it's hard to get back."

Asked about the NESN survey of 2010, Epstein said, "It played into part of the reason why I thought it was time to move on."

In the hours before the Red Sox–Cubs finale at Wrigley (another ESPN Sunday night game), Francona wandered into the cramped visitors' clubhouse with his ESPN crewmates, greeted Youkilis with "Mazel tov," hugged McCormick and Bogar, and managed to cajole a smile out of Beckett. Chewless, wearing a gray suit and purple tie, the ex-manager looked slightly awkward in the familiar setting.

"It was a little uncomfortable," he said later. "I'd been with them for so long, and all of a sudden I was a visitor."

Spotting coach Alex Ochoa holding a cup of Gatorade, Francona peered into the container and said, "Is that beer? Where's the chicken?"

Across the room, Pedroia, sitting next to Daisuke Matsuzaka, spotted his old cribbage partner, burst out of his chair, and screamed, "Tito, you should have seen it, man. Dice came back and pitched for the first time last week. He struck out a guy in the first inning, and we were throwing the ball around the infield, and I caught it and went over to Dice and said, 'Aaaaaaaahhhhhhhhhh!' He should pitch with a samurai sword, man."

It was as lively as the room had been all season. Hours later, Francona's ESPN teammate Buster Olney characterized the Sox clubhouse as "toxic.'"

Youkilis, who'd been miserable around Valentine for four months, was traded to the White Sox in late June. He immediately went on a tear, hitting .478 with three homers and 10 RBI in the week leading into the All-Star break. The Red Sox lost eight of 11 before the break. Lucchino sent a pandering letter to season-ticket holders, reminding fans that the Sox "look forward to the return of the varsity."

Francona came back to Fenway Park with the ESPN *Sunday Night* crew July 8 and participated in the standard manager's briefing session with Shulman and Hershiser. It was the ex-manager's first visit to his old office since he'd been fired. On orders from Henry, the ancient space had been totally renovated for Valentine. Plush red carpet covered the floor, the Pesky couch had been replaced with a new model, and a privacy wall had been erected to separate the skipper's desk from the latrine.

The former Sox manager found himself surprisingly unmoved by the renovations and said little while Shulman and Hershiser peppered Valentine with questions.

"My pictures were gone, the couch was gone, and it looked so different," said Francona. "I can't believe they put that wall up. That's where I did some of my best work. I conducted a lot of meetings from that bathroom. The place was completely changed and didn't even feel like I had worked there.

". . . For eight years I had asked them to redo that office," he said with a chuckle. "It was the first thing I asked for at every one of those roundtable meetings. It was kind of a running joke. I'd say it, and Larry would write it down on his yellow legal pad. I guess we should have won a third World Series."

In late July, when the Sox had an off day after flying from Texas to New York, Sox owners agreed to meet at the Palace Hotel with players who wanted to complain about Valentine. The meeting was requested via text by someone using Adrian Gonzalez's cell phone. Players were angry that Lester had been left on the mound to take an 11-run beating against Toronto at Fenway on July 22. One player complained that Valentine had been harsh with rookie third baseman Will Middlebrooks, saying, "Nice inning, Will," after Middlebrooks had a tough inning in the field. Players were unhappy with Valentine's limited communication skills.

When the clubhouse mutiny was first reported by Yahoo Sports in August, the Sox were in Baltimore, the same place where everything imploded at the end of 2011. In an effort to stop the bleeding, Henry, Werner, and Lucchino made an emergency trip to Camden Yards. They tried to explain the meeting as another in their series of "roundtable" discussions, but according to Francona, in his eight years on the job there had never been a roundtable without the manager or outside of Fenway Park. A fuming Lucchino disputed Francona's contention.

Beloved Sox ambassador Johnny Pesky died August 13. Fans reacted angrily when it was learned that only four Red Sox players took the time to attend Pesky's funeral. That same week, the Sox fell to seven games under .500, effectively eliminating themselves from playoff consideration for a third straight season.

A poll conducted by Channel Media & Market Research revealed that 70 percent of respondents said the Red Sox had changed for the worse over the last five years. In the same poll, Boston's baseball owners were ranked least popular among owners of New England sports teams.

Lucchino blamed the Boston media for exaggerating the ball club's dysfunction. Werner burst into the NESN broadcast booth defending the beleaguered Sox ownership group. Henry issued a few "votes of confidence" for Valentine, then withdrew almost completely.

Rarely answering questions from reporters, the Sox principal owner also ignored multiple emails from Francona requesting cooperation for this book.

Francona's final email to John Henry, sent in August 2012, read: "Hello John. I can't tell you how disappointed I was that after 8 years together and what I thought was mutual respect you chose not to even respond to my email. I guess I know now where I stand with you. Good luck. Tito."

Francona enjoyed his year in the broadcast booth, but never lost his desire to return to the clubhouse and the dugout. He loved the game too much. Throughout the summer of 2012, every time he'd walked into a big league clubhouse wearing his gray suit and purple tie he'd felt the urge to get back in uniform and peel open a can of Lancaster. When ESPN granted his request to cover the Little League World Series in Williamsport, Pennsylvania, Francona was invigorated watching freckle-faced 12-year-olds from Goodlettsville, Tennessee, and Kearney, Nebraska.

"I love the Little League World Series," he said. "It was always on TV in the clubhouse when I was around the big leagues. It was something I'd always wanted to attend. Everybody has got the right attitude about the game. It's where county fair meets baseball. It's the joy we all had when we first started playing. It's the way baseball is *supposed* to be."

On Saturday, August 25, the Red Sox shocked the baseball world when they traded Gonzalez, Crawford, Beckett, and Nick Punto to the Dodgers for first baseman James Loney and four prospects. The Dodgers assumed $261 million in future contract payments. It was the biggest Boston baseball trade since Babe Ruth was dealt to the Yankees in 1920, and it signaled the end of a failed era that had begun when the Sox tried throwing money at their problems after they were swept by the Angels in the 2009 playoffs. From 2009 to 2012, the Sox spent $629 million in player payroll and won zero playoff games.

The day the mega-deal was announced, Red Sox GM Ben Cherington said, "We are not who we want to be."

The Sox lost their final eight games and 12 of their last 13. They went 7–22 in September-October. On the night the season ended at Yankee Stadium, a 14–2 loss to the Yankees, Valentine said he'd been undermined by his own coaches during the season.

Valentine went to Lucchino's home in Brookline the following morning and met with Henry, Werner, Lucchino, and Cherington. At 12:47 PM on Thursday, October 4, exactly 14 hours and 14 minutes after the last out at Yankee Stadium, the Red Sox issued a statement announcing that Valentine would not return for the 2013 season. This time Lucchino admitted that his manager was fired. The Red Sox hired Toronto manager John Farrell — Francona's pitching coach in Boston from 2007 to 2010 — to succeed Valentine. The Sox had to part with infielder Mike Aviles to acquire Farrell, who had one year remaining on his contract with the Blue Jays.

On Monday, October 8, Terry Francona was named the 42nd manager of the Cleveland Indians. Tito Francona, who hit .363 for the Tribe in 1959, attended his son's introductory press conference at Progressive Field.

The circle of the baseball life was complete. Only 100 miles from where he grew up in New Brighton, Pennsylvania, Terry Francona was back in a big league clubhouse, reading scouting reports, and making friends with the clubbies.

"Managing the Red Sox was the hardest job I ever had," he said. "And it was the best job I ever had. Now it's time to try it somewhere else."

# ACKNOWLEDGMENTS

### Terry Francona

I want to thank my dad, who was always there for me, even when he was away playing in the major leagues. His phone calls home were always the same. He'd say, "Did you try your best? Did you have fun?" And then, at the very end of the phone call, he'd sneak in, "How many hits did you get?"

I would like to thank every coach or manager I ever had the opportunity to play for. You might have thought I was not paying attention, but I watched and listened to everything! I'd also like to thank:

Greg Fazio, New Brighton High School baseball coach and lifelong friend: For allowing me to skip study hall so I could drag the infield on rainy days so we could always play.

Jerry Kindall: For teaching me not only to play the game correctly and with respect but also to respect the people in the game.

Larry Bearnarth, Memphis Chicks: For teaching me to *never* refer to a manager as "coach."

Felipe Alou: For making me feel fearless on the playing field.

Dick Williams: Even though your stare terrified me, I knew and respected that you were always three innings ahead of the game.

Jim Fanning: For demonstrating that you can be a nice guy and a major league manager.

Bill Virdon: Honest as the day is long, and a sense of humor that flew way under the radar.

Buck Rodgers: For having a way of making the 25th player, me, feel just as important as one of his regulars.

Jim Frey: For giving me a second chance to play in the major leagues.

Gene Michael: For being able to view the game and see the big picture.

Pete Rose: I would have run through a wall for him, and the way I swung the bat, he probably wishes I did.

Steve Swisher: For believing that a 28-year-old with two rickety knees could fight his way back to the big leagues.

Doc Edwards: For allowing his players the freedom to play the game.

Tom Trebelhorn: For teaching me how important it is to communicate, be enthusiastic, and be organized.

Gaylen Pitts: For making Triple A baseball — after ten seasons in the big leagues — feel like a wonderful experience.

I want to thank Buddy Bell for believing in me and giving me an opportunity not only to manage in the minor leagues but also to serve under him as a major league coach. Without him, I would probably still be retaking the test for my real estate license.

Thanks to Brad Mills, John Farrell, DeMarlo Hale, and many other coaches who not only are some of my best friends, but help babysit me through thick and thin.

Thanks to the Philadelphia Phillies and Lee Thomas for going out on a limb and hiring a 36-year-old first-time manager. And to Ed Wade, who fired me, but who I enjoyed working with and respect to this day. Some people in this game become almost like family, and Bill Giles is one of those special people.

I want to thank Mark Shapiro and Chris Antonetti for showing me friendship and direction when I needed it most. And Ken Macha for allowing me to serve as his bench coach and inspiring me to want to manage again.

Thanks to the Boston Red Sox for allowing me seven years and five months of the hardest but best years of my life. Special thanks to Theo

Epstein for believing in me and trusting me with the responsibility of being the manager. Thanks to the hardworking men of Boston baseball ops, who will be friends for life.

If you had told me on September 1, 2011, that by November of 2011, I would be jobless and writing a book with Dan Shaughnessy, I would have told you as eloquently as only I can do that this would happen as soon as a 200-pound hog jumps out of my ass. It turned out to be not only fun but very healthy for me to look back at the eight years of whirlwind ups and downs.

Thanks to Joe Buck for being gutsy enough and crazy enough to talk me into working two playoff games in the broadcast booth, which opened a whole new door for me, and a wonderful year with ESPN.

# Dan Shaughnessy

Nobody predicted this pairing of authors. I am a sports columnist for the *Boston Globe* and Terry was manager of the Boston Red Sox for eight amazing seasons. Terry was mad at me for a good portion of his Red Sox years, and a couple of times per season I'd get a call from a team publicist telling me that Terry wanted a word. This usually resulted in a mildly heated exchange. I liked the fact that the manager could engage in a spirited debate or clarification, then move forward as if nothing had happened. I loved his daily press conferences with the Boston media. I always thought the two of us could have had a strong relationship if our jobs, by definition, hadn't placed us in a perpetual position of conflict. This turned out to be true.

In a sense, I started working on this book when my dad drove my brother and me to my first Red Sox game at Fenway Park against the Baltimore Orioles in 1961. Dad saved up enough S&H Green Stamps to get me a Tito Francona mitt in the summer of '62.

Houghton Mifflin Harcourt editor Susan Canavan and my agent, David Black, both sent me emails on September 30, 2011, the night Terry was fired by the Red Sox owners. They wanted me to see if Terry would like to work on a book about his eight seasons in the Sox dugout. I sent Terry an email the next day, and though he was hesitant, he told me he would meet with David Black.

On November 29, 2011, Terry picked me up at my Newton home in his Cadillac Escalade. He was packed for our trip to New York, where publishers wanted to talk to the ex-manager of the Red Sox about a book project. Still skeptical, ever-hilarious, Terry's opening remark when I got in his car was, "Our first stop is going to be someplace where we can get these windows tinted so nobody'll see me driving you around!"

Our interview process for the book was exhaustive, thorough, and fun. We met dozens of times—in hotel lobbies and coffee shops in Boston, Brookline, Bristol, Fort Myers, Detroit, Chicago, and New York. We met at the McDonald's at the Charlton rest stop on the Massachusetts Turnpike. I had my Sony digital recorder, and Terry had five decades of baseball stories, including tales of two World Series winners in a town that hadn't won a World Series since 1918.

A word about the process of this book: The narrative represents Terry's perspective and his recollections of his eight years in the Red Sox dugout, fortified by my own thousands of hours behind the scenes with the 2004–2012 Boston Red Sox and almost four decades of baseball coverage dating back to 1977 when I was a cub reporter covering the Orioles for the *Baltimore Evening Sun*.

We had a lot of help. The great Tito Francona was always available when I had a question about the Francona family or Terry's youth. Terry's trusted lieutenants Brad Mills, DeMarlo Hale, and John Farrell submitted to lengthy interviews, as did Sox players (and ex-players) David Ortiz, Josh Beckett, Jon Lester, Dustin Pedroia, and Kevin Youkilis. Brian O'Halloran gave me an hour at spring training. In briefer sessions, I also spoke with Derek Lowe, Johnny Damon, Darnell McDonald, Bud Black, Twig Little, Derek Jeter, and Mark Teixeira. Sox traveling secretary Jack McCormick was a tremendous source on all things Tito. Clubbies Pookie Jackson and Steve Murphy embraced the project. Veterans Tommy McLaughlin and Joe Cochran were terrific, as always. A big thank-you goes out to Sox publicist Pam Kenn (she was Pam Ganley before August 2012), a Tito favorite. Thanks to Sox employees Ken Nigro, Dr. Larry Ronan, Ben Cherington, Sam Kennedy, Jon Shestakofsky, Leah Tobin, Abby DeCiccio, Peter Cohenno, Dick Bresciani, Debbie Matson, Sarah Narracci, Sheri Rosenberg, Brita Meng Outzen, Dr. Charles Steinberg, Billy Broadbent, Sarah

McKenna, John Carter, Larry Cancro, Kevin Doyle, and Guy Spina. Thanks to Jenn Katz, Bernadette Serrette, John Perolito, John Keenan, Bob Mosher, Holly Munroe, Caitlin Neves, Abby Taylor, John McDermott, and Bob Allen on the EMC level. Bill and Alli Achtmeyer and Jan Aughe get big props for their help behind the scenes. Thanks to Jim Kaat, Tim McCarver, Dan Shulman, and Orel Hershiser. Thanks to Bobby Valentine for letting me inspect the "new" manager's office at Fenway. Thanks to Sox CEO Larry Lucchino and chairman Tom Werner for their cooperation during a difficult season. Both agreed to be interviewed for this project.

The *Boston Globe* has been great to me. Thanks to Chris Mayer, Marty Baron, Caleb Solomon, and Joe Sullivan. Big thanks to Pete Abraham, Nick Cafardo, and Mike Vega for all they do on the Sox beat. Thanks to Bill Tanton, Dave Smith, Vince Doria, Don Skwar, John Lowe, Tim Kurkjian, Laurel Prieb and Wendy Selig-Prieb, Phyllis Merhige, Bob Ryan, Chris Gasper, Amalie Benjamin, Jonny Miller, Steve Buckley, Gordon Edes, Sean McAdam, Rob Bradford, Bill Ballou, Mike Fine, Lenny Megliola, Mike Silverman, Ian Browne, Gary Tanguay, Tony Massarotti, Bill Bridgen, Joe Amorosino, Lou Merloni, Jessica Moran, Mike Barnicle, Wendi Nix, Tom Verducci, and Joel Feld. Books like this are impossible to write today without a website like baseball-reference.com. Michael Holley's *Red Sox Rule* was a terrific source.

Owing to his enduring respect for Terry Francona, Theo Epstein submitted to a three-hour interview (at Starbucks in Brookline) with multiple follow-ups.

David Black is master of the universe and has great taste in whiskey and New York eateries. Susan Canavan steered the ship from the first day, believed in us at every stop, and paid the tab at Post 390. Friends Paul Comerford and Christy Lemire were terrific early-version editors. The ever-patient computer whiz Sean Mullin gets his own thank-you page. Thanks to Jeremiah Manion and Charlie Smiley in the *Globe* library. Stan Grossfeld brought his soul and perspective to the project. Thanks to Ashley Gilliam for taking my calls and printing a million versions of the book as we went through the summer. Copyeditor Cindy Buck had more saves than Mariano Rivera. Thanks to Gary Gentel,

Linda Zecher, Eric Shuman, and Bruce Nichols at Houghton Mifflin Harcourt. Thanks to Jeremy Kapstein, Ed Kleven, Lesley Visser, Kevin Dupont, Stephen Stills, Steve Sheppard, Sue Lodemore, and everybody else who tolerated a cranky author. Thanks to Megan Wilson, the best publicist any author could have.

As ever, thanks to Marilou, Sarah, Kate, Rob, and Sam for their help and support at home in Newton.

# INDEX